Building a Successful Biodiesel Business

Building a Successful Biodiesel Business

Technology Considerations — Developing the Business — Analytical Methodologies

Second Edition

Jon Van Gerpen
University of Idaho

Rudy Pruszko
Iowa State University

Davis Clements
Renewable Products Development Laboratory

Brent Shanks
Iowa State University

Gerhard Knothe
USDA/NCAUR

Publisher: Biodiesel Basics

ISBN: 0-9786349-0-X

First Edition: 2005
Second Edition: 2006

Biodiesel Basics .com

Acknowledgements

We want to thank everyone who has helped make our workshop program and this book possible. This second edition, even more than the first, has benefited from the feedback we have received from the almost 400 alumni of our workshops. Their questions, ideas, proposals and other contributions have helped us to fine-tune the book to better address the needs of this exciting new industry. Iowa State University and the Iowa Energy Center have continued to be important supporters of the program, and they have been joined by the University of Idaho. Gregg Cameron of Iowa State University's Continuing Education and Communication Services department has always done a great job of providing logistical support for the workshops. The graduate students who have helped with the laboratories for the workshops have become too numerous to mention by name, but without their contributions we would not have been able to provide the hands-on experiences for the workshop attendees. We wish everyone looking to get into the biodiesel field the best of luck and hope that you find this book to be helpful in reaching your goal.

The book was edited by Helen K. Randall, HKR Communications & Marketing. Cover design and interior layout were provided by Tim S. Randall.

Jon Van Gerpen
L. Davis Clements
Rudy Pruszko
Gerhard Knothe
Brent Shanks

Table of Contents

Introduction

We started the first edition of this book by noting that it was an exciting time for everyone in the biodiesel industry. Little did we know how exciting it would become! Interest in biodiesel is exploding. Biodiesel is selling as fast as producers can make it. In many cases, it is priced below petroleum-based diesel fuel. Investors and entrepreneurs are eager to get an early start in this growth opportunity.

We have been offering three- to five-day workshops on various aspects of biodiesel since 2002. We have had almost 400 total attendees with representatives from 42 states and 21 countries outside of the United States. Our goal has always been to provide comprehensive and unbiased information about biodiesel. The notes from these workshops were the origin of this book. This second edition incorporates revisions and updates to material from the first edition and includes three completely new chapters. Feedback from our students has been invaluable in improving the material and making it more useful to readers.

Recent developments in the industry have re-emphasized the importance of fuel quality to the continued growth of biodiesel. We want to call the reader's attention to the chapters on biodiesel specifications, test methods and engine requirements. Everyone should constantly repeat the mantra, "If it doesn't meet the ASTM specification, it's not biodiesel." There is no greater threat to this industry than a public perception that the fuel cannot be trusted.

We wish everyone the best of luck in their ventures; however, we want to add a note of caution to temper the enthusiasm that currently surrounds biodiesel. New industries tend to go through periods of rapid growth followed by corrections, where those businesses that are well-managed survive and those that are uncompetitive fail. We don't expect that biodiesel will be any different. It will not be long before the rapid expansion of the biodiesel industry produces a shortage of feedstocks causing price increases that will challenge the profitability of biodiesel production. Companies that have controlled costs, built efficient plants, stayed abreast of new developments and captured a supply of feedstock will have the best chance for survival. We hope this book helps your project be one of the successful companies.

Jon Van Gerpen
L. Davis Clements
Rudy Pruszko
Gerhard Knothe
Brent Shanks

SECTION ONE

Introduction to Biodiesel Fuels

CHAPTER 1

Background

Biodiesel is an alternative fuel for diesel engines that is receiving great attention worldwide. Although it attracts the most attention because it is renewable, other attributes include that it can be used either pure or in blends with diesel fuel in unmodified diesel engines, and it reduces some exhaust pollutants. It is also attractive because it can be produced easily from common feedstocks. However, the relative simplicity of biodiesel production can disguise the importance of maintaining high quality standards for any fuel supplied to a modern diesel engine. It is essential to the growth of the biodiesel industry that all fuel produced and sold meet these quality standards.

What Is Biodiesel?

Biodiesel is defined as the mono-alkyl esters of fatty acids derived from vegetable oils or animal fats. In simple terms, biodiesel is the product obtained when a vegetable oil or animal fat is chemically reacted with an alcohol to produce fatty acid alkyl esters. A catalyst such as sodium or potassium hydroxide is required. Glycerol is produced as a co-product. The approximate proportions of the reaction are:

100 lbs of oil + 10 lbs of methanol → 100 lbs of biodiesel + 10 lbs of glycerol

Soybean oil is the most popular feedstock in the U.S. Soybeans are a major U.S. crop and government subsidies are available that make the fuel economically attractive to consumers who need or want to use a nonpetroleum-based fuel. Biodiesel from soybeans is sometimes called soydiesel, methyl soyate or soy methyl esters (SME). In Europe, most biodiesel is made from rapeseed oil, and it is known as rapeseed methyl esters (RME). The University of Idaho has done considerable work with rapeseed to produce rapeseed ethyl esters (REE).
(Website: **http://www.uidaho.edu/bae/biodiesel)**

Biodiesel can also be made from other feedstocks:

- Other vegetable oils such as corn oil, canola (an edible variety of rapeseed) oil, cottonseed oil, mustard oil, palm oil, etc.
- Restaurant waste oils such as frying oils
- Animal fats such as beef tallow or lard
- Trap grease (from restaurant grease traps), float grease (from wastewater treatment plants), etc.

All vegetable oils and animal fats consist primarily of triglyceride molecules such as that shown schematically in **Figure 1.**

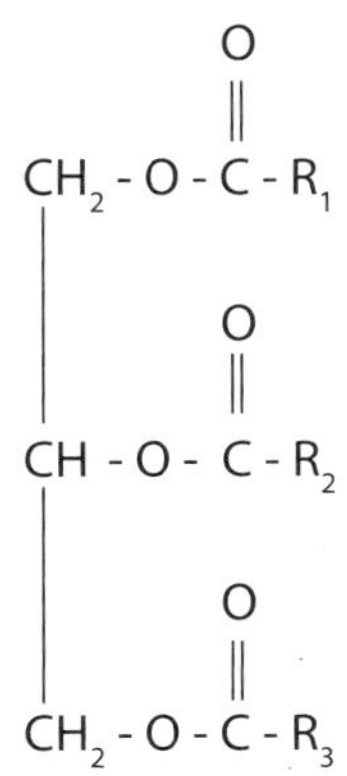

Figure 1. Triglyceride.

R_1, R_2 and R_3 represent the hydrocarbon chains of the fatty acyl groups of the triglyceride. In their free form, fatty acids have the configuration shown in **Figure 2**, where R is a hydrocarbon chain > 10 carbon atoms.

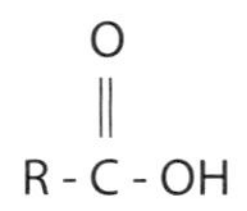

Figure 2. Fatty Acid.

Transesterification is the process of reacting a triglyceride molecule with an excess of alcohol in the presence of a catalyst (KOH, NaOH, $NaOCH_3$, etc.) to produce glycerol and fatty esters. The chemical reaction with methanol is shown schematically in **Figure 3**.

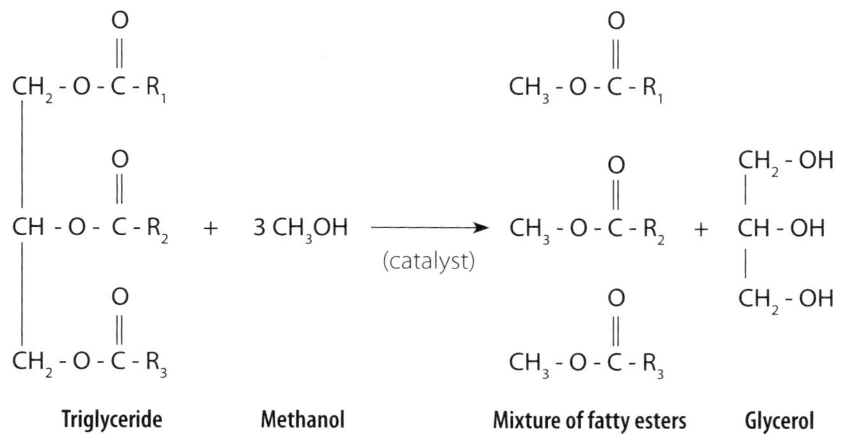

Figure 3. Transesterification.

The mixture of fatty esters produced by this reaction is referred to as *biodiesel.* At the present time, a large portion of the biodiesel sold in the U.S. is produced from the methyl esters of soybean oil, so biodiesel, soydiesel and methyl soyate are often used interchangeably. In general, biodiesel may also include fatty esters derived from other vegetable or animal sources. Other alcohols besides methanol may be used so it will be necessary to distinguish between the three terms.

The fuel properties of biodiesel are determined by the amounts of each fatty acid in the feedstock used to produce the esters. Fatty acids are designated by two numbers: the first number denotes the total number of carbon atoms in the fatty acid chain and the second is the number of double bonds present in the chain. For example, 18:1 designates oleic acid, which has 18 carbon atoms and one double bond. **Table 1** lists the fatty acid profiles of a number of common vegetable oils and animal fats.

Oil or fat	14:0	16:0	18:0	18:1	18:2	18:3	20:0	22:1
Soybean		6-10	2-5	20-30	50-60	5-11		
Corn	1-2	8-12	2-5	19-49	34-62	trace		
Peanut		8-9	2-3	50-65	20-30			
Olive		9-10	2-3	73-84	10-12	trace		
Cottonseed	0-2	20-25	1-2	23-35	40-50	trace		
Hi linoleic Safflower		5.9	1.5	8.8	83.8			
Hi Oleic Safflower		4.8	1.4	74.1	19.7			
Hi Oleic Rapeseed		4.3	1.3	59.9	21.1	13.2		
Hi Erucic Rapeseed		3.0	0.8	13.1	14.1	9.7	7.4	50.7
Butter	7-10	24-26	10-13	28-31	1-2.5	.2-.5		
Lard	1-2	28-30	12-18	40-50	7-13	0-1		
Tallow	3-6	24-32	20-25	37-43	2-3			
Linseed Oil		4-7	2-4	25-40	35-40	25-60		
Tung Oil		3-4	0-1	4-15		75-90**		
Yellow Grease*	1.3	17.4	12.4	54.7	8.0	0.7	0.3	0.5

* Typical analysis: listed in Tat, M.E. and J.H. Van Gerpen, "Fuel Property Effects on Biodiesel," ASAE Paper 036034, American Society of Agricultural Engineering Annual Meeting, Las Vegas, NV, July 27-30, 2003.

** The dominant fatty acid in tung oil is a conjugated isomer of linolenic acid called eleostearic acid. The three double bonds in eleostearic acid are located at 9:10, 11:12, and 13:14 instead of at 9:10, 12:13 and 15:16 as in linolenic acid.

Peterson, C.L., "Vegetable Oil as a Diesel Fuel: Status and Research Priorities," ASAE Transactions, V. 29, No. 5, Sep.-Oct. 1986, pp. 1413-1422.
Linstromberg, W.W., Organic Chemistry, Second Edition, D.C. Heath and Company, Lexington, Mass., 1970.

Table 1. Composition of Various Oils and Fats.

The common names of the fatty acids given in **Table 1** are as follows:

14:0 Myristic Acid (tetradecanoic acid)
16:0 Palmitic Acid (hexadecanoic acid)
18:0 Stearic Acid (octadecanoic acid)
18:1 Oleic Acid
18:2 Linoleic Acid
18:3 Linolenic Acid
20:0 Arachidic Acid (eicosanoic acid)
22:1 Erucic Acid

How Much Biodiesel Can Be Made?

It is sometimes claimed that biodiesel could be the answer to the world's dependence on fossil fuels. It is enlightening to look at the potential impact of biodiesel on diesel fuel consumption.

Using the rough guideline that a pound of oil or fat will give a pound of biodiesel, we can use the total production of fats and oils in the U.S. to estimate the potential impact of biodiesel on total diesel consumption. **Table 2** lists the annual U.S. production figures for vegetable oils and animal fats.

Vegetable oil production (billion pounds/yr)		Animal fats (billion pounds/yr)	
Soybean	19.550	Edible tallow	1.753
Peanuts	0.161	Inedible tallow	3.657
Sunflower*	0.305	Grease	2.617
Cottonseed	0.990	Yellow grease	1.367
Corn	2.413	Other grease	1.250
Canola	0.626	Lard	0.265
Safflower*	0.056	Poultry fat	1.019
Total veg. oil	**23.659**	**Total animal fat**	**11.928**

Source: Jacobsen Publishing Company for the Animal Fats and Vegetable Oil unless otherwise noted.

* Bureau of the Census and Agricultural Marketing Service, USDA for Oct. 2004 to Sep. 2005

Table 2. Total Annual 2005 Production of U.S. Fats and Oils.

The combined U.S. vegetable oil and animal fat production is 36.0 billion pounds per year. Soybean oil dominates the vegetable oil market comprising over 80 percent of the total vegetable oil volume. Animal fats total almost 50 percent of the vegetable oil market. At about 7.6 pounds per gallon of oil, total fat and oil production would equal 4.74 billion gallons of biodiesel. **Table 3** lists the total on-highway consumption of diesel fuel from 1996 to 2004.

	On-highway diesel	
	(billion gallons/yr)	(billion pounds/yr)
1996	26.96	191.1
1997	28.61	202.9
1998	30.15	213.8
1999	32.06	227.3
2000	33.13	234.9
2001	33.22	235.5
2002	34.31	243.2
2003	37.10	263.0
2004	37.13	263.2

(from the Energy Information Administration, **www.eia.doe.gov**)

Table 3. Sales of On-highway Diesel Fuel.

It is obvious that biodiesel is not going to completely replace petroleum-based diesel fuel in the near future. If all of the vegetable oil and animal fat were used to produce biodiesel, we could only replace about 13 percent of the current demand for on-highway diesel fuel.

So, why bother with biodiesel? There are five primary reasons for encouraging the development of biodiesel in the U.S.

1. ***It provides a market for excess production of vegetable oils and animal fats.*** There is increasing demand around the world for soybean meal to provide protein for human and animal consumption. If new markets are not found for the soybean oil, then the price of soybeans will be low and farmers will have even more difficulty making a profit. The animal by-products industry also has a problem with more supply than the current market demands. This is compounded by the potential for even greater restrictions on the use of animal fats in animal feeds because of concerns about the spread of BSE (Bovine Spongiform Encephalopathy - Mad Cow Disease).

2. ***It decreases the country's dependence on imported petroleum.*** Obviously, this reason should not be overemphasized since the percentage of the country's fuel supply that can be replaced with biodiesel will be small. However, petroleum markets tend to be sensitive to small fluctuations in supply, so an additional source of fuel can have a surprising impact on stabilizing fuel prices.

3. ***Biodiesel is renewable and does not contribute to global warming due to its closed carbon cycle.*** Because the primary feedstock for biodiesel is a biologically-based oil or fat that can be grown season after season, biodiesel is renewable. And, since the carbon in the fuel was originally removed from the air by plants, there is no net increase in carbon dioxide levels. Again, this reason should not be overemphasized because biodiesel does not have the potential to make a major impact on total carbon dioxide production. It should also be noted that the primary alcohol used to produce biodiesel is methanol. Methanol makes up about 10 percent of the feedstock input; and since most methanol is currently produced from natural gas, biodiesel is not completely renewable.

4. ***The exhaust emissions from biodiesel are lower than with regular diesel fuel.*** Biodiesel provides substantial reductions in carbon monoxide, unburned hydrocarbons and particulate emissions from diesel engines. While the carbon monoxide and unburned hydrocarbons from diesels are already low compared with gasoline engines, biodiesel reduces them further. Particulate emissions, especially the black soot portion, are greatly reduced with biodiesel. Unfortunately, most emissions tests have shown a slight oxides of nitrogen (NOx) increase with biodiesel. This increase in NOx can be eliminated with a small adjustment to an engine's injection timing while still retaining a particulate reduction.

5. ***Biodiesel has excellent lubricating properties.*** Even when added to regular diesel fuel in an amount equal to 1-2 percent (volume percent), it can convert fuel with poor lubricating properties, such as ultra-low-sulfur diesel fuel, into an acceptable fuel. This could be a major use for biodiesel. Starting in June 2006, on-highway diesel fuel will be required to contain less than 15 ppm of sulfur. This ultra-low-sulfur diesel fuel needs to have a lubricity additive to be used successfully in modern diesel engines. While many lubricity additives are available, U.S. pipeline operators have announced that they intend to ban diesel fuel with lubricity additives from their pipelines because of concern about possible contamination of jet fuel. This means the lubricity additives will need to be added at the distribution terminals where biodiesel could be one of the options if the local market asks for it.

CHAPTER 2 | Diesel Engine Fuel Requirements

Modern diesel engines are known for their durability, reliability and excellent fuel economy. When the fuel quality is high, diesel engines routinely last for tens of thousands of hours and provide fuel economy that is 30-40 percent better than gasoline-fueled engines. Diesel engines are used almost exclusively in the trucking, construction, agricultural and logging industries. In Europe, over half of the new passenger cars purchased are powered by diesel engines. In the United States, diesel engines have not gained widespread consumer acceptance in light-duty vehicles, except in pick-up trucks. There is some indication that diesels will gain popularity in the sport utility market as automakers are pressured to improve the fuel economy of these vehicles.

This chapter discusses the requirements that the diesel engine places on the fuel. Any alternative fuel must meet these requirements.

Background on Diesel Combustion

Diesel combustion is the process that occurs when a hydrocarbon fuel, chosen for its ability to auto-ignite, is injected into a volume of air that has been compressed to a high temperature and pressure. When fuel is injected into the turbulent compressed air inside the engine cylinder, it does not ignite immediately. There is a time period called the *ignition delay*, during which the fuel must heat up, vaporize, mix with air and undergo chemical precombustion reactions that produce the radicals necessary for spontaneous ignition or *auto-ignition*. The classical notion of ignition delay calls the heating, vaporization and mixing processes the "physical delay" and the pre-reactions the "chemical delay." This can be deceptive in that all of the processes can be, and probably are, occurring simultaneously.

After sufficient time has elapsed, ignition will occur spontaneously at multiple locations in regions of fuel-air mixture that have fuel-air ratios close to the chemically correct ratio or the *stoichiometric ratio*. Combustion proceeds very rapidly due to the backlog of prepared or nearly prepared fuel-air mixture formed during the ignition delay period. The rapidly rising temperatures and pressure in the cylinder accelerate the combustion in an uncontrolled manner until the backlog is depleted. This portion of the combustion process is usually called *premixed combustion*. The remainder of the fuel in the spray core is still too rich to burn, so combustion slows and is controlled by the rate at which the air is entrained and a combustible mixture formed. This portion of the combustion process is called *mixing controlled* or *diffusion burning*. Thus, while chemical kinetics dominates the ignition delay, the high temperatures and pressures of the post-ignition gases promote very fast reaction rates that make fuel-air mixing the rate determining process.

The Fuel Must Ignite in the Engine

One of the most important properties of a diesel fuel is its readiness to auto-ignite at the temperatures and pressures present in the cylinder when the fuel is injected. The laboratory test that is used to measure this tendency is the Cetane Number Test (ASTM D 613). Fuels with a high *cetane number* will have short ignition delays and a small amount of premixed combustion since little time is available to prepare the fuel for combustion. The cetane number has been criticized in recent years for not accurately reflecting the auto-ignition conditions in modern turbocharged engines, particularly with alternative fuels.

The Fuel Must Release Energy When It Burns

The energy content of a fuel is usually characterized by the amount of heat that can be extracted when the fuel, starting at ambient conditions, is burned and the products are cooled to ambient conditions. This is defined as the *enthalpy* of combustion (or the heat of combustion). Since the combustion process will produce water and the energy will be different depending on whether the water is liquid or vapor, two separate cases are considered. First, if all of the water in the exhaust products is assumed to be condensed, the heat extracted will be the *higher heating value* or the *gross heating value*. If all of the water is considered to be vapor, then the heat extracted is the *lower heating value* or the *net heating value*. Since engines do not have the ability to condense water in the exhaust, the lower heating value is the most commonly used measure of fuel energy content.

The energy content of the fuel is not controlled during manufacturing. The actual value of lower heating value for diesel fuel will vary depending on the refinery in which it was produced, the time of year and the source of the petroleum feedstock. Beyond these differences, the heating value will depend on the composition of the fuel. Diesel fuels with high percentages of aromatics tend to have high energy contents per gallon even though the aromatics tend to have low heating values per pound. Their high density more than compensates for their lower energy content on a weight basis. This is of special importance for diesel engines because fuel is metered to the engine volumetrically. A fuel with lower energy content per gallon will cause the engine to produce less peak power.

Biodiesel fuels do not contain aromatics, but they contain fatty acids with different levels of saturation. Fuels with more saturation tend to have lower energy contents and those with greater unsaturation tend to have higher energy content.

As can be seen in the following table, biodiesel has a lower energy content (lower heating value) than No. 2 diesel fuel. On a weight basis, the energy level is 12.5 percent less. Since biodiesel is denser than the diesel fuel, the energy content is only 8 percent less on a per gallon basis.

	Btu/lb	Btu/gal
No. 2 diesel	18,300	129,050
Biodiesel (soybean)	16,000	118,170
	(12.5% less)	(8% less)

Table 4. Energy Content.

Since diesel engines will inject equal volumes of fuel, most diesel engine operators see a power loss of about 8 percent with B100. In some cases, the power loss may be less than this. Biodiesel's higher viscosity can decrease the amount of fuel that leaks past the plungers in the diesel fuel injection pump. This increases the volume of fuel injected and offsets a portion of the power loss.

Tests have shown that the actual efficiency at which the energy in the fuel is converted to power is the same for biodiesel and petroleum-based diesel fuel. Therefore, the brake specific fuel consumption (BSFC), the parameter most often used by engine manufacturers to characterize fuel economy, will be at least 12.5 percent higher for biodiesel.

$$BSFC = \frac{\dot{m}_{fuel}}{P_b}$$

The specific fuel consumption (SFC) is similar to efficiency in that it measures how little fuel may be required to do a certain quantity of work, hence the smaller the SFC, the more efficient the engine. However, it does not contain information about the amount of energy that may be available from the fuel, so it cannot be used to make comparisons between engines burning different fuels.

The thermal efficiency is usually defined as the ratio of the power produced by the engine to the energy in the fuel consumed, as indicated by the lower heating value (LHV) of the fuel. As with the SFC, both indicated and brake values of the thermal efficiency can be defined.

$$\textit{Brake Thermal Efficiency} = \eta_{bt} = \frac{P_b}{\dot{m}_{fuel}\,(LHV)}$$

The LHV can be measured by calorimetry, or it can be estimated from equations based on the fuel composition.

There is some ambiguity in the definition of the thermal efficiency based on whether the higher or lower heating value should be used. Furnace efficiencies tend to be based on the higher heating value. Natural gas is sold on the basis of higher heating value. However, for engines, the normal procedure is to use the lower heating value since the engine cannot condense the water vapor in the exhaust gases.

Example: An engine produces 230.4 kW of brake power while burning 50 kg/hr of fuel. The lower heating value of the fuel is 42,567 kJ/kg. What are the BSFC and brake thermal efficiency?

$$BSFC = \frac{\dot{m}_{fuel}}{P_b} = \frac{50\ ^{kg}/_{hr}}{230.4\ \text{kW}} \times \frac{1000\ g}{\text{kg}} = 217.0\ ^{g}/_{kW\text{-}hr}$$

$$\eta_{bt} = \frac{P_b}{\dot{m}_{fuel}\,(LHV)} = \frac{230.4\ kW}{50\ ^{kg}/_{hr} \times 42{,}567\ ^{kJ}/_{kg}} \times \frac{1\ kJ}{1\ kW - s} \times \frac{3600\ \text{s}}{hr} = 0.390\ (39.0\%)$$

The Fuel Must Provide a Large Amount of Energy per Gallon

The density of the fuel is not itself an important parameter for the diesel engine, but it is generally related to the fuel's energy content. As a first approximation, hydrocarbon diesel fuels all have about the same energy content per pound. The differences between fuels relate mostly to the density. Denser fuels provide greater energy per gallon and since fuel is sold volumetrically, the higher the density, the greater the potential energy.

The Fuel Must Not Limit the Operability of the Engine at Low Temperatures

Diesel fuel contains a small amount of long chain hydrocarbons, called waxes, which crystallize at temperatures within the normal diesel engine operating range. If temperatures are low enough, these wax crystals will agglomerate and plug fuel filters and prevent engine operation. At a low enough temperature, the fuel actually will solidify.

Additives, known as *pour point* depressants, can be used to inhibit the agglomeration of the wax crystals, which then lowers the point at which fuel filter plugging occurs. It is also common to add No. 1 diesel fuel to No. 2 diesel fuel to lower its operating point. No. 1 diesel fuel has a very low level of waxes, so it dilutes the waxes in No. 2 diesel fuel and lowers the temperature at which they cause the fuel to solidify.

ASTM D 2500 and D 97 are used to determine the *cloud point* and *pour points* of the fuels, respectively. Other tests are used to measure the tendency of the fuel to plug fuel filters.

The Fuel Must Not Contribute to Corrosion

Many of the parts in the diesel fuel injection system are made of high-carbon steels and thus are prone to corrosion when in contact with water. Water damage is the leading cause of premature failure of fuel injections systems. Many diesel engines are equipped with water separators that cause small water droplets to coalesce until they are large enough to drop out of the fuel flow where they can be removed. Diesel fuel containing excessive water that enters the injection system can cause irreversible damage in a very short time.

ASTM D 2709 is used to measure the total amount of water and sediment in a diesel fuel sample. The method uses a centrifuge to separate the water so it can be measured volumetrically. The ASTM specification for biodiesel limits the amount of water and sediment to 0.05 percent.

Some compounds in diesel fuel, especially sulfur compounds, can be corrosive. Since copper compounds are particularly susceptible to this type of corrosion, copper is used as an indicator of the tendency of the fuel to cause corrosion. ASTM D 130 uses polished copper strips soaked in the fuel to characterize the tendency to corrode metals. Some tarnish is typically allowed, but corrosion causes the fuel to fail the test.

The Fuel Must Not Contain Sediment that Could Plug Orifices or Cause Wear

Diesel fuel filters are designed to capture particles that are larger than 10 microns in size. Some newer engines are even equipped with filters that capture particles as small as 2 microns. These filters should stop foreign materials from entering the fuel injection system. However, when fuels are exposed to high temperatures and the oxygen in air, they can undergo chemical changes that form compounds that are insoluble in the fuel. These compounds form varnish deposits and sediments. These deposits and sediments can plug orifices and coat moving parts causing them to stick. Several test procedures have been developed that attempt to measure the tendency of diesel fuels to produce these sediments, such as ASTM D 2274, but none have gained the acceptance needed to be included in the diesel fuel specification (ASTM D 975).

When diesel fuel burns, it should be converted entirely to carbon dioxide and water vapor. Inorganic materials present in the fuel may produce ash that can be abrasive and contribute to wear between the piston and cylinder. ASTM D 482 is used to characterize ash from diesel fuels. The ASTM specification for biodiesel, D 6751, requires that ASTM D 874 be used. This method measures sulfated ash, which is specified because it is more sensitive to ash from sodium and potassium. These metals are likely to be the main sources for ash in biodiesel.

When fuel is exposed to high temperatures in the absence of oxygen, it can pyrolyze to a carbon-rich residue. While this should not occur in the cylinder of a properly operating engine, some injection systems have the potential to create a region within the injection nozzle where this residue can collect and limit the range of motion of moving parts. Various test procedures such as ASTM D 189, D 524 and D 4530 have been developed as an attempt to predict the tendency of a fuel to form in-cylinder carbon deposits. Unfortunately, it is very difficult to reproduce in-cylinder conditions in a test, so the correlation of these procedures to actual engine deposits is limited.

Diesel fuel injection systems have very closely fitting parts that are subjected to high loads. These parts require lubrication to prevent rapid wear. All diesel injection systems rely on the fuel itself to provide this lubrication. While the mechanism is not fully understood, it is known that as refiners reduce the sulfur content of diesel fuel, the ability of the fuel to provide the necessary lubrication decreases. The property that characterizes the ability of the fuel to lubricate is the lubricity. Several tests are commonly used to measure *lubricity*, but they have not been accepted or required as part of the ASTM specification for diesel fuel.

The Fuel Should Not Cause Excessive Pollution

As mentioned earlier, when conditions are ideal all of the carbon in the diesel fuel will burn to carbon dioxide and all of the hydrogen will burn to water vapor. In most cases, virtually all of the fuel does follow this path. However, if sulfur is present in the fuel, it will be oxidized to sulfur dioxide and sulfur trioxide. These oxides of sulfur can react with water vapor to form sulfuric acid and other sulfate compounds. The sulfates can form particles in the exhaust and elevate the exhaust particulate level. In 1993, the Environmental Protection Agency (EPA) mandated that diesel fuel should contain no more than 500 ppm of sulfur. This was a factor of 10 reduction in sulfur level and greatly reduced sulfur as a source of exhaust particulate. In 2006, the EPA has mandated a new reduction in sulfur to 15 ppm. This will eliminate sulfur as a component of exhaust particulate and allow the introduction of catalytic after-treatment on diesel engines. Sulfur is a powerful catalyst poison and limits the options available for controlling emissions on future engines.

Aromatics are a class of hydrocarbon compounds that are characterized by stable chemical structures. They are usually present in diesel fuel at levels between 30-35 percent. They are desired by diesel engine operators because they provide greater energy per gallon of fuel, but they are believed to contribute to higher emissions of particulate and NOx, and they tend to have lower cetane numbers. In the early 1990s, the California Air Resources Board (CARB) implemented standards that limited the aromatic content of diesel fuels sold in California to 10 percent. CARB later allowed the aromatic content to be higher if fuel producers could show that their fuels produced equivalent or lower emissions than the low aromatic fuel.

The Fuel Should Not Deviate from the Design Fuel

Fuel viscosity is specified in the standard for diesel fuel within a fairly narrow range. Hydrocarbon fuels in the diesel boiling range easily meet this viscosity requirement. Most diesel fuel injection systems compress the fuel for injection using a simple piston and cylinder pump called the plunger and barrel. In order to develop the high pressures needed in modern injection systems, the clearances between the plunger and barrel are approximately one ten-thousandth of an inch. In spite of this small clearance, a substantial fraction of the fuel leaks past the plunger during compression. If fuel viscosity is too low, the leakage will correspond to a power loss for the engine. If fuel viscosity is too high, the injection pump will be unable to supply sufficient fuel to fill the pumping chamber. Again, the effect will be a loss in power. If fuel viscosity is extremely excessive, as is the case with vegetable oils, there will be a degradation of the spray in the cylinder causing poor atomization and the production of black smoke.

The Fuel Should Be Intrinsically Safe

Diesel engine operators are accustomed to treating diesel fuel as if it were nonflammable. The volatility of both No. 1 and No. 2 diesel fuel are low enough that the air-vapor mixture above the fuel is below the flammability limit. The property that characterizes this behavior is the *flash point*. The temperature at which the fuel will give off enough vapor to produce a flammable mixture is the point. This temperature is 125°F to 150°F for diesel fuel and below -40°F for gasoline.

CHAPTER 3

Diesel Fuel Specification

ASTM Specifications for Diesel Fuel Oils (D975-05)

Diesel fuel is characterized in the U.S. by the ASTM standard D 975. This standard identifies seven grades of diesel fuel; three grades under the classification No. 1-D, three grades under No. 2-D, and No. 4-D grade. These are described as follows.

Grade No. 1-D (S15, S500, and S5000): A light distillate fuel for applications requiring a higher volatility fuel for rapidly fluctuating loads and speeds as in light trucks and buses. The specification for this grade of diesel fuel overlaps with kerosene and jet fuel, and all three are commonly produced from the same base stock. One major use for No. 1-D is to blend with No. 2-D during winter to provide improved cold flow properties. S15 grade fuel contains 15 ppm sulfur, S500 grade fuel contains up to 500 ppm sulfur and S5000 may contain up to 5000 ppm (0.5 percent) sulfur. Starting in 2006, S15 is required for on-highway vehicles.

Grade No. 2-D (S15, S500, and S5000): A middle distillate fuel for applications that do not require a high volatility fuel. Typical applications are high-speed engines that operate for sustained periods at high load. As with 1-D fuel, S15 grade fuel contains 15 ppm sulfur, S500 grade fuel contains up to 500 ppm sulfur and S5000 may contain up to 5000 ppm (0.5 percent) sulfur. Starting in 2006, S15 is required for on-highway vehicles. 2-D, or No. 2 diesel fuel, is more dense than 1-D so it provides more energy per gallon. This makes it the preferred fuel for on-highway trucks.

Grade No. 4-D: A heavy distillate fuel that is viscous and may require fuel heating for proper atomization of the fuel. It is used primarily in low and medium speed engines.

ASTM D975 specifies the property values shown in **Table 5** for these grades of diesel fuel. The surprising aspect about ASTM D 975 is how few requirements are actually included. The standard says nothing about the composition of the fuel or its source. It only defines some of the property values needed to provide acceptable engine operation and safe storage and transportation. Each of the properties has an ASTM standard that defines the methods used to measure the property. So, the individual standards define the measurement methods and ASTM D 975 defines the allowable value. The individual properties that are defined for diesel fuel are described below along with some properties that perhaps should be included. The next chapter will describe the actual test procedures for the measurement of each property.

Flashpoint – Diesel fuel flashpoints are measured with ASTM D93. The flash point is the lowest temperature at which a combustible mixture can be formed above the liquid fuel. It is dependent on both the lean flammability limit of the fuel and the vapor pressure of the fuel constituents. The flash point is determined by heating a sample of the fuel in a stirred container and passing a flame over the surface of the liquid. If the temperature is at or above the flash point, the vapor will ignite and an easily detectable flash can be observed. No. 2 diesel fuel is required to have a flashpoint above 52°C (125°F) and No. 1 diesel fuel must have a flashpoint above 38°C (100°F). Some special applications, such as marine fuels, may require higher flash points. Biodiesel's flash point is usually well above 150°C.

Property	Grade No.1-D	Grade No.2-D	Grade No. 4-D
Flashpoint °C, min.	38	52	55
Water and sediment, % vol, max.	0.05	0.05	0.50
Distillation temp. °C , 90%			
Min.	–	282	–
Max.	288	338	–
Kinematic viscosity, mm^2/s at 40°C			
Min.	1.3	1.9	5.5
Max.	2.4	4.1	24.0
Ramsbottom carbon residue, on 10%, %mass, max.	0.15	0.35	–
Ash, %mass, max.	0.01	0.01	0.10
Sulfur, %mass, max.	S15: 0.0015 S500: 0.05 S5000: 0.5	S15: 0.0015 S500: 0.05 S5000: 0.5	2.00
Copper strip corrosion, Max. 3 hrs at 50°C	No. 3	No. 3	–
Cetane number, min.	40	40	30
One of following properties must be met:			
(1) cetane index	–	–	–
(2) Aromaticity, % vol., max.	–	–	–
Lubricity	520	520	–
Cloud point, °C, max.	Determined by local climate.		
	Should be 6°C higher than the tenth percentile minimum ambient temperature for the region. For Iowa:		
	Month	10th % min. temp.	
	Oct.	-2°C	
	Nov.	-13	
	Dec.	-23	
	Jan.	-26	
	Feb.	-22	
	Mar.	-16	

Table 5. Requirements for Diesel Fuel Oils (ASTM D 975-05).

Water and sediment – Water and sediment in diesel fuel are measured using ASTM D 2709. This method used to measure water and sediment is only sensitive to free water. It uses a centrifuge to separate small water droplets and particles to be sure they do not exceed 0.05 percent (500 ppm). An earlier method (D 1796) specified that a solvent should be added to allow the measurement to include dissolved water but since diesel fuel will only dissolve about 50 ppm of water, this has little impact on whether the fuel exceeds the specification value. This test is particularly important when working with biodiesel because biodiesel is usually water-washed to remove traces of soap and free glycerol. Vacuum drying is usually needed to remove residual water following the washing process.

Distillation curve – The distillation curve is determined by relating the fraction of a fuel sample that is removed by heating a fuel sample to progressively higher temperatures. The procedure is specified in ASTM Standard D 86. Typically, the curve is characterized by the initial point, the temperature at which the first drop of liquid leaves the condenser, the temperatures at each 10 percent of the liquid, and the end point. ASTM D 975 indicates that No. 1 diesel fuel should have a 90 percent point (90 percent of the fuel components boil below this temperature) that is no higher than 288°C. No lower limit is specified for No. 1 fuel. No. 2 diesel fuel has a lower limit of 282°C for the 90 percent point and an upper limit of 338°C. Since diesel fuel consists of hundreds of different compounds, a distillation curve provides important information about the composition of the fuel. However, biodiesel usually only contains 4 to 5 major compounds that all boil at about the same temperature. In addition, the boiling temperature is so high at atmospheric pressure that the biodiesel compounds usually decompose (crack) during the distillation test. Distillation tests following ASTM D 86 are not appropriate for biodiesel. The ASTM standard D 6751 specifies a distillation test although it recommends ASTM D 1160, which is conducted under vacuum. While this test will allow the biodiesel to be distilled without decomposing, the procedure specified in the technique for converting the distillation curve back to atmospheric pressure is only valid for petroleum products and should be used with caution for biodiesel.

Kinematic viscosity – Viscosity is a measure of a fluid's resistance to flow. The greater the viscosity, the less readily the liquid flows. The viscosity of petroleum oils is a strong function of temperature with the viscosity decreasing as the temperature increases. ASTM D 445 is a standard test procedure for determining the kinematic viscosity of liquids. Biodiesel is more viscous than No. 2 diesel fuel but only by a small amount. Depending on feedstock and amount of oxidation, biodiesel viscosity will vary between 4.0 and 6.2, while No. 2 diesel fuels tend to fall in the a narrower range of 2.4 to 2.6.

Ash – Ash consists of the residue left when the fuel is heated to a sufficiently high temperature that combustible material burns and leaves as CO_2 and H_2O. The procedure is described in ASTM D 482. It primarily consists of inorganic compounds and their oxides. These materials can be abrasive and contribute to engine wear. Diesel fuels are generally very low in ash and ASTM D 975 allows no more than 0.01 percent by weight.

Sulfur – Before 1993, No. 2 diesel fuel was allowed to contain up to 0.5 percent (5000 ppm) sulfur. This level is still acceptable for off-highway fuels, but after 1993, on-highway fuels were required to contain less than 0.05 percent (500 ppm) sulfur. In 2006, the sulfur specification for on-highway fuels will be lowered to 15 ppm. This is an extremely low level of sulfur and will be difficult to measure. The current method, D 2622, will need to be changed to a new method when the sulfur level drops.

Copper strip corrosion – Many of the compounds in diesel fuel can be corrosive. Copper and copper compounds tend to be particularly susceptible to chemical attack. As an indicator of the tendency of a fuel to cause corrosion, ASTM D 130 describes placing polished copper strips in the fuel for 3 hours at 50°C. Then the strips are washed in a solvent and compared to standard descriptions of tarnish and corrosion.

Cetane number – The cetane number is an engine-based test defined by ASTM D 613. It is based on a special engine produced by Waukesha Engine Company that is similar to the Octane Test Engine used for rating gasolines. The cetane number relates to the readiness of the fuel to self-ignite when exposed to the high temperatures and pressure in the diesel engine combustion chamber. It is not directly related to the energy content of the fuel. On-highway diesel fuels in the U.S. are required to have cetane numbers of 40 or higher. Most will be in the range of 40 to 45. In Europe, cetane numbers are higher and tend to be in the 50-55 range.

Cloud point – The cloud point is the temperature at which a cloud of wax crystals first appears in a fuel sample that is cooled under conditions described by ASTM D2500.

ASTM D 975 does not specify a specific value for the cold flow requirements of diesel fuel. Instead, it suggests that the cloud point be no more than 6°C higher than the 10th percentile minimum ambient temperature for the month the fuel will be used. The 10th percentile temperature corresponds to the minimum temperature that would be reached no more than 3 days out of 30 for the month. **Figure 4** shows the values of 10th percentile temperatures during the month of January in the U.S. ASTM D 975 contains similar maps for other low temperature months.

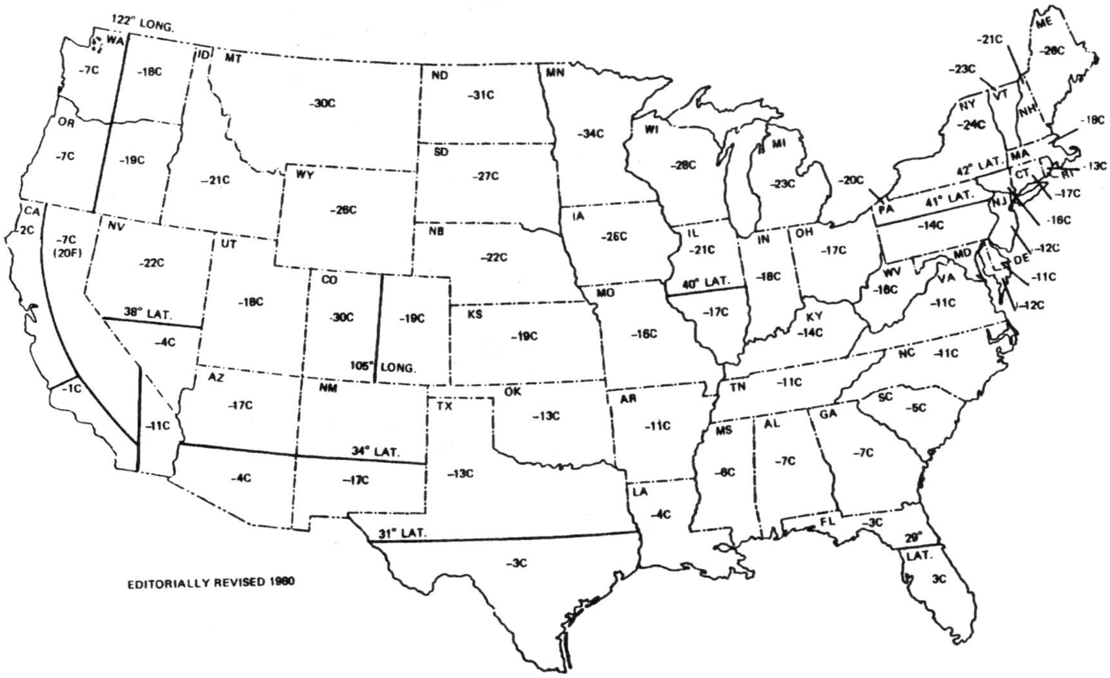

Figure 4. 10th Percentile Minimum Air Temperatures for January (ASTM D975).

Lubricity – The need for diesel fuel lubricity has been recognized for many years. Most early concern focused on the use of No. 1 diesel fuel in place of No. 2 diesel fuel under cold weather conditions. Higher wear rates with No. 1 diesel fuel would be aggravated by No.1 diesel fuel's lower viscosity. However, when the U.S. Environmental Protection Agency (EPA) mandated that the sulfur content of on-highway diesel fuel be lowered from 5000 ppm to 500 ppm in 1993, fuel lubricity captured national attention. There is still disagreement about what specific fuel changes are caused by the sulfur reduction that result in lubricity reduction. Some have suggested that sulfur compounds themselves provide lubricity, others have suggested that nitrogen compounds or naphthenic hydrocarbons are responsible. In any case, there is general agreement that the severe hydrotreating process used by petroleum refineries to remove sulfur results in lower fuel lubricity. Recent EPA regulations will lower the sulfur content of diesel fuel to 15 ppm by 2007. This change is expected to worsen fuel lubricity. The addition of small amounts of biodiesel (0.25-2 percent) to diesel fuel has a dramatic effect on the lubricity of that fuel. Pure biodiesel and high level blends have excellent lubricity.

Lubricity can be defined as: "The property of a lubricant that causes a difference in friction under conditions of boundary lubrication when all the known factors except the lubricant itself are the same. The lower the friction the higher the lubricity." (Kajdas, C., S.S.K. Harvey, and E. Wilusz, *Encyclopedia of Tribology*, Elsevier, New York, 1990.)

Other important fuel properties that are not included in ASTM D 975 (but maybe should be) include:

Fuel stability – A fuel is considered unstable when it undergoes chemical changes that produce undesirable consequences such as deposits, acidity or a bad smell. There are three different types of stability commonly described in the technical literature. *Thermal stability* addresses fuel changes that occur due to elevated temperature. These changes may occur at conditions encountered in modern fuel injection systems as fuel is recirculated through the engine cylinder head and back to the fuel tank. *Oxidative stability* refers to the tendency of fuels to react with oxygen at temperatures near ambient. These reactions are much slower than those that would occur at combustion temperatures, and they produce varnish deposits and sediments. *Storage stability* is also a frequently used term and refers to the stability of the fuel while it is in long-term storage. These terms are not necessarily exclusive terms. For example, oxidative attack is probably one of the primary concerns of storage stability, but storage stability might also involve issues of water contamination and microbial growth. While there is no stability test procedure specified in ASTM D 975, ASTM D 2274 is included in some private purchase requirements.

Heating value, net and gross – There are actually two heating values in common use, the higher, or gross, heating value and the lower, or net, heating value. Both quantities are measured using a calorimeter where the heat transfer from the hot gases resulting from combustion of the fuel with air is measured as the gases are cooled to the initial temperature of the reactants. The procedure is described in ASTM D 240. The higher heating value assumes that all of the water in the products is condensed liquid while the lower heating value assumes all of the water is present as vapor, even though the product temperature may be below the dew point temperature. The lower heating value is the most common value used for engine applications. It is used as an indicator of the energy content of the fuel. In general, the higher the heating value of the fuel, the less fuel that will be required to do a given amount of work. Biodiesel generally has a lower heating value that is 12 percent less than No. 2 diesel fuel on a weight basis. Since the biodiesel has a higher density, the lower heating value is only 8 percent less on a volume basis.

CHAPTER 4

Basic Organic Chemistry

As demonstrated in the preceding chapter, biodiesel production involves many chemical processes. In order to facilitate comprehension of the organic chemistry underlying biodiesel, its feedstocks, production and analysis, we will introduce some basic chemistry definitions. Some or parts of some definitions given here are taken from *Webster's Third New International Dictionary*, published by Merriam-Webster, Springfield, MA, 1993.

Basic Chemistry

Chemistry is the science that deals with the composition, structure and properties of substances and of the transformations that these substances undergo. In most cases, those transformations are called *reactions*.

More than 100 *elements* are now known, and most of them occur naturally. An element is a substance that cannot be separated into simpler substances by ordinary chemical means. Elements are arranged in a systematic fashion based on their atomic numbers (see the following definition) and properties in the *periodic table*.

An *atom* is the smallest particle of an element that can exist either alone or in combination with similar particles (other atoms) of the same or of a different element. It is the smallest particle of an element that enters into the composition of molecules (which leads us to the following definition of a molecule). An element is comprised exclusively of atoms having the same *atomic number*. The atoms of each element have a specific *atomic weight*, which is the average relative weight of an element as it occurs in nature referred to some element taken as standard. Atoms are composed of *protons, electrons and neutrons*, which are responsible for the atomic weight. The nucleus of an atom consists of the protons and neutrons. The *nucleus* is the bulk of the weight of the atom. Electrons are smaller and lighter, and they orbit the nucleus. It is <u>very important</u> that a proton has a positive charge and an electron has a negative charge. The number of protons, neutrons and electrons that comprise an atom give the atom its properties and ultimately the properties of the element. Thus, the element with one proton (and one electron) is hydrogen. Adding another proton (and electron as well as neutrons) gives the next element (helium), etc. Carbon has six protons and six electrons, while oxygen has eight protons and electrons.

A *molecule* is the smallest particle of an aggregate of at least two connected atoms of either the same or different elements and has a combination of properties (chemical and physical) that is specific to it. Each molecule has a *molecular weight*, which is the weight of the molecule calculated as the sum of the atomic weights of its constituent atoms.

A *compound* is a chemically distinct substance formed by the union of two or more ingredients (as elements) in definite proportion by weight and with definite structural arrangement.

A *mole* is the quantity of a chemical substance that has a weight in mass units (in grams) numerically equal to the molecular weight or that, in the case of a gas, has a volume occupied by such a weight under specified conditions (as 22.4 liters at $0^{\circ}C$ and a pressure of 760 millimeters of mercury). Thus, the number of molecules (6.022×10^{23}) in one mole of any given substance is equal to the number of molecules in one mole on any other substance. The concept of mole is very important in chemistry for calculating the amounts of compounds that undergo a reaction with each other. Often reactions are influenced by the amounts of compounds reacted with each other, the *molar ratio* (a related term is *stoichiometry*) of a reaction.

Elements are in most cases comprised of the same kind of elementary particles known as atoms. Atoms can form aggregates known as molecules. Molecules can consist of atoms of many different elements or of atoms of one element. If a molecule is comprised of atoms of different elements, it is also known as a compound.

Each element has a symbol, for example H for hydrogen (hydrogen is the lightest element, each atom consisting only of one proton and one electron, see earlier notation; a hydrogen atom with the electron removed is referred to as a proton that can be written as H^+) and O for oxygen, so that a *chemical formula* for each compound can be written. If more than one atom of an element is in the compound, a subscript number indicates that. Thus water, as a compound comprised of two hydrogen atoms and one oxygen atom, has the molecular formula H_2O. In other words, each water molecule is composed of two hydrogen atoms and one oxygen atom.

Organic Chemistry

Although there are many compounds that bridge the gap, most compounds are still classified either as *inorganic* or *organic*. The reason goes far back into the history of chemistry, when it was assumed that all compounds came either from inanimate matter (*inorganic compounds*) or living matter (*organic compounds*). With some exceptions (an example is carbon dioxide, CO_2) that will not concern us here, *organic chemistry* is the chemistry of the element carbon (symbol C). If organic chemistry deals only with the chemistry of one element, then why is it distinguished from *inorganic chemistry*? The reason is that most known compounds (approximately 90 percent) are organic compounds. If you think of the astounding variety of plant and animal life and that, on a basic level, organic compounds are responsible for that life, it becomes clear why there are so many organic compounds. More scientifically, carbon atoms are very versatile in forming compounds with atoms of other elements. So why are we going to deal with some organic chemistry here? The answer is that both biodiesel and conventional diesel fuel (derived from petroleum) are mixtures of organic compounds.

Besides carbon, several other elements are very important to organic chemistry. Three especially important elements are hydrogen, oxygen and nitrogen. Other important elements are phosphorus, sulfur and the so-called halogens (fluorine, chlorine, bromine and iodine). The following table contains some basic information on common elements (except halogens) in organic chemistry.

Name	Symbol	Atomic Number	Atomic Weight
Carbon	C	6	12.011
Hydrogen	H	1	1.008
Nitrogen	N	7	14.007
Oxygen	O	8	15.9994
Phosphorus	P	15	30.974
Sulfur	S	16	32.06

Table 6. Some Elements of Importance to Organic Chemistry.

For biodiesel purposes, carbon, hydrogen and oxygen are especially important. It is a good idea to memorize their atomic weights. For (rough) estimates, it suffices to just use the values H = 1, C = 12, O = 16, etc.

When atoms form molecules, they connect via **bonds**. For organic chemistry, it is extremely important that each carbon atom in a molecule forms four bonds to other atoms. These other atoms can be other carbon, hydrogen or oxygen or atoms of other elements. The bonds can be formed to a mixture of other atoms; for example, one carbon atom can form a bond to one hydrogen, one oxygen and two carbon atoms to meet the requirement of four bonds. When a carbon atom bonds to *four* other atoms, regardless of the nature of these atoms, these bonds are *single bonds*. Hydrogen can form only one bond to other atoms and oxygen can form two bonds to other atoms. It is also extremely important that carbon atoms can form double bonds or triple bonds. Compounds that have double or triple bonds are often called *unsaturated compounds*. (Can you explain why?) Thus, one carbon atom can be connected to an oxygen atom by a single bond, another carbon by a single bond and another carbon by a double bond to meet the requirement of four bonds. Two carbon atoms can also be connected by triple bonds. A carbon atom and an oxygen atom can also be connected by a double bond. (Obviously then, the oxygen atom cannot form a bond with another atom.) This versatility in bonding is responsible for the many different, often complex structures of organic compounds.

As in the case of water discussed earlier, organic molecules can be written as formulas, for example, CH_4 for methane. For more complex molecules, it is a good idea and indeed customary to write structural formulas because of the many possible structures for a given combination of carbon atoms with other atoms (a feature known as *isomerism*). That way, features like double or triple bonds can also be indicated by the corresponding number of lines between carbon atoms or carbon and other atoms. Examples of some simple organic molecules containing only carbon, hydrogen and oxygen that also show the diversity of organic compounds (check to see that there are always four bonds per carbon) are shown in **Figure 5**.

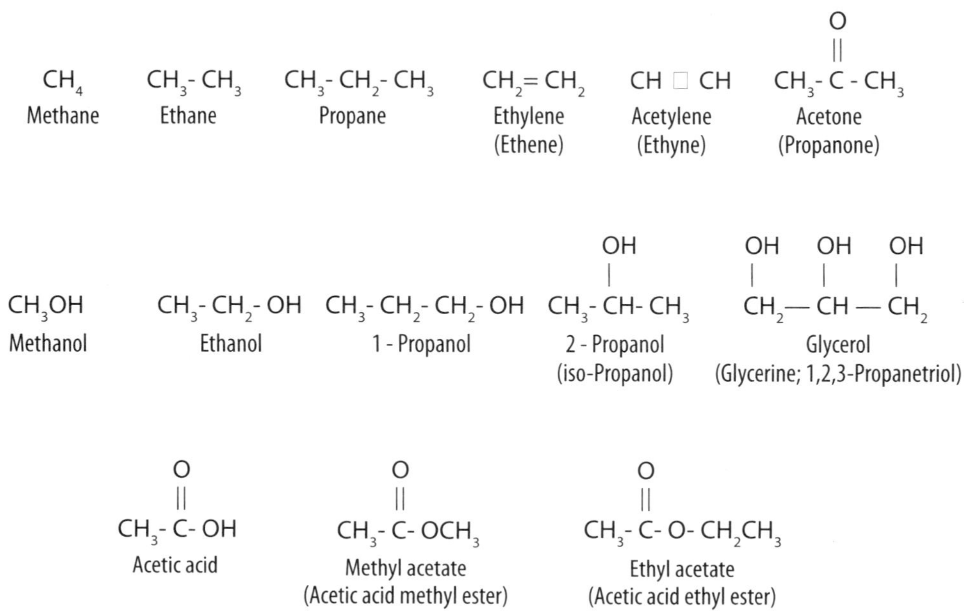

Figure 5. Simple Organic Molecules.

Organic compounds can be categorized into various *classes of compounds*. In the earlier examples, note that methanol, ethanol (the alcohol contained in alcoholic beverages) and 1- and 2-propanol belong to the class of compounds known as *alcohols*. (Note the difference in 1- and 2-propanol; what does the difference in nomenclature indicate?) Their common feature is the presence of an OH group (also known as hydroxy group). A feature such as an OH group is also known as a *functional group*. Besides OH, there are many more functional groups. A functional group usually imparts specific properties and the propensity to undergo certain reactions to a molecule. There are innumerable compounds that belong to the class of compounds known as alcohols. Many organic molecules contain the characteristics of several or even many classes of compounds, usually because they contain several or many different functional groups.

Note how the names of alcohols are derived from the compounds in the top row known as hydrocarbons (hydrocarbons are compounds containing carbon and hydrogen) by dropping the -e and adding the suffix –ol. (Why is the name "glycerol" preferable to "glycerine" for the same compound?) Similarly, note that the CH3 moiety as it is found in methyl esters is called a *methyl* group and CH_2CH_3 is called *ethyl*, again a systematic change in nomenclature. Such changes in nomenclature, especially rational nomenclature, are common in organic chemistry. For some of the compounds, more than one name is given. This results from many compounds having so-called trivial names in addition to their rational ones (historically, the trivial names are older than the rational names). Trivial names are important, because for many organic compounds of more complex structure the rational names become unwieldy, so that the trivial names are much easier to learn and use.

When reading (or writing) the structural formulas of organic compounds, you will often encounter R or R_1, R_2 or R_3, etc. "R" means *organic rest* and is an abbreviation for a part of the molecule which then has to be defined. General formulas for some classes of compounds can be written as shown in **Figure 6.**

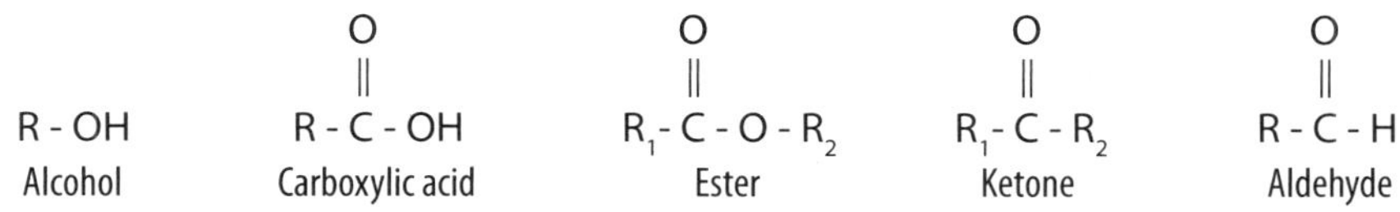

Figure 6. Classes of Organic Molecules.

The specific structural features of organic compounds (such as the hydroxy groups in alcohols) are responsible for the fact that many compounds can react with other compounds to form new compounds. The new may then belong to a different class of compounds. These structural features are also responsible for the differences in physical properties (melting point, boiling point, etc.) of the organic compounds. For example, acetic acid can react with methanol to form methyl acetate (which belongs to the ester class of compounds; esters are formed by the reaction of an acid with an alcohol), in which case water is also formed as a byproduct:

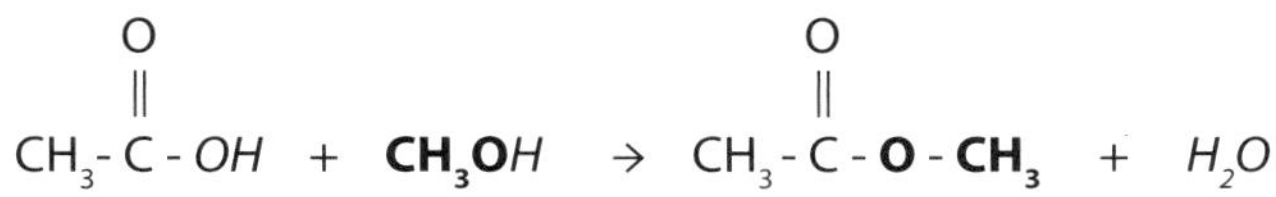

Figure 7. Chemical Reaction to Form Methyl Acetate.

You can also perform the back reaction in which methyl acetate reacts with water to form acetic acid and methanol. When methanol is the alcohol participating in the reaction, the ester formed is a *methyl ester*, when ethanol is the alcohol participating in the reaction, an *ethyl ester* is formed.

Vegetable Oils and Biodiesel

Now we can start to deal with biodiesel. As you know, biodiesel is derived from vegetable oils. The major components of vegetable oils are *triglycerides*. The term *triacylglycerols* is being used more and more, but we will use the classical term in this discussion. *Triglycerides* are esters of *glycerol* (as noted earlier, an alcohol with a hydroxy group on each of its three carbon atoms) with long-chain acids, commonly called fatty acids. **Tables 7 and 8** list the most common *fatty acids* and their corresponding methyl esters.

Note from the comparison of the rational names of the fatty acids with their structural formulas how the position of the double bonds is defined by numbers. The number of carbon atoms is counted by beginning with the first carbon having the functional group defining the fatty compound as acid or ester. As you can see from the previous example (1-propanol and 2-propanol), this way of counting holds for other functional groups as well. The trivial names of fatty acids and their esters are far more commonly used than their rational names.

Fatty acid (trivial name / rational name);	Structure [a)]	Common acronym	Methyl ester (trivial name / rational name)
Palmitic acid / Hexadecanoic acid;	R-$(CH_2)_{14}$-CH_3	C16:0	*Methyl palmitate / Methyl hexadecanoate*
Stearic acid / Octadecanoic acid	R-$(CH_2)_{16}$-CH_3	C18:0	*Methyl stearate / Methyl octadecanoate*
Oleic acid / 9(Z)-octadecenoic acid	R-$(CH_2)_7$-CH=CH-$(CH_2)_7$-CH_3	C18:1	*Methyl oleate / Methyl 9(Z)-octadecenoate*
Linoleic acid / 9(Z),12(Z)-octadecadienoic acid	R-$(CH_2)_7$-CH=CH-CH_2-CH=CH-$(CH_2)_4$-CH_3	C18:2	*Methyl linoleate / Methyl 9(Z),12(Z)-octadeca-dienoate*
Linolenic ac id / 9(Z),12(Z),15(Z)-octadecatrienoic acid	R-$(CH_2)_7$-(CH=CH-$CH_2)_3$-CH_3	C18:3	*Methyl linolenate / Methyl 9(Z),12(Z),15(Z)-octadecadienoate*

a) R = COOH (CO_2H) or $COOCH_3$ (CO_2CH_3); $(CH_2)_7$ = CH_2- CH_2- CH_2- CH_2- CH_2- CH_2- CH_2, etc.

Table 7. Chemical Structure of Common Fatty Acids and Their Methyl Esters.

Fatty acid *Methyl ester*	Formula	Molecular weight	Melting point (°C)
Palmitic acid *Methyl palmitate*	$C_{16}H_{32}O_2$ $C_{17}H_{34}O_2$	256.428 270.457	63-64 30.5
Stearic acid *Methyl stearate*	$C_{18}H_{36}O_2$ $C_{19}H_{38}O_2$	284.481 298.511	70 39
Oleic acid *Methyl oleate*	$C_{18}H_{34}O_2$ $C_{19}H_{36}O_2$	282.465 296.495	16 -20
Linoleic acid *Methyl linoleate*	$C_{18}H_{32}O_2$ $C_{19}H_{34}O_2$	280.450 294.479	-5 -35
Linolenic acid *Methyl linolenate*	$C_{18}H_{30}O_2$ $C_{19}H_{32}O_2$	278.434 292.463	-11 -52 / -57

Table 8. Characteristics of Common Fatty Acids and Their Methyl Esters.

A glance at the differences in the nomenclature will indicate why. Often fatty acids and their esters are referred to, especially in writing, by the abbreviations given in the last column of **Table 7**. (Question: How do those abbreviations arise?)

All double bonds in the rational nomenclature of fatty acids and esters also are defined by a letter (Z), which is the result of the possibility of two kinds of double bonds. These two kinds of double bonds are a result of how the other parts of the molecule are attached in space to the carbon atoms forming the double bond. Besides this technical aspect, we will not discuss any further details of such arrangements of molecules in space except nomenclature. In rational nomenclature, the counterpart to Z is defined by the letter E. However, in the literature the terms *cis* and *trans* also are often used. In organic chemistry, a *cis* double usually, but not always, corresponds to a Z double bond and a trans double bond usually, but not always corresponds to an E double bond. In the case of the fatty acids we are interested in here, cis and Z correspond to each other as do trans and E. However, each of these fatty acids has a counterpart in which the double bonds are trans (or E), but their occurrence (in nature) is far less common. (Have you heard health reports on trans fatty acids?) Under certain conditions, the double bonds can change from Z to E. These conditions include heating and types of reactions called *hydrogenations* (a hydrogenation reaction is the production of more saturated fats from unsaturated ones by adding hydrogen to double bonds).

The two kinds of double bonds cause the compounds to have different physical properties such as melting point. For example, while the melting point of oleic acid is 15°C (compare to that of methyl oleate earlier; note that fatty acids have higher melting points than their corresponding methyl esters), that of elaidic acid (the acid in which the Z double bond in oleic acid is replaced by E; the methyl ester would be called methyl elaidate) is 45°C. The existence of different compounds for the same formula (both oleic acid and elaidic acid are $C_{18}H_{34}O_2$; their methyl esters are $C_{19}H_{36}O_2$) is called *isomerism*. The corresponding compounds are isomers. There are many different kinds of isomerism. A more detailed discussion of isomerism is beyond the scope of this book.

It is extremely important to realize that vegetable oils are mixtures of triglycerides from various fatty acids. The composition of vegetable oils varies with the plant source. Often the terms *fatty acid profile* or *fatty acid composition* are used to describe the specific nature of fatty acids occurring in fats and oils. Several examples were described in earlier chapters, and profiles for common vegetable oils and animal fats were given earlier. The chemical and physical properties of fats and oils and the esters derived from them vary with the fatty acid profile. For example, imagine how mixtures containing larger amounts of methyl palmitate or methyl stearate will solidify at higher temperatures than those containing lesser amounts (see melting points in **Table 8**).

The corresponding esters of those fatty acids in **Tables 7 and 8** with glycerol are known as tripalmitin, tristearin, triolein, trilinolein and trilinolenin. (Exercise: Calculate the molecular weights of these triglycerides.)

In a variation of the formation of esters from acids and alcohols as discussed above, an ester can react with another alcohol. In that case, the new alcohol is derived from the original ester and the new ester is derived from the original alcohol. Thus, an ethyl ester can react with methanol to form a methyl ester and ethanol. This process is called *transesterification*. Transesterification is extremely important for biodiesel. Biodiesel as it is defined today is obtained by transesterifying the triglycerides with methanol. Methanol is the preferred alcohol for obtaining biodiesel because it is the cheapest (and most available) alcohol. However, for the reaction to occur in a reasonable time, a substance called a *catalyst* (catalysts are substances that, often present in small amounts, accelerate the speed of a reaction; in many cases virtually no reaction would occur without a catalyst) must be added to the mixture of the vegetable oil and methanol. The transesterification reaction given in **Figure 8** is a more generic form.

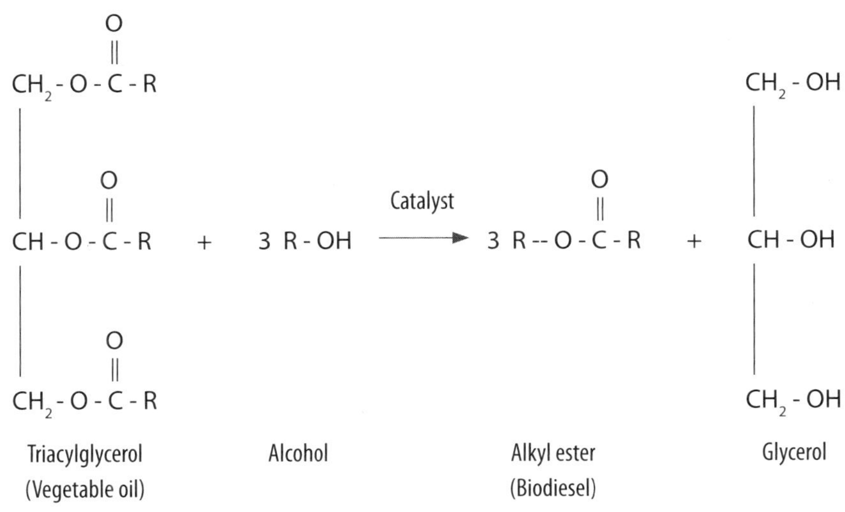

Figure 8. The Transesterification Reaction.

R is a mixture of various fatty acid chains. The alcohol used for producing biodiesel is usually methanol (where the R- corresponds to CH_3).

As noted in the figure caption, R can be any of the fatty acid chains. The number 3 indicates the number of moles needed to satisfy the formation of the methyl esters. When one mole is necessary, the number 1 is usually not written. It is also important to keep it mind that this only formally states the molar ratios of starting materials and products. In practice, the molar ratios may need to vary to obtain a nearly complete reaction.

For the transesterification to occur, usually 6 moles of alcohol are used for every mole of triglyceride, which is more than the equation indicates. The reason is that the reaction is desired to proceed in the direction of the arrow, i.e., to the right. In other terms, the *equilibrium* of the reaction needs to be shifted toward the right side of the equation. As the term "equilibrium" indicates, not all reactions easily proceed to completion, and after some time, the starting materials and reaction products are present in constant amounts (the equilibrium has been attained). In many cases, the fact that a reaction can proceed in the reverse fashion (from right to left in the equation) also plays a role in formation of the equilibrium. To force the equilibrium in the direction of the products (as is almost always desired), one or more parameter(s) of the reaction may need to be changed. Such parameters include the molar ratio as well as others such as temperature, pressure and use of a catalyst.

In accordance with the names of fatty acids and their esters (see **Tables 7 and 8**), the methyl ester of soybean oil is often called *methyl soyate*. The term *soybean oil methyl ester* (SME) is also very common. The same holds for the esters of other vegetable oils. Another common abbreviation that is commonly used is FAME (fatty acid methyl ester).

Besides triglycerides, *mono-* and *diglycerides* can also exist. They are formed as intermediates during the transesterification reaction. This is one of the problems when conducting chemical reactions in general, not only the transesterification reaction. It is almost always the goal of chemical reactions to obtain products that are as pure as possible. However, as discussed, hardly any chemical reaction proceeds to full completion. Therefore, intermediates (in the case of transesterification, the intermediates are partially reacted triglycerides, i.e., the mono- and diglycerides) often can contaminate the final product. Other materials that can contaminate biodiesel are residual methanol (or other alcohol), glycerol and catalyst.

When the transesterification reaction is conducted, you can visually observe that not all materials can be readily mixed with each other. A term often used for this phenomenon is that there are two *phases*. At the start of the reaction, you can notice that methanol and vegetable oil do not readily mix. At the end of the reaction, you can see there are two layers (phases), one consisting mainly of glycerol, the other of the methyl esters. Obviously, glycerol and methyl esters do not mix readily. How readily one compound will dissolve in another depends on the structural features of the compounds, for example, the existence of OH groups. Thus, compounds containing OH groups and those not containing OH groups often will not readily mix. (Question: Why is water effective in washing biodiesel after the transesterification reaction?)

The catalyst used for carrying out the transesterification is usually sodium hydroxide (NaOH) or potassium hydroxide (KOH). These compounds belong to a class of materials known as bases and also are inorganic compounds (inorganic compounds are often used in organic chemistry for carrying out or catalyzing reactions). Other bases are also suitable for the transesterification reaction. The counterparts of bases are known as *acids*. Many acids can also be used as catalysts in the transesterification reaction. However, the base-catalyzed reaction has advantages, such as a higher reaction rate. More details are covered in the production part of this book.

In the soap formation reaction described in **Figure 7**, a fatty acid and base reacted to form a new compound, which was called soap, and water. Compounds such as soap, in which the hydrogen (*proton*) of an acid has been replaced with a metal ion, are often called *salts*. The reason that such compounds exist is that materials such as NaOH (or KOH) can split apart (dissociate) in a fashion that gives Na^+ and OH^- (or K^+ and OH^-) in which the protons and electrons are not evenly distributed, leading to charged particles. Thus, having the same charge, Na^+ or K^+ can replace H^+ here.

Another important aspect of the chemistry of fatty acids and their esters is that the more unsaturated fatty acids and their esters (linoleic, linolenic) can relatively easily react with air (more specifically, the oxygen in air) and form degradation products with time (the time depends also on other factors such as temperature). This is due to the existence of CH_2 between two carbons double-bonded to other carbons (see the structural formulas in **Table 7**).

Why Are Methyl Esters of Fatty Acids Suitable as Diesel Fuel?

Earlier, we briefly discussed compounds such as methane, ethane and propane. These compounds belong to a class of compounds called *alkanes*. Alkanes are compounds consisting only of carbon and hydrogen, and they only contain single bonds. Therefore, the general formula for straight-chain alkanes is CH_3-$(CH_2)_n$-CH_3. Other alkanes (for example, with branching in the chain or with the carbons arranged in a ring) can also exist. However, in each case the carbons are connected only with single bonds.

When dealing with diesel fuels, you may have heard the term *cetane number*, which derives its name from hexadecane (trivial name: cetane), a straight-chain alkane with 16 carbons ($C_{16}H_{34}$). The cetane number is a dimensionless indicator of the ignition quality of a diesel fuel and is conceptually similar to the better-known octane number used for gasoline. Hexadecane is the high-quality standard on the cetane scale and has been assigned a cetane number of 100. A highly branched alkane, 2,2,4,4,6,8,8-heptamethylnonane (HMN; nine carbons in a chain with seven CH_3 groups attached to it besides the two terminal CH_3 groups, thus it also has 16 carbon atoms; try drawing its structure) is the low-quality compound on the cetane scale and has a cetane number of 15. Therefore, long, unbranched alkanes comprise an "ideal" petrodiesel fuel. The analogy to hexadecane as ideal petrodiesel component shows why biodiesel is suitable as an alternative diesel fuel. The fatty acids whose methyl esters are now used as biodiesel also are long-chain compounds similar to long-chain alkanes, such as hexadecane, which make good petrodiesel.

Petrodiesel consists of many components. Besides hydrocarbons, petrodiesel often contains significant amounts of compounds known as *aromatics*. Aromatics are cyclic compounds such as benzene or toluene (see structures in **Figure 9**). The carbon atoms are not shown here explicitly in the structures. They are assumed to be at the intersections of the lines that show the bonds between the carbons.

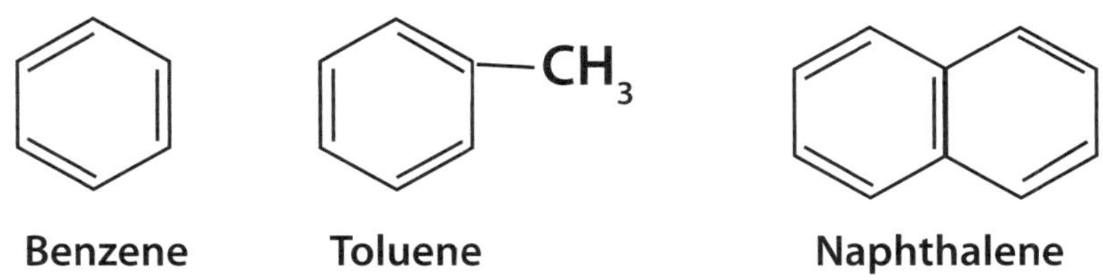

Figure 9. Aromatic Molecules.

Thus, benzene has the formula C_6H_6, and toluene has the formula C_7H_8. The common feature of all aromatic compounds is such a ring structure. There are many aromatic compounds in which the ring structures are fused, for example, naphthalene (formula $C_{10}H_8$). They have low cetane numbers and are undesirable components of petrodiesel. However, they have high densities and thus help elevate the energy contained in a gallon of the fuel. Biodiesel's lack of aromatic compounds is often cited as an advantage.

Compounds in which several benzene rings are fused together (even more than in naphthalene) are termed *polyaromatic hydrocarbons* (PAHs). They are found in exhaust emissions of petrodiesel and, in reduced amounts, of biodiesel fuel.

Why are vegetable oils transesterified to produce biodiesel? This question will mainly be dealt with in other parts of the book, but briefly, vegetable oil methyl esters have lower viscosities (resistance to flow of a liquid) than the parent vegetable oils. Think of honey or syrup, which have high viscosities and flow with difficulty, vs. water or milk, which have low viscosities and flow easily. Compared to the viscosities of the parent vegetable oils, the viscosities of vegetable oil methyl esters are much closer to that of petrodiesel. High viscosity causes operational problems in a diesel engine, such as poor quality fuel injection and the formation of deposits.

The Concept of Acids and Bases

Many compounds, both organic and inorganic, can be classified as acids and bases. The most useful concept of acids and bases as these materials relate to biodiesel is that of *proton donors* (acids) or *proton acceptors* (bases), also often called *Brønsted* (or Lowry) acids and bases. Transferring a proton from an acid to a base is an acid-base reaction. A proton is a hydrogen atom stripped of its electron. Thus, a proton carries a positive charge and is often written H^+.

Whether a compound can behave as an acid or base depends on some of the atoms contained in it. Thus, organic compounds containing nitrogen often have base character, while those containing oxygen in form of OH (hydroxy) groups often have acid character. There are compounds that have the characteristics of both, but that will not concern us here. Also, there are acids and bases in which a proton has been replaced by another particle carrying a positive charge called a *cation*, derived from another element. Note that a charged particle, be it negatively or positively charged, is called an *ion*. Without going into further detail, that acids and bases split (*dissociate*) into ions instead of atoms results from the position of the atoms in that compound in the periodic table of the elements and the corresponding *electron configuration* of those atoms.

Thus, in water (H_2O), one proton (H^+) can be replaced by the positively charged cation of sodium (Na^+) or potassium (K^+) to give NaOH (sodium hydroxide) or KOH (potassium hydroxide). Obviously, the remaining hydroxy part (OH) is then negatively charged to give OH^- (hydroxide). As a result, NaOH and KOH are strong bases as they have a tendency to take up a proton. Similarly, in an alcohol such as methanol (CH_3OH), the proton attached to the oxygen atom can be replaced by, for example, Na^+. The resulting CH_3ONa (sodium methylate; sodium methoxide; sodium methanolate) is a strong base, $CH3O^-$ being the reason.

Similarly, but in the reverse fashion, there are acids, the best known being sulfuric acid (H_2SO_4), hydrochloric acid (HCl) and nitric acid (HNO_3). Acids have a tendency to give off protons, leaving a negatively charged particle called an anion. For example, the anion of sulfuric acid is $HSO4^-$.

It is not too surprising that water can act as an acid and/or a base as water can be seen as H^+ and OH^- giving H_2O. Actually, water reacts with itself as acid and base in the following fashion:

$$H_2O + H_2O \rightleftharpoons H_3O^+ + OH^-$$

The double arrow indicates that a reaction can go in both directions. The concentrations of both H_3O^+ and OH^- in neutral water are very low. In chemical terminology it is then said that the *equilibrium* of the reaction is on the left side of the equation. (Also $H_2O + H_2O$ would usually be written as $2\ H_2O$.)

The example reaction forms the basis of the important concept of the *pH value*. The pH value is defined as the concentration of H_3O^+ (hydronium ion) in an aqueous solution.

It is <u>very important</u> to know that the pH of neutral water is 7, which results from the concentration of H_3O^+ in neutral water. (Question: How does the concentration of OH^- in neutral water compare to that of H_3O^+?) An acidic solution has a pH below 7 while a basic solution has a pH above 7.

The pH value of an aqueous solution can be conveniently measured directly in a matter of seconds using a *pH meter*. There are also strips of paper (*indicator paper*; a simpler version is *litmus paper*) that when held into a solution change color due to an acid-base reaction with the *indicator* on the paper. Comparison of the obtained color with a chart on the box in which the indicator paper is stored then gives the pH.

Thus, an acid can be seen as a compound that produces H_3O^+ when dissolved in water beyond the H_3O^+ present in neutral water anyway. Conversely, a base produces OH^- when dissolved in water beyond the OH^- present in neutral water. (Exercise: Try writing the above equation for dissolving an acid or base in water. For an acid use the generic formula HA and for a base A^-.)

Another important part of the acid-base chemistry is that of *neutralization*. Many chemical reactions, including the transesterification reaction that yields biodiesel, require the use of *catalysts*. Catalysts are compounds that are often present in small amounts that cause a reaction to proceed at a faster rate. Often such catalysts can be acids or bases. That means that even when the reaction is over, the reaction solution will still be acidic or basic, depending on the kind of catalyst used (and the other materials present). Then it is usually necessary to *neutralize* the solution, i.e., the solution needs to attain the neutral pH. Not surprisingly, a basic solution can be neutralized with an acid and, vice versa, an acidic solution can be neutralized with a base. In many cases, a reaction of an acid and a base forms a *salt*, i.e., a product of part of the acid and base (the anion and cation). Actually, common table salt (NaCl) is an example as it can be seen as arising from sodium hydroxide and hydrochloric acid: $NaOH + HCl \rightarrow NaCl + H_2O$.

Many other salts are possible. When producing biodiesel under basic (more common) or acidic (less common) conditions, neutralization is required. That neutralization may lead to a salt that, although a side product of the process, may even be sold. It may be worthy of consideration to choose the neutralizing acid (or base) for not only its cost but also for the market potential of the salt. However, attention is required to possible contamination of the glycerol by the salt, which may influence the purification, applicability and market potential of the glycerol. The salts may possess different properties, including retention, relative to glycerol.

For application of the acid-base concept to the organic chemistry of fats and oils, biodiesel and the transesterification reaction, it is important to recall that fatty acids are acids (organic acids are often termed *carboxylic acids*), although rather weak ones. The important functionality in the fatty acid molecule is -COOH. The proton can dissociate to give $-COO^-$ and H^+. Thus fatty acids can also form salts, in which the H^+ is replaced by some cation such as Na^+ or K^+. Such salts of fatty acids form the basis of *soaps*, as described earlier.

Fatty acids can also form esters, which react with an alcohol, for example, methanol:

$$\mathrm{R{-}{-}}\overset{\displaystyle \mathrm{O} \atop \|}{\mathrm{C}}\mathrm{-}\mathbf{OH} + \mathit{CH_3O}\mathbf{H} \rightarrow \mathrm{R{-}{-}}\overset{\displaystyle \mathrm{O} \atop \|}{\mathrm{C}}\mathrm{-}\mathit{OCH_3} + \mathbf{H_2O}$$

The italicized and bolded parts of the equation indicate which parts are exchanged. Actually in the esterification of the acid with methanol, it is OH^- that is exchanged for CH_3O^-. Thus, under the right conditions, a carboxylic acid can give off either H^+ or OH^-.

It is also possible to transform one ester into another:

$$\mathrm{R{-}{-}}\overset{\displaystyle \mathrm{O} \atop \|}{\mathrm{C}}\mathrm{-}\mathit{OCH_3} + \mathbf{C_2H_5O}\mathit{H} \rightarrow \mathrm{R{-}{-}}\overset{\displaystyle \mathrm{O} \atop \|}{\mathrm{C}}\mathbf{OC_2H_5} + \mathit{CH_3OH}$$

However as most carboxylic acids, including fatty acids, and alcohols are very weak acids and bases, stronger acids or bases need to be added to their mixture to achieve a reaction. These acids or bases function as the catalysts.

Catalysts

Two of the most commonly used catalysts for transesterification are NaOH and KOH. These catalysts operate by reacting with the alcohol according to the reaction given here (written using methanol and NaOH, but other alcohols and catalysts could be substituted).

$$CH_3OH + NaOH \rightarrow CH_3O\text{-}Na + H_2O \qquad \textbf{(Eq. 1)}$$

Similar to H_2O "consisting" of H^+ and OH^-, CH_3O-Na can be seen as consisting of CH_3O^- (alkoxide; alkylate) and Na^+. CH_3O^- is the species that attacks the ester moieties in the glycerol molecule in the following fashion:

(Eq. 2)

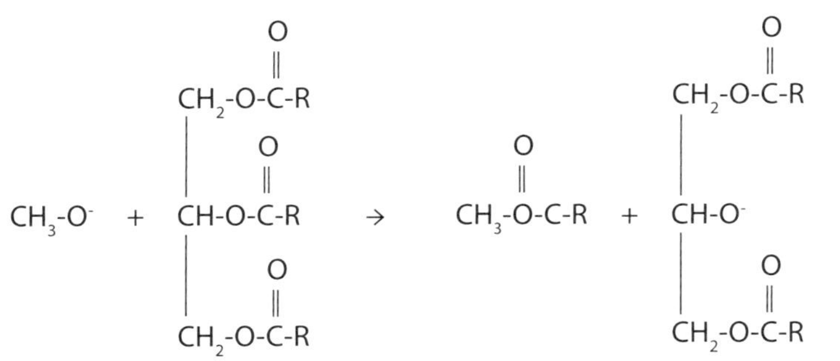

Alkoxide Triglyceride Methyl ester Triglyceride anion

While the ester molecule is complete after this reaction, the anion of the triglyceride needs to pick up a proton to give a stable product (a diglyceride in this case). If this proton is taken from methanol, then the alkoxide catalyst will be recovered as shown in **Eq. 3**.

(Eq. 3)

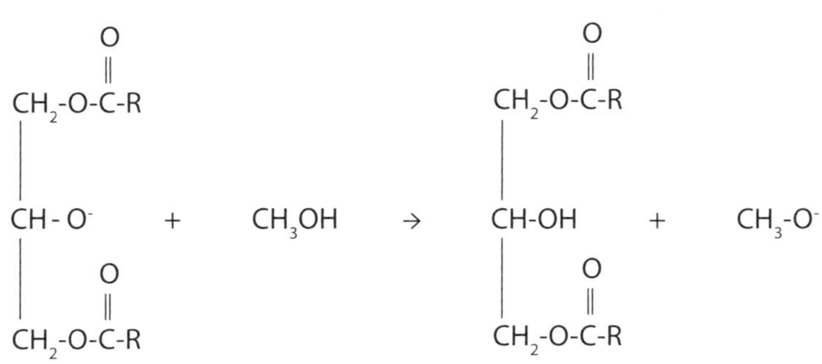

Triglyceride anion **Methanol** **Diglyceride** **Alkoxide**

Several other reactions could be written that would allow the anion of the triglyceride to pick up a proton, such as reaction with free fatty acids and water. The presence of water affects the transesterification negatively because the triglyceride anion in **Eq. 3** will react with the water to form OH^-, which can behave in a fashion similar to $CH_3\text{-}O^-$ but a free fatty acid will result instead of a methyl ester.

(Eq. 4)

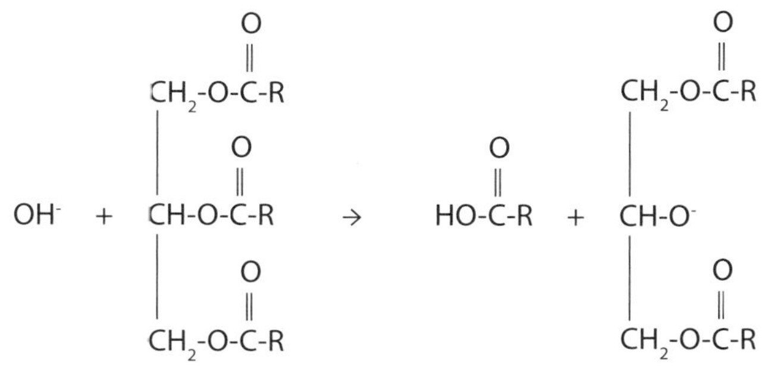

Hydroxide Triglyceride Free fatty acid Triglyceride anion

The free fatty acid can react with the Na^+ to form soap. Some water in the system can be tolerated, because $R\text{-}O^-$ is a stronger base than OH^-, so the transesterification reaction occurs at a higher rate than the "saponification" of glycerol leading to free fatty acids. The equilibrium of the following reaction (formation of free fatty acids from the ester) is far on the left side of the reaction equation:

(Eq. 5)

Alkyl ester Free fatty acid

As a result of the slow reaction rate, only very minor amounts of free fatty acids are formed during transesterification if the reaction is free of water at the beginning. Another aspect of the weaker basicity of OH^- vs. $R\text{-}O^-$ is that the ester moieties in the triglyceride molecules will react less with OH^- than $R\text{-}O^-$. However, when too much catalyst (or water) is present, **Eq. 5** becomes more prevalent and causes enhanced formation of mono- and diglyceride molecules instead of reactions with all positions in the glycerol backbone. Thus, the effect of too much catalyst or too much water leads to the same result, namely, enhanced formation of undesirable mono- and diglycerides as well as free fatty acids. Too much catalyst can lead to soap formation when conditions encourage free fatty acid production.

Because of the possibility of the reactions described above leading to free fatty acids and mono- and diglycerides, direct use of sodium or potassium alkylate (R-ONa or R-OK; the alkylate moiety must correspond to the R- moiety in the alcohol) as catalysts is becoming of greater interest. The reaction of alcohol and XOH given above (**Eq. 1**) cannot occur in this case (assuming, of course, that the reaction system is free of water). Instead, the transesterification according to **Eq. 2** can occur directly.

CHAPTER 5

The ASTM Specification for Biodiesel Fuel

The ASTM system is the basis for defining product specifications and measurement methods for most segments of the fuels and industrial products market in the U.S. ASTM D 6751 – 03a sets forth the specifications as shown in **Table 9** that must be met for a fatty acid ester product to carry the designation "biodiesel fuel" or "B100." Products that meet the specification, by implication, will perform properly as a compression ignition fuel either as B100 or in blends with any petroleum-derived diesel fuel defined by ASTM Specification D 975 Grades 1 -D, 2-D, and low sulfur 1 -D and 2-D.

In this chapter, you will:

1. Become familiar with the structure of ASTM Specification and Methods documents.
2. Learn the Specifications for B100 fuel.
3. Introduce the methods used to measure the performance parameters for B100 fuel.

ASTM Specification Structure

All ASTM Specifications have the same basic format. The "Scope" section describes the specific materials and applications that are covered by the Specification. The "Referenced Documents" section lists all of the ASTM Standards and Methods that are used in conducting the tests to verify that the material meets the specifications of the Standard. The methods provide detailed equipment requirements and experimental methodologies to be used for each parameter measured.

The "Terminology" section consists of a set of definitions explicit to the Standard. The methods to be used for obtaining and protecting samples, test methods, the required property values to meet the standard (typically in the form of limits) and the required units for reporting results are presented as "Requirements."

"Test Methods" is a listing of the required test methods specified for each property to be measured. Often a given test method will involve a number of additional ASTM Specifications and/or Methods to correctly carry out the procedure. In some cases, such as ASTM 6751, there will be appendices to the Standard that specify additional equipment requirements or methods that are not currently a part of the ASTM approved methods.

"Workmanship" describes additional precautions to be followed in conducting the testing. The "Key Words" are included as an aid to cross-referencing and for database searches.

Definition of Biodiesel

Biodiesel is defined in the standard as a fuel comprised of mono-alkyl esters of long chain fatty acids derived from vegetable oils or animal fats, designated B100.

A "mono-alkyl ester" is the product of the reaction of a straight chain alcohol, such as methanol or ethanol, with a fat or oil (triglyceride) to form glycerol (glycerin) and the esters of long chain fatty acids. Biodiesel can be used as B100 (neat) or in a blend with petroleum diesel. A blend of 20 percent biodiesel with 80 percent petrodiesel, by volume, is termed "B20." A blend of 2 percent biodiesel with 98 percent petrodiesel is "B2," and so on.

Property	Method	Limits	Units
Flashpoint, closed cup	D 93	130 min	°C
Water and sediment	D 2709	0.050 max	% volume
Kinematic viscosity, 40°C	D 445	1.9 – 6.0	mm^2/s
Sulfated ash	D 874	0.020 max	wt. %
Total sulfur	D 5453	0.05 max	wt. %
Copper strip corrosion	D 130	No. 3 max	
Cetane number	D 613	47 min	
Cloud point	D 2500	Report to customer	°C
Carbon residue	D 4530	0.050 max	wt. %
Acid number	D 664	0.80 max	mg KOH/g
Free glycerin	D 6584	0.020	wt. %
Total glycerin	D 6584	0.240	wt. %
Phosphorus	D 4951	0.0010	wt. %
Vacuum distillation end point	D 1160	360°C max, at T-90	% distilled
Storage stability	To be determined	To be determined	To be determined

Table 9. ASTM D 6751 – 03a Requirements, for Grade S 15.

The flashpoint is defined as the "lowest temperature corrected to a barometric pressure of 101.3 kPa (760 mm Hg), at which application of an ignition source causes the vapors of a specimen to ignite under the specified conditions of test." This test, in part, is a measure of residual alcohol in the B100. The flashpoint is also a determinant for flammability classification of materials. B100's typical flashpoint is > 200°C, classifying it as "nonflammable."

Water and sediment is a test that determines the volume of free water and sediment in middle distillate fuels having viscosities at 40°C in the range 1.0 to 4.1 mm2/s and densities in the range of 700 to 900 kg/m3. This test is a measure of cleanliness of the fuel. For B100, it is particularly important because water can react with the esters, making free fatty acids, and can support microbial growth in storage tanks.

Kinematic viscosity is the resistance to flow of a fluid under gravity. (Equal to the dynamic viscosity/density.) The kinematic viscosity is a basic design specification for the fuel injectors used in diesel engines. Too high a viscosity and the injectors do not perform properly.

Dynamic viscosity is the ratio between applied shear stress and rate of shear of a liquid.

Density is the mass per unit volume of a substance at a given temperature.

Sulfated ash is the residue remaining after a sample has been carbonized, and the residue subsequently treated with sulfuric acid and heated to a constant weight. This test monitors the mineral ash residual when a fuel is burned. For biodiesel, this test is an important indicator of the quantity of residual metals in the fuel that came from the catalyst used in the esterification process.

Total sulfur method covers the determination of total sulfur in liquid hydrocarbons, boiling in the range from approximately 25-400°C, with viscosities between approximately 0.2 and 20 cSt (mm2/s) at room temperature. Vegetable oil feedstocks typically have very little sulfur, but this test is an indicator of contamination of protein material and/or carry-over catalyst material or neutralization material from the production process.

Copper strip corrosion relates to the detection of the corrosiveness to copper of fuels and solvents. This test monitors the presence of acids in the fuel. For B100, the most likely source of a test failure would be excessive free fatty acids, which are determined in accordance with an additional specification.

Cetane number is a measure of the ignition performance of a diesel fuel obtained by comparing it to reference fuels in a standardized engine test. Cetane for diesel engines is analogous to the octane rating in a spark ignition engine. It is a measure of how the fuel will ignite in the engine. For B100, the cetane number can be very accurately predicted using the ester's composition. It is unlikely that an individual producer will ever run cetane tests on-site.

Cloud point is defined as the temperature at which a cloud of wax crystals first appears in a liquid when it is cooled down under conditions prescribed in this test method. The cloud point is a critical factor in cold weather performance for all diesel fuels. The chemical composition of some biodiesel feedstocks leads to a B100 that may have higher cloud points than desired. The cloud point, however, is another parameter that can be predicted accurately with knowledge of the esters composition, but producers are advised to be able to perform this test.

Carbon residue in petroleum products is the part remaining after a sample has been subjected to thermal decomposition. The carbon residue is a measure of how much residual carbon remains after combustion. This is particularly important in diesel engines because of the possibility of carbon residues clogging the fuel injectors. The most common cause of excess carbon residues in B100 is an excessive level of total glycerin. Total glycerin is also measured directly using an additional procedure specified in this Standard.

Acid number tells the quantity of base, expressed as milligrams of potassium hydroxide per gram of sample, required to titrate a sample to a specified end point. The acid number is a direct measure of free fatty acids in B100. The free fatty acids can lead to corrosion and may be a symptom of water in the fuel.

Free glycerin is the glycerin present as molecular glycerin in the fuel. Free glycerin results from incomplete separation of the ester and glycerin products after the esterification reaction. This can be a result of incomplete washing and may be accompanied by incomplete alcohol removal and a lowered flashpoint. The free glycerin is a source of carbon deposits in the engine because of incomplete combustion.

Total glycerin is the sum of free and bonded glycerin. Bonded glycerin is the glycerin portion of the mono-, di- and triglyceride molecules." Elevated total glycerin values are indicators of incomplete esterification reactions and predictors of excessive carbon deposits in the engine. The terms "glycerin" and "glycerol" are used interchangeably.

The phosphorus test covers the quantitative determination of barium, calcium, copper, magnesium, phosphorus, sulfur and zinc in unused lubricating oils and additive packages. In the case of B100, phosphorus can come from incomplete refining of the vegetable oil and from bone and proteins encountered in the rendering process.

Vacuum distillation end-point testing covers the determination, at reduced pressures, of the range of boiling points for petroleum products that can be partially or completely vaporized at a maximum liquid temperature of 400°C.

Petroleum fractions have tens to hundreds of individual compounds mixed together. The distillation curves are used to characterize the broad chemistry of a given crude oil source. In B100, there are at most 10 different esters present, and they can be identified using gas or liquid chromatography. The same chromatography that determines free and total glycerin can determine the esters composition in the B100. This composition allows calculation of the T-90 point for the fuel, without having to perform the testing for every batch of product. The T-90 point is the temperature at which 90 percent of the liquid volume has been distilled.

Storage stability testing is done because all fuels are subject to degradation over time when they are stored. This degradation may be due to microbial action, water intrusion, air oxidation, etc. The standard and the test methods for determining storage stability for B100 are still in the development stage within the ASTM process, however, ASTM D 6751 - 03a provides a number of guidelines for the long-term storage of biodiesel.

ASTM Testing Procedures

ASTM Standards are based upon the use of a number of ASTM Methods to perform the property measurements. The Methods provide very explicit directions with respect to the equipment, the measurements and the data analysis. Standardization of procedure and reproducibility of results among laboratories is the goal of the process. Each test method outlines a specific procedure to perform a property measurement, and specifies the equipment to be used and the data analysis techniques to be followed. The goal of the Methods is to define procedures that will give identical results for the same sample in any laboratory that follows the ASTM procedures in making the measurement.

It is common for a Method to reference a number of additional Standards and Methods. In order to conform to ASTM requirements, all elements incorporated into a given test must meet all of the Standards and Methods of that test.

CHAPTER 6

An Introduction to the Liquid Fuels Production, Marketing and Regulatory Systems

Biodiesel producers and users are inextricably tied to the petroleum industry. Biodiesel fuel is mixed with petroleum derived petrodiesel, marketed through the conventional petroleum marketing system and used in engines designed to operate on petrodiesel. The biodiesel blends, whether B2 to B5 (2-5 percent biodiesel in 98-95 percent petrodiesel), B20 or B100, are subject to the same engine performance and emissions expectations as petrodiesel. The fact that the original compression ignition (diesel) engines were designed to operate on peanut oil is historically interesting, but the reality of today's liquid fuel business is that as biodiesel producers, we are competing in a petroleum dominated market.

In this chapter, you will learn to:

1. Describe the five essential characteristics of the liquid fuels industry and discuss how they came to be.
2. Describe the production and marketing pathway followed by petrodiesel and contrast it with the production of biodiesel.
3. Describe the performance factors and regulations that drive the quality and use of diesel fuels.
4. Examine briefly diesel fuel markets and how biodiesel fits into these markets.

Origins of the Petroleum Liquid Fuels Industry

The liquid fuels industry has its roots in the production of lighting for households, industry and towns. Lamps that used animal fats, vegetable oils and, rarely, crude petroleum as fuels have a history more than 5,000 years old. The lamps provided poor light and were smoky, but were better than torches and less expensive than candles for common use.

Petroleum has an equally long history of use by humans, but more for medicinal properties, as a lubricant, a sealant for ships, fuel for torches and other applications. The oil was derived from natural seeps that occur in many parts of the world.

The first true liquid fuels industry was the recovery of sperm whale oil that began in the 1700s. The industry flourished, particularly in the United States. However, by about 1850, the population of sperm whales had declined significantly, and the whaling trade was becoming less profitable. With increasingly urban, industrial-age populations, a new source of liquid fuel was needed.

With the decline in availability of whale oil, there was a search for new sources of lighting fuels. About this same time, chemists in Galacia developed a method for refining the crude petroleum from the seepage springs into a clear liquid that burned with a bright, non-smoky flame. This fraction was called "kerosene." The race for oil was on.

In August 1859, Edwin Drake completed the first oil well in western Pennsylvania. The result was a chaotic rush to drill, refine and market this new lighting fuel that quickly spread throughout the world.

The rush for oil was largely dominated by Americans, with the British, Russians and Romanians also significant players. However, one man, John D. Rockefeller, gathered enough power to define the characteristics of the industry that continue until today. These characteristics are an industry that is:

- Vertically integrated
- Capital intensive
- Competitive/collaborative
- Standards driven
- Multinational

Industry Characteristics - *Vertically Integrated*

The oil boom came shortly after the Gold Rush of 1849 and had many similarities to it. With each new strike, hundreds of seekers drilled as many holes as they could, as fast as they could. As much oil was spilled as was produced, refined and shipped. The new fields often played out very quickly, because there was no understanding of how the oil was situated underground or how to most effectively recover it. The most important persons in the Oil Patch were the "wildcatters" who found techniques to identify promising locations for drilling.

At the same time, the refining techniques were not much more advanced than the distillation of moonshine whiskey. In fact, some stills were pressed into service to recover kerosene. A kerosene product that had not been refined enough contained too much of the more flammable fraction, gasoline, and too often exploded in the lamps of unknowing customers.

The kerosene product was, at first, transported in modified water wagons and marketed door-to-door. As the industry grew, there was a growing partnership with the railroads that defined big business in the late 1800s and early 1900s. One outcome of this alliance was the Sherman Anti-Trust Act of 1890 that limits the level of consolidation within an industrial sector. An additional innovation for transport was the use of pipelines to move crude oil from the oil field to larger, more sophisticated refineries some distance away.

Rockefeller created and dominated the petroleum industry. His influence continues today. He was first a marketer, with the promise that Standard Oil products would have a high, consistent quality for the consumer. This meant improvements in refining techniques and control of the refining operations. Higher quality refining meant that the refineries were more expensive, so they were consolidated, requiring a system for transporting crude oil to the refinery and refined product to the consumer.

In order to minimize the price and availability boom-bust pattern, control of production sites and control of the amount of crude produced was necessary. Thus, Standard Oil was one of the first, and definitely the largest, fully vertically integrated company in history.

American oil interests were not alone in the world. The British were still an empire, and they actively pursued integrated production, refining and marketing concessions throughout the world. In this era of WWI and the 1920s, Shell Oil, Royal Dutch Oil and several national oil companies were begun.

With the advent of the electric light and centralized power production, coincident with the invention of spark ignition and compression ignition engines, petroleum entered the transportation fuel business. A key figure in establishing this shift was a little-known Secretary of the Admiralty (British) by the name of Winston Churchill. He insisted that the British Navy be petroleum powered instead of coal powered. World War I affirmed his decision.

Industry Characteristics - *Capital Intensive*

Locating new oil fields has become a highly technical science, but the odds of success in a new area are still small. Even drilling in a known field is expensive. Increasingly, new fields are remote, deep and either wet or cold, or both.

Pipeline systems have been built throughout the world to move oil from the source to the refinery or to oil tankers for further transport. Building, maintaining and, occasionally, defending these transport systems is the most expensive single business in the world.

Refining has transformed into refining and petrochemical processing. World-class refineries in the capacity range of 200,000 to more than 500,000 barrels per operating day represent capital investments in the billions of U.S. dollars at a single site.

With the rise of the petrochemical complex, the refiners can sell a fraction of their product within the organic chemicals market at very reasonable profit margins and recover the cost of the raw material from fuel sales. This is one factor in the large variations in fuel price with changes in oil availability or the spot price of a barrel of oil.

Industry Characteristics - *Competitive/Collaborative*

At first, anyone with a drill, a still and enough resources to open a well could become an independent producer. As the industry became more sophisticated, and as Standard Oil grew, the smaller independents were pushed out of refining and distribution. There are still a number of independent wildcatter operations seeking new oil fields, but the cost of leases to explore is prohibitive. The most recent rounds of large corporation consolidations have nearly re-created organizations similar to the original Standard Oil.

The large companies' goal has long been stability in terms of market share, quantity and price. In the early years, the drive for stability led to a number of agreements dividing the world into market regions and, in more recent years, into production regions.

In the early days, the small companies' goal was quick profits. Many of the smaller, but successful, companies merged to create names like Texaco, Sinclair, Conoco and Phillips that have persisted until recently.

The costs of developing and maintaining large production operations overseas led to a number of collaborations among the larger companies to fund the exploration and development of the fields and the transport infrastructure, and to deal with different and occasionally unstable governments.

The costs of developing the scientific foundation for the refining and, later, the petrochemical business led to a unique collaboration. The American Petroleum Institute (API) was founded by the large oil companies. The Institute was, and is, used as a forum to represent the industry, but also as a way of combining resources to discover basic chemical and physical properties and principles that form the foundation for the petrochemical industry worldwide.

The smaller oil companies, specifically in Texas, created another collaboration with long lasting significance. The small companies were developing the rich Texas oil fields in a manner that was destructive for all concerned. The producers united and asked the State of Texas to step in and manage the allocation of production of oil in a manner that was equitable and would minimize the boom/bust trends. The result was the Texas Railroad Commission that continues to function today. The impact of the Commission is felt at the international level, as it was the model for the creation of the Organization of Petroleum Exporting Countries (OPEC) that now represents multi-national interests in petroleum production.

Industry Characteristics - *Standards Driven*

The twin concepts of using quality as the basis for marketing and verifiable standards as a measure of quality largely began with the petroleum industry. The technical orientation of the industry from its early days allowed the industry to respond fairly easily when the market shifted from kerosene for lighting to gasoline, diesel and fuel oil for transportation. The use of specifications based on properties of the fuels that were tied to engine specifications made it possible to use chemical property predictions to predict engine performance.

In 1934, the API began Project 44, a complete characterization of the physical and chemical properties of all the compounds in a "standard" barrel of oil. The project continues today. The results are the foundation of modern physical-organic chemistry.

The need for standardized, repeatable testing methods for the petroleum industry was responsible, in part, for the establishment of the American Society for Testing Materials (ASTM). Today, ASTM continues in its role as a forum for interested parties to establish specifications for the properties and performance of a wide range of materials. An additional function of ASTM is to establish and validate methods to measure the properties used in the standards.

The current trend toward placing emissions and health standards on fuels and engines is a logical result of the standards-driven liquid fuels industry.

Industry Characteristics - *Multinational*

The initial efforts to develop markets in the Far East led to competition based upon European colonial lines, but with U.S. companies and expertise as key participants. For some companies, going international was a way to expand market share and profits without disturbing the stability of the market within the U.S. For the British, expansion around the world was necessary to provide a secure supply of fuel oil for the British Navy.

As the international markets developed, and competitors, such as Russia, Romania, Mexico and later the Middle East, became producers, the issue became one of balancing supply and demand. There were a number of meetings among the industry leaders that established market territories and, in the Middle East, national boundaries.

In a way, the formation of OPEC completed the circle of supply balancing attempts by using the Texas Railroad Commission model to structure its collaboration of producers. At first, OPEC was not particularly effective, but the Oil Embargo of 1973 restructured the international economy, as well as the international petroleum industry by shifting market dominance away from the U.S. and European major companies.

Production and Marketing of Petroleum Products

The path of production of all petroleum-derived liquid fuels is essentially the same. First, there is a phase of exploration to find new fields where oil (or natural gas) can be recovered. The resource is accessed through drilling into the formation where it is trapped, and the resource is produced.

In the early life of an oil or gas field, the naturally occurring pressure in the formation pushes the fluid into and up the well bore. Oil is pumped, in part to control the flow rate to the surface. As a field ages, the pressure drops and the recovery rate slows. Some older wells may only produce a few barrels per day. There are a number of formations where secondary means of recovery, such as water flooding, are used to recover additional oil. There is no risk in finding the wells, but the recovery operation is not always cost effective.

Refining and petrochemical complexes vary from small refiners, with low capacity and few products, to massive complexes. In some areas, such as the Texas Gulf Coast, the chemical complexes are interconnected, sharing products and raw materials to make fuels, polymers and basic chemicals for a wide range of end uses.

For those of us involved in biodiesel, the interface with the liquid fuel industry is at the level of the fuel distributor. Bulk fuel product can be transported to the distribution terminal by pipeline, boat or truck. The biodiesel blends are most commonly created for sale at the terminal.

Fuel distributors usually operate under a franchise with the petroleum company. Their business is to provide fuel to company stations, fleet purchasers and customers on the spot market. Some of the large truck stop chains have their own distributor network, but usually in collaboration with a specific refiner.

It is most typical that the biodiesel producer will deliver product as B100 to the distribution terminal. At the terminal, the biodiesel is added on top of the petrodiesel fuel. Commonly, this will be in the tanker used to move the fuel to market. The two fuels will mix due to the motion if the tanker. The biodiesel is slightly heavier than petrodiesel, so they can be mixed based only on their density differences.

The Refiner's View of Diesel Fuel

In the refinery, the different products from the sequence of distillation towers at the beginning of the operation are re¬ferred to as "cuts." The lightest cut is a mixture of hydrocarbon gases, such as methane, ethane and propane. Butane is an intermediate cut that can be sold as a fuel, converted into other chemical products or added to gasoline in winter. The gasoline and kerosene cuts are next in order. Diesel is a cut below kerosene and above the valuable gas oil cut. The gas oil cut can be reformed into gasoline, increasing the total production.

Sulfur is a poison to the catalysts used to control harmful emissions from internal combustion engines. Crude oil with high sulfur content can be treated to remove the sulfur, but in many cases the sulfur compounds will naturally distill out in the diesel cut. Until recently, there have been minimal restrictions on the sulfur content in diesel fuel. However, the market for diesel fuel is extremely large and mostly for heavy duty truck engines. The emissions from these engines have been increasingly identified as sources of air pollution and possible health risks. New regulations on emissions from diesel engines are key factors in creating a market niche for biodiesel.

Factors Creating the Market for Biodiesel

Increasingly stringent enforcement of the Clean Air Act in the U.S. has led to requirements of lowered sulfur content and fewer par¬ticulate emissions from diesel fuel. There are more questions regarding the long-term health effects of exposure to diesel particulates. Particularly in Air Quality Non-attainment areas, there are questions of how to meet these more stringent requirements. Biodiesel blends add lubricity to low-sulfur diesel fuels and significantly reduce particulate emissions.

The market potential for B2 to B5 blends has been impacted by the requirements that all on-road diesel fuel must now contain less than 15 parts per million sulfur (Ultra-low Sulfur or ULS) diesel. This requirement applies to the United States, Canada and the European Union. ULS diesel has very poor lubricating properties that can be compensated for by using B2 to B5 biodiesel blends.

Diesel Fuel Marketing

The dominant market for diesel fuel is over-road trucking. The fuel supply system for the trucking industry is predomi¬nantly through company-serviced truck stops. The same jobber that delivers gasoline to the truck stop will likely also deliver diesel fuel. Unless the petroleum company and the local jobber approve of the blending and sale of biodiesel blends at their facilities, there will be no biodiesel sales.

The task of the biodiesel producer is to deliver clean, on-specification B100 to the distributor. Strategies to ensure the delivery of top quality biodiesel to the blender include:

- Clean, dry transport tanks.
- Clean, dry storage tanks.
- Making and certifying the specified blend levels.
- Minimizing storage time before transport and sales.
- Requiring high quality petrodiesel blend stock (ASTM 975).

CHAPTER 7

Patent Discussion

Although transesterification is an old and well-known process in organic chemistry, much of the modern chemical techniques were described in patents issued to chemists for DuPont and Colgate-Palmolove-Peet during and immediately after World War II. A careful examination of these patents shows that most of the techniques used since that time were anticipated and described in these patents.

Prior to the 1940s, most soap was made following a traditional process of saponifying fats through the addition of sodium or potassium hydroxide and water. The glycerol freed by this process often remained with the soap and provided some moisturizing benefit. Techniques were available to separate the glycerol, such as salt addition, but they provided a water-contaminated crude product and were not suitable for use with potassium soaps. Because of the high demand for glycerol to produce explosives for the war effort, there was a strong motivation to remove the glycerol during soap production and provide a high-quality, dry product. George Bradshaw of E.I duPont received a patent (#2,271,619) for a process that added about 1.6 times the theoretical amount of an alcohol, such as methanol, which contained 0.1 to 0.5 percent sodium or potassium hydroxide, to an oil or fat. When performed at 80°C, this process provided 98 percent conversion to a high-quality, water-free glycerol and alcohol esters. The esters were then saponified with additional caustic soda to produce soap, and the alcohol was recovered by heating. The residual glycerol and methanol in the alcohol esters could be removed by two or three washes with a small amount of water.

Bradshaw recognized that this process could be conducted in stages:

> "... *the amount (of alcohol) can be reduced substantially by working in steps. For instance, if the theoretical quantity of methyl alcohol is used, the conversion will be about 80 per cent. The glycerine can then be removed and more alcohol and catalyst added. If the alcohol is added in three steps the amount required can be reduced, for instance, from one and six-tenths times theory to one and two-tenths times theory.*"

In his second patent (#2,360,844) Bradshaw says that with three or four steps the amount of alcohol can be reduced to as low as 1.1 times theory.

Bradshaw noted that other alcohols beside methanol could be used for the process including ethanol, propanol, isopropanol, the butanols and the pentanols. Bradshaw further describes how fats containing small amounts of water and free fatty acids could still be processed with this technique by adding additional catalyst. He mentions that a drawback of this approach is that soap will be formed that can inhibit the separation of the glycerol. He suggests acidification of the reaction mixture to split the soaps back to free fatty acids and salts to facilitate the separation. His suggestions for acceptable acids include formic, acetic, propionic, hydrochloric, sulfuric, sulfamic and phosphoric.

Bradshaw's work was followed by an extensive set of patents from researchers at the Colgate-Palmolive-Peet company. These patents describe the same essential process as Bradshaw but add special features to improve the quality or quantity of the product. Each of these patents will be discussed with emphasis given to the unique contributions they make to the art.

H.D. Allen described modifications to Bradshaw's process that were intended to make it easier to implement as a continuous flow process. He raised the temperature of the reaction to the range 80°C to 160°C under enough pressure to keep the alcohol in the liquid phase. Allen noted that either acid or alkali catalysts could be used and that acid catalysts were better suited to feedstocks with high free fatty acid levels since the free fatty acids would be esterified during the process. At the higher temperatures, the reaction times were shortened to 3 to 4 minutes at 135°C with an alkali catalyst and 45 minutes with an acid catalyst. Allen's proposed system for transesterification is shown in **Figure 10** and includes a reaction coil (plug flow reactor) to provide the time needed to complete the reaction.

Although previous researchers had noted the problem with using alkali catalysts with high free fatty acid feedstocks and some had commented on the suitability of acid catalysts for this purpose, G.I. Keim appears to have been the first to propose a sequence of an acid catalyzed reaction to convert the free fatty acids to esters followed by an alkali catalyzed step to convert the triglycerides. Keim notes that the alkali catalysts are 10 to 50 times faster with triglycerides than the acid catalysts. However, the alkali reacts with the free fatty acids to form soap, which causes gelling problems and difficulty separating the glycerol. By starting with the acid to convert the FFAs to esters until the FFA level is at a tolerable amount for the alkali catalyst, then switching to the alkali catalyst, the reaction time is kept short without undesirable soap formation. Keim recognizes that the conversion of the FFAs to esters produces water. He says that removal of this water might be necessary for the reaction to proceed to completion and he says this can be accomplished by absorption with sodium sulfate or by heating. He also observed that transesterification reactions with higher alcohols, beyond methanol, are especially sensitive to water.

Arrowsmith makes a number of additional observations about the alkyl ester production process. He notes that it is desirable to minimize the amount of alkali catalyst used because the amount of soap formed increases with increasing catalyst. He also observed that increasing the amount of excess alcohol could decrease the amount of catalyst required. However, excess alcohol can cause problems with separation of the esters and glycerol. This problem is particularly severe for ethanol and higher alcohols. The separation process can be enhanced by adding water, which will pull the glycerol out of the ester phase. However, now the lower phase is a mixture of glycerol, alcohol and water. Considerable extra energy is required to separate and purify these three compounds, especially if they involve azeotropes as is the case with ethanol and water. Arrowsmith proposes to solve this problem with a water-free process.

Instead of using a washing process to purify the esters, he recommended using vacuum distillation. According to Arrowsmith, the esters and glycerol form an azeotrope with a boiling temperature that is lower than the boiling temperature of either pure component. Distillation is performed at pressures of 1 to 15 mm Hg. Arrowsmith states that the transesterification reaction can be conducted at temperatures as high as 250°C, but that at temperatures above about 140-160°C, the soap can catalyze the reverse reaction and break up the esters to free fatty acids and alcohol. By staying at 130°C, it was possible to attain 30 second reaction times and achieve 97 percent complete reaction.

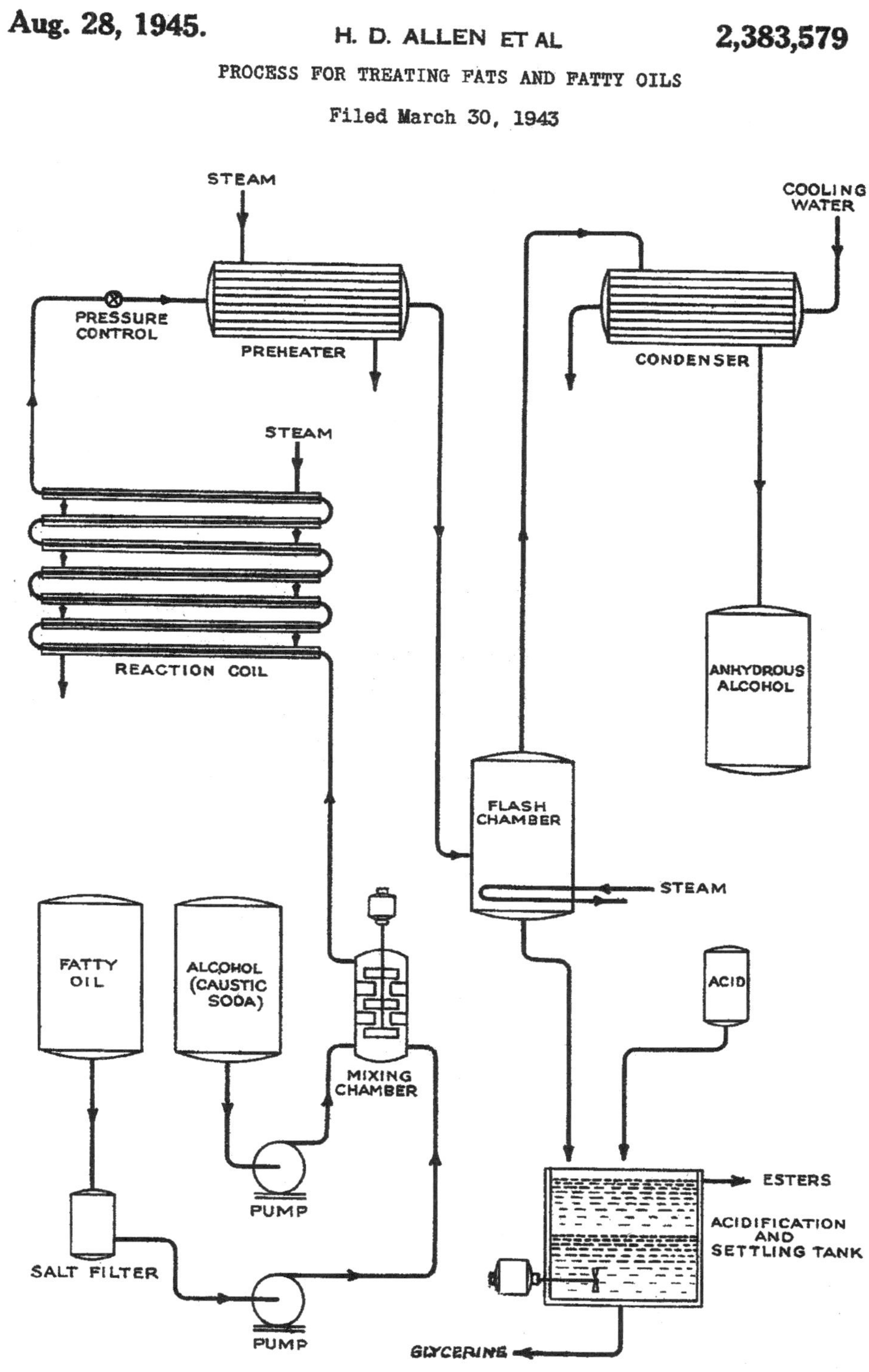

Figure 10. Process Flow Diagram by H.D. Allen, Patent No. 2,383,579.

Trent also noted the tendency of excess alcohol to serve as a cosolvent and inhibit the separation of the ester and glycerol. Trent developed an apparatus for removing the alcohol and separating the esters and glycerol in a single device. Two variations of the device are shown in **Figure 11**. The first device (left side) shows a system where the oil and alcohol/catalyst mixture are mixed and passed through a coil of sufficient length to allow the reaction to reach completion before entering a chamber that is heated by a steam coil. The entire process is at or near atmospheric pressure. The temperature of the chamber is high enough that the alcohol boils off leaving the mixture of esters and glycerol. Acid is added to neutralize the mixture and split the soap into free fatty acids and salts. Then, the ester/glycerol mixture passes through a tube into the lower half of the chamber where it separates. Esters are removed from the top of the lower chamber, and the glycerol is removed from the bottom.

The second variation (right side) is similar except that the reaction is conducted at higher temperature and pressure. Instead of a steam coil to boil off the alcohol, the pressurized mixture is sprayed into the upper half of the container, which acts as a flash chamber. Vaporized alcohol is removed and the ester/glycerol enters the bottom half of the container where it separates as with the other device.

Trent mentions other novel aspects of the alcoholysis process: Trent says that esters produced with higher alcohols (ethanol and above) have a greater tendency to gel from soap formation than with methanol. He claims that better yields are found if reactants are mixed cold and then heated, instead of preheated and then mixed. He says that early increases in temperature increase saponification and recommends using a step-wise increase in temperature from 40-50°C then up to 110-123°C. He also mentions water washing using either batch, concurrent or countercurrent flow systems. In addition to water washing, Trent claims that a liquid-liquid extraction process using a selective solvent that is at least partially immiscible with the esters can be used. He specifically suggests polar solvents such as furfural, sulfur dioxide, nitromethane, methanol, ethanol, ethylene glycol, allyl alcohol, ethyl sulfate, acetaldehyde, acetamide, dichlorodiethyl ether and methyl carbitol. Trent notes that alcohol can be used as a washing agent for the oil to remove free fatty acids and water.

Percy points out that since alcohol is immiscible with the triglyceride feedstock and the final glycerol product is immiscible with the ester layer, this requires a high level of agitation to prevent saponification (soap formation) and separation of the glycerol layer and consequent removal of the alcoholysis catalyst. However, when mono- and diglycerides are present, the reaction mixture will be homogeneous. Percy advocates a process where only enough alcohol or catalyst is added to form mono- and diglycerides. This reaction mixture will be homogeneous. This partially reacted mixture can be held and a stream bled off to which more alcohol and catalyst are added to complete the reaction. Percy claims that if the reaction is conducted at 10 atm and 175°C, the reaction will proceed to completion without catalyst. When using higher alcohols, if ester/glycerol separation is a problem, adding methanol can help encourage the separation.

Dreger observes that ethanol and higher alcohols allow the use of higher reaction temperatures, but he says that this increases the tendency toward soap formation. He advocates lower reaction temperatures of around 40°C. In addition, he recommends the use of small amounts of methanol to ensure complete reaction and to promote separation of glycerol when using ethanol and higher alcohols. Adding 10-30 percent methanol to whatever alcohol is being used will encourage complete reaction and ester/glycerol separation.

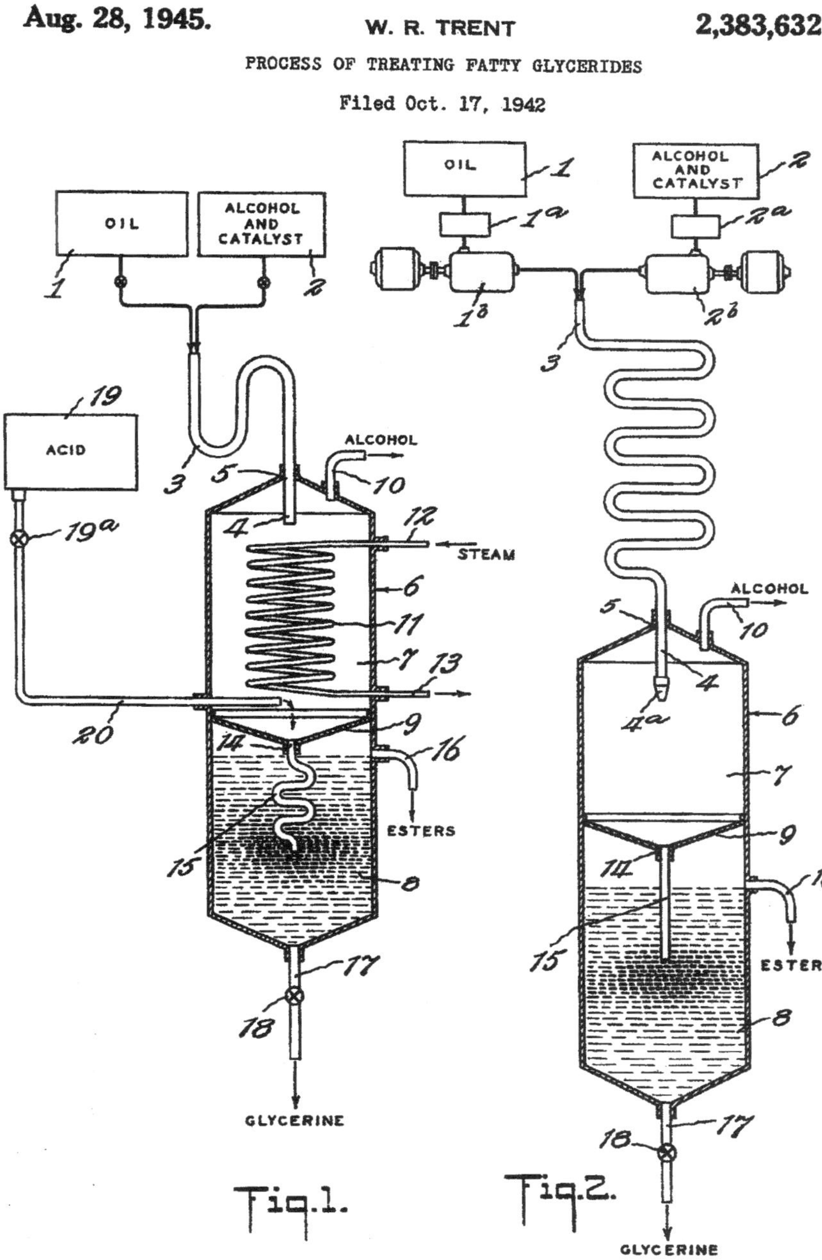

Figure 11. Trent's Combined Reactor and Glycerol Separator.

Patents Referenced

Bradshaw (1942, 1944): E.I. duPont de Nemours & Company, 2,271,619 and 2,360,844.
Allen (1945): Colgate-Palmolive-Peet, 2,383,579.
Arrowsmith (1945): Colgate-Palmolive-Peet, 2,383,580 and 2,383,581.
Percy (1945): Colgate-Palmolive-Peet, 2,383,614.
Keim (1945): Colgate-Palmolive-Peet, 2,383,601 and 2,383,602.
Trent (1945): Colgate-Palmolive-Peet, 2,383,632 and 2,383,633.
Dreger (1945): Colgate-Palmolive-Peet, 2,383,596.

Patent List for Biodiesel

Pre-1940s

Bollman, H. 1921. Process for the Removal of Fatty Acids, Resins, Bitter and Mucilaginous Substances from Fats and Oils. U.S. Patent 1,371,342.

1940s

Bradshaw, G.B., "New Soap Process," Soap, V. 18, May 1942, pp. 23-24, 69-70.

Bradshaw, G.B. and W.C. Meuly, Process of Making Pure Soaps, U.S. Patent 2,271,619. 1942.

Goss, W.H. and H.F. Johnstone, "Process for Producing Fatty Acid Polyhydric Esters from Glycerides," U.S. Patent No. 2,290,609. 1942.

Bradshaw, G.B. and W.C. Meuly, "Preparation of Detergents," U.S. Patent No. 2,360,844. Awarded Oct. 24, 1944.

Buxton, L.O., "Process for Treating Fatty Oils," U.S. Patent 2,345,097. 1944.

Allen, H.D. and W. A. Kline, "Process for Treating Fats and Fatty Oils," U.S. Patent 2,383,579. 1945.

Trent, W.R., "Process of Treating Fatty Glycerides," U.S. Patent 2,383,632. 1945.

Trent, W.R., "Process for Treating Fatty Glycerides," U.S. Patent 2,383,633. 1945.

Glossop, G.A., "Treating Fatty Glycerides," U.S. Patent 2,383,533. 1945.

Glossop, G. A., "Treating Fatty Glycerides," U.S. Patent 2,383,599. August 28, 1945.

Arrowsmith, C.J. and J. Ross, "Treating Fatty Materials," U.S. Patent 2,383,580. 1945.

Arrowsmith, C.J. and J. Ross, "Process for Preparing Fatty Materials," U.S. Patent 2,383,581. 1945.

Keim, G.I., "Treating Fats and Fatty Oils," U.S. Patent 2,383,601. Awarded Aug. 28, 1945.

Keim, G.I. and J. Ross, "Process for Treatment of Fatty Glycerides," U.S. Patent 2,383,602. 1945.

Dreger, E.E., "Method of Treating Fatty Glycerides," U.S. Patent 2,383,596. 1945.

Percy, J.H., "Treatment of Fatty Glycerides," U.S. Patent 2,383,614. 1945.

Colbeth, I. M., "Process of Reacting Glyceride Oils," U.S. Patent 2,469,371. May 10, 1949.

1950s

Sprules, F. J. and D. Price, "Production of Fatty Esters," U.S. Patent 2,494,366. January 10, 1950.

Price, D. and F.J. Sprules, "Transesterification Process," U.S. Patent 2,543,421. 1951.

Kuceski, V. P., "Purification of Alpha, Omega Alkanedioic Acids by Partial Esterification," U.S. Patent 2,824,122. February 18, 1958.

1970s

Hay, R. G., J. G. McNulty, W. L. Walsh, "Process for Preparing Esters," U.S. Patent 3,692,822. September 19, 1972.

Schmerling, L., "Transesterification of Carboxylic Acids," U.S. Patent 4,112,235. 1978.
(tin halide catalyst)

Kawahara, Y. and T. Ono, "Process for Producing Lower Alcohol Esters of Fatty Acids," U.S. Patent 4,164,506. Awarded Aug. 14, 1979.

Mims, S.S., "Esterification and Extraction Process," U.S. Patent No. 4,082,788. 1978.

1980s

Tanaka, Y., A. Okabe and S. Ando, "Method for the Preparation of a Lower Alkyl Ester of Fatty Acids," U.S. Patent 4,303,590, December 1, 1981.

Reierson, R.L., "Process for Removing Glycerin," U.S. Patent 4,360,407. 1982.

Matsukura, T. and Y. Nakagawa, "Method for Manufacturing High Quality Fatty Acid Esters," U.S. Patent 4,371,470. February 1, 1983.

Feldman, J. and J.M. Hoyt, "Extractive Distillation of Alcohol-ester Mixtures," U.S. Patent 4,473,444. 1984.

Stage, H., "Process for Deodorizing and/or Physical Refining of High-boiling Organic Edible Oils, Fats, and Esters," U.S. Patent 4,599,143. July 8, 1986.

Lepper, H. and L. Friesenhagen, "Process for the Production of Fatty Acid Esters of Short-chain Aliphatic Alcohols from Fats and/or Oils Containing Free Fatty Acids," U.S. Patent 4,608,202. August 26, 1986.

Lepper, H. and L. Friesenhagen, "Process for the Production of Fatty Acid Alkyl Esters," U.S. Patent 4,652,406. March 24, 1987.

Stage, H., "Process for Degassing, Dehydrating, and Precut Separation in Straight-run Distillation of Crude Fatty Acids," U.S. Patent No. 4,680,092. July 14, 1987.

Stern, R., G. Hillion, P. Gateau, and J.C. Guibet, "Process for Manufacturing a Composition of Fatty Acid Esters Useful as Gas Oil Substitute Motor Fuel with Hydrated Ethyl Alcohol and the Resultant Esters Composition," U.S. Patent 4,695,411. Awarded Sept. 22, 1987.

Billenstein, S., B. Kukla, and H. Stuhler, "Process for the Preparation of Fatty Acid Esters of Short-chain Alcohols," U.S. Patent 4,668,439. May 26, 1987.

Jeromin, L., E. Peukert, and G. Wellmann, "Process for the Pre-esterification of Free Fatty Acids in Fats and Oils," U. S. Patent 4,698,186. October 6, 1987.

Maruzeni, S., W. Matsumoto, and N. Yasuda, "Reaction Method for Transesterifying Fats and Oils," U.S. Patent 4,874,699. 1989.

1990s

Jeromin, L., E. Peukert, B. Gutsche, G. Wollman, and B. Schleper, "Process for the Continuous Transesterification of Fatty Acid Lower Alkyl esters," U.S. Patent 4,976,892. December 11, 1990.

Matsumoto, W., E. Nakai, T. Nezu, and K. Suzuki, "Process for the Transesterification of Fat and Oil," U.S. Patent 5,089,404. 1992.

Klok, R and H. H. Verveer, "Process for Producing Fatty-acid Lower-alkyl Mono-esters," U.S. Patent 5,116,546. May 26, 1992.

Venter, J. J. and C. McDade, "Transesterification Catalyst," U.S. Patent 5,183,930. February 2, 1993.

Connemann, J., A. Krallmann, and E. Fischer, "Process for the Continuous Production of Lower Alkyl Esters of Higher Fatty Acids," U.S. Patent 5,354,878. October 11, 1994.

Kesling Jr., H. S., L. J. Karas, and F. J. Liotta Jr., "Diesel Fuel," U.S. Patent 5,308,365. May 3, 1994.

Wimmer, T., "Process for the Production of Fatty Acid Esters of Lower Alcohols," U.S. Patent 5399731. March 21, 1995.

Wimmer, T., "Process for Preparing Fatty Acid Esters of Short-chain Monohydric Alcohols," U.S. Patent 5,434,279. 1995.

Bam, N., Drown, D.C., Korus, R., Hoffman, D.S., Johnson, T.G., and Washam, J.M., "Method for Purifying Alcohol Esters," U.S. Patent 5424467. June 13, 1995.

Demmering, G., C. Pelzer and L. Friesenhagen, "Process for the Production of Fatty Acid Lower Alkyl Esters," U.S. Patent 5,455,370, October 3. 1995.

Stern, R., G. Hillion and J. J. Rouxel, "Improved Process for the Production of Esters from Fatty Substances Having a Natural Origin," U.S. Patent 5,424,466. June 13, 1995.

Basu, H.N. and M.E. Norris, "Process for Production of Esters for Use as a Diesel Fuel Substitute Using a Non-Alkaline Catalyst," U.S. Patent 5,525,126. June 11, 1996.

Assmann, G., G. Blasey, B. Gutsche, L. Jeromin, J. Rigal, R. Armengaud and B. Cormary, "Continuous Process for the Production of Lower Alkyl Esters," U.S. Patent # 5,514,820. May 7, 1996.

Johnson, L.A. and E.G. Hammond, "Soybean Oil Ester Fuel Blends," U.S. Patent 5,520,708. May 28, 1996.

Bayense, C. R., H. Hinnekens and J. Martens, "Esterification Process," U.S. Patent 5,508,457. April 16, 1996.

Bradin, D.S., "Biodiesel Fuel," U.S. Patent 5578090. November 26, 1996.

Hunt, T. K., L. Jeromin, W. Johannisbauer, B. Gutsche, V. Jordon and H. Wogatzki, "Recovery of Tocopherols," U.S. Patent 5,646,311, July 8, 1997.

Foglia, T.A., L.A. Nelson, and W.N. Marmer, "Production of Biodiesel, Lubricants and Fuel and Lubricant Additives," U.S. Patent 5,713,965. 1998.

Mittelbach, M., "Method for the Preparation of Fatty Acid Alkyl Esters," U.S. Patent 5849939. Dec. 15, 1998.

Stoldt, S.H. and D. Harshida, "Esters Derived from Vegetable Oils Used As Additives for Fuels," U.S. Patent 5,730,029. 1998.

Granberg, E. P., R. G. Schafermeyer and J. A. Letton, "Method for Purifying an Inert Gas While Preparing Lower Alkyl Esters," U.S. Patent 5,844,111. December 1, 1998.

Hayafuji, S., T. Shimidzu, S. Oh, and H. Zaima, "Method and Apparatus for Producing Diesel Fuel Oil from Waste Edible Oil," U.S. Patent 5,972,057. 1999.

Stern, R., G. Hillion, J. J. Rouxel, and S. Leporq, "Process for the Production of Esters from Vegetable Oils or Animal Oils Alcohols," U.S. Patent 5,908,946. June 1, 1999.

2000s

Nourreddini, H., "Process for Producing Biodiesel Fuel with Reduced Viscosity and a Cloud Point below 32 deg. Fahrenheit," U.S. patent 6,015,440. January 18, 2000.

Stern, R., G. Hillion, and M.N. Eisa, "Process for the Production of Ethyl Esters," U.S. Patent 6,013,817. Awarded Jan. 11, 2000.

Stidham, W. D., D. W. Seaman, and M. F. Danzer, "Method for Preparing a Lower Alkyl Ester Product from Vegetable Oil," U. S. Patent 6,127,560. October 3, 2000.

Noureddini, H., "System and Process for Producing Biodiesel Fuel with Reduced Viscosity and a Cloud Point Below Thirty-two (32) Degrees Fahrenheit," U.S. Patent 6,174,501. January 16, 2001.

Boocock, D.G.B., "Single-Phase Process for Production of Fatty Acid Methyl Esters from Mixtures of Triglycedrides and Fatty Acids," Canadian Patent 2,381,394. 2001.

Peter, S., R. Ganswindt and E. Weidner, "Method for Producing Fatty Acid Esters," U.S. Patent 6,211,390. April 3, 2001.

Ergun, N. and P. Panning, "Method for Producing Fatty Acid Methyl Ester and Equipment for Realizing the Same," U.S. Patent 6,440,057. 2002

Haas, M. J., S. Bloomer and K. Scott, "Process for the Production of Fatty Acid Alkyl Esters," U.S. Patent 6,399,800. June 4, 2002.

Waller, F.J. and R. Quinn, "Method of Producing a Diesel Fuel Blend Having a Pre-determined Flash-Point and Pre-determined Increase in Cetane Number," U.S. Patent application 20020134008. 2002.

Matsumura, M. and S. Murakami, "Method and Equipment of Refining Plant Oil and Waste Vegetable Oil into Diesel Engine Fuel," U.S. Patent 6,364,917. 2002.

Wu, W. T. and J. W. Chen "Method of Preparing Lower Alkyl Fatty Acids Esters and in Particular Biodiesel," U. S. Patent 6,398,707. June 4, 2002.

Yeh, L. I., R. H. Schlosberg, R. C. Miller, J. R. Bateman, and R. F. Caers, "Diesel Fuel Composition," U.S. Patent 6,447,557. September 20, 2002.

Yeh, L. I., R. H. Schlosberg, R. C. Miller, J. R. Bateman, and R. F. Caers, "Diesel Fuel Composition," U.S. Patent 6,458,176. October 1, 2002.

Yeh, L. I., R. H. Schlosberg, R. C. Miller, J. R. Bateman, and R. F. Caers, "Diesel Fuel Composition," U.S. Patent 6,447,558. September 10, 2002.

Barnhorst, J. A., M D. Staley and D. A. Oester, "Transesterification Process," U.S. Patent 6,489,496. December 3, 2002.

Fleisher, C.A., "Continuous Transesterification Process," U.S. Patent application 20030229238. 2003.

Koncar, M., M. Mittlebach, H. Gossler, and W. Hammer, "Method for Preparing Fatty Acid Alkyl Esters," U.S. Patent 6,696,583. 2004.

Boocock, D.G.B., "Process for Production of Fatty Acid Methyl Esters from Fatty Acid Triglycerides," U.S. Patent 6,712,867. 2004.

Branson, J. D., "System and Method for Extracting Energy from Agricultural Waste," U.S. Patent 6,824,682. 2004.

Ginosar, D.M. and R.V. Fox, "Process for Producing Biodiesel. Lubricants, and Fuel and Lubricant Additives in a Critical Fluid Medium," U.S. Patent 6,887,283. 2005.

SECTION TWO

Types of Production

CHAPTER 8

Production Processes

This chapter provides an overview of the steps in the production of biodiesel from preparation of the feedstock to the recovery and purification of the fatty acid esters (biodiesel) and the co-product glycerol (also called glycerin). We will review several chemistries used for transesterification and esterification and different approaches to product preparation and purification.

The emphasis is on the choices to be made in matching feedstock selection, capacity and operating mode with the selection of the basic process chemistry and layout for a specific location. While no specific process technology is favored here, an effort has been made to describe the major approaches currently in use and in development in the industry.

Feedstocks Used in Biodiesel Production

The primary raw materials used in the production of biodiesel are vegetable oils, animal fats and recycled greases. These materials contain triglycerides, free fatty acids and other contaminants, depending on the degree of pretreatment they have received prior to conversion. Since biodiesel is a mono-alkyl fatty acid ester, the primary alcohol used to form the ester is the other major feedstock.

Most processes for making biodiesel use a catalyst to initiate the transesterificaton and esterification reactions. The catalyst is required as a solubilizer because the alcohol is sparingly soluble in the oil phase and because non-catalyzed reactions are extremely slow. The catalyst promotes an increase in alcohol solubility to allow the reaction to proceed at a reasonable rate. The most common transesterification catalysts are strong mineral bases such as sodium hydroxide, potassium hydroxide and sodium or potassium methoxide. After the transesterification is complete, the base catalyst acccumulates primarily in the glycerol co-product phase. Mineral acid catalysts used in homogenous esterification reactions tend to accumulate in an acid-alcohol-water phase.

Typical proportions for the chemicals used to make biodiesel by transesterification are in the following **Table 10**.

Reactants:	Fat or oil (e.g. 100 kg soybean oil) Primary alcohol (e.g. 10 kg methanol)
Catalyst:	Mineral base (e.g. 0.3 kg sodium hydroxide)
Neutralizer:	Mineral acid (e.g. 0.25 kg sulfuric acid)

Table 10. Typical Chemical Proportions.

Fats and oils: Choice of the fats or oils to be used in producing biodiesel is both a process chemistry decision and an economic decision. With respect to process chemistry, the greatest difference among the choices of fats and oils is the amount of free fatty acids that are associated with the triglycerides. Other contaminants, such as color and odor bodies, can reduce the value of the glycerin produced and reduce the public acceptance of the fuel if the color and odor persist in the fuel.

Most vegetable oils have a low percentage of associated free fatty acids. Crude vegetable oils contain some free fatty acids and phospholipids. The phospholipids are removed in a "degumming" step, and the free fatty acids can be removed in a "refining" step. Oil can be purchased as crude, degummed or as refined oil. The selection of the type and quality of the oil affects the production technology that is required.

Animal tallows and recycled (yellow) grease have much higher levels of free fatty acids. Yellow grease is limited to 15 percent free fatty acids and is a traded commodity that is typically processed into animal and pet food. Specifications for yellow grease are described in the chapter on feedstock preparation. Trap greases come from traps under kitchen drains. These greases can contain between 50 and 100 percent free fatty acids. There is no market for these greases at this time; most are landfilled. Trap grease is not yet used for biodiesel production and has some technical challenges that have not yet been fully resolved, such as difficult to break emulsions (gels), fine silt that will cause equipment wear, high water contents and very strong color and odor bodies that affect biodiesel and glycerin products. There are also unresolved questions about small quantities of other contaminants such as pesticides that might be present in the fuel.

The options for the triglyceride choice are many. Among the vegetable oils sources are soybean, canola, palm and rape. Animal fats are products of rendering operations. They include beef tallow, lard, poultry fat and fish oils. Yellow greases can be mixtures of vegetable and animal sources. There are other less desirable but also less expensive triglyceride sources, such as brown grease and soapstock. The free fatty acid content affects the type of biodiesel process used, and the yield of fuel from that process. The other contaminants present can affect the extent of feedstock preparation necessary to use a given reaction chemistry to produce an on-specification product.

Alcohol: The most commonly used primary alcohol in biodiesel production is methanol, although other alcohols, such as ethanol, isopropanol and butanol are used. A key quality factor for the alcohol is the water content (< 0.08 wt%). Water interferes with transesterification reactions and can result in poor or no ester yields and high levels of soap, free fatty acids and triglycerides in the final product. Unfortunately, all the lower alcohols are hygroscopic and, therefore, are capable of absorbing water from the air.

As long as the product esters meet ASTM 6751, it does not make any chemical difference which alcohol is used in the process. Other issues, such as cost of the alcohol, amount of alcohol needed for the reaction, ease of recovering and recycling the alcohol, fuel tax credits and global warming, influence the choice of alcohol. Some alcohols also require slight technical modifications to the production process, such as higher operating temperatures, lower mixing speeds, or longer separation times.

Since the reaction to form the esters is on a molar basis and alcohol is purchased on a volume basis, their properties make a significant difference in raw material costs. It takes three moles of alcohol to react completely with one mole of triglyceride. In April, 2006, one gallon of methanol costs $1.07. That gallon contains 93.56 gram-moles of methanol at a cost of $0.0114 per gram-mole. By contrast, a gallon of ethanol at the current price of $2.40 per gallon for fuel-grade ethanol costs $0.0370 per gram-mole, or 3.2 times more per mole.

In addition, a base catalyzed process typically uses an operating mole ratio of 6:1 mole of alcohol rather than the 3:1 ratio required by the reaction. The reason for using extra alcohol is that it uses the Law of Mass Action to "drive" the reaction closer to the 99.7 percent yield needed to meet the total glycerol standard for fuel-grade biodiesel. The unused alcohol must be recovered and recycled back into the process to minimize operating costs and environmental impacts. Methanol is considerably easier to recover than ethanol because ethanol forms an azeotrope with water, making it expensive to purify the ethanol to the required dryness during recovery. If the water is not removed, it will interfere with the reactions, as noted above. Methanol recycles more easily because it does not form an azeotrope and is vaporized at a lower temperature.

These two factors are the reason that even though methanol is more toxic, it is the preferred alcohol for producing biodiesel. Methanol has a flashpoint of 10°C, while the flashpoint of ethanol is 8°C, so both are considered highly flammable. Methanol should never come into contact with skin, eyes or through respiration, as it can be readily absorbed. Excessive exposure to methanol can cause blindness, other health effects and death.

Methanol does have a somewhat variable pricing history. When the production of MTBE was mandated for the reduction of emissions from gasoline engines in the winter, there was a significant expansion in world capacity for the material. The excess capacity and a crash in demand led to methanol prices of $0.31 per gallon in early 2002. However, in late July 2002, the production/consumption levels regained equilibrium, and the methanol price doubled back to the more typical value of $0.60± per gallon. The more recent ban on the use of MTBE and the mandated use of fuel oxygenates during certain seasons has led to a mid-2006 price of $1.07 per gallon.

For the use of ethanol as a feedstock, the alcohol quality requirements are that it be pure (~ 99.8 wt%) and anhydrous. Since chemical grade ethanol is typically denatured to prevent its abuse, purchase ethanol that has been denatured with methanol. The product esters will be a mix of methyl and ethyl esters, that can meet ASTM D 6751.

Catalysts and neutralizers: Catalysts may either be base, acid or enzyme materials. The most commonly used catalyst materials for converting triglycerides to biodiesel are sodium hydroxide, potassium hydroxide and sodium methoxide. Most base catalyst systems use low FFA materials as a feedstock. Even refined fats and oils contain small amounts (<2 percent) of free fatty acids. These FFA will be titrated almost instantanousely by the dissolved catalyst to form soaps. Most of the soaps will end up in the crude glycerin co-product.

The base catalysts are highly hygroscopic and they form chemical water when dissolved in the alcohol reactant. They also absorb water from the air during storage. If too much water has been adsorbed, the catalyst will perform poorly because of excess water in the reacting system. The ester product likely will not meet the total glycerin standard in ASTM D 6751.

Although acid catalysts can be used for transesterification, the reaction rates are generally too slow for industrial processing. Acid catalysts are more commonly used for the esterification of free fatty acids. The typical choices include sulfuric acid and phosphoric acid. There are a number of solid acid catalyst systems that are finding application for the esterification of free fatty acids and feedstocks that contain high levels of free fatty acids. The free fatty acids are converted into esters. In the case of homogeneous (dissolved) acid catalysts, the acids must be neutralized when this process is complete, but this can be accomplished using the base catalyst that is added for the transesterification step. Use of heterogeneous (solid) acid catalysts does not require neutralization before the transesterification. However, the products of acid esterification must be dried before beginning the transesterification step. Both esterification and neutralization produce "chemical" water as a product, resulting in a reaction mixture that is too wet to yield biodiesel quality esters.

There is continuing interest in using lipases as enzymatic catalysts for the production of alkyl fatty acid esters. Some enzymes work on the triglyceride, converting them to methyl esters, while others work on the fatty acids. The commercial use of enzymes is currently limited to the production of specialty chemicals from specific types of fatty acids. The commercial use of enzymes is limited because costs are high, the rate of reaction is low and yields of methyl esters are typically less than the 99.7 percent required for fuel-grade biodiesel. Enzymes are being considered for fatty acid conversion to biodiesel as a pretreatment step, but this system is not commercial at this time.

Neutralizers are used to remove the base or acid catalyst from the product biodiesel and glycerol. If you are using a base catalyst, the neutralizer is typically an acid, and vice versa. If the biodiesel is being washed, the neutralizer can be added to the wash water. While sulfuric acid is a common choice to neutralize base catalysts, if phosphoric acid is used, the resulting salt has value as a chemical fertilizer.

Catalyst selection: **Base catalysts** are used for essentially all vegetable oil processing plants. The initial free fatty acid con-tent and the water content are generally low. Tallows and greases with free fatty acid contents greater than about 2 percent should be pretreated to either remove the FFA or convert the FFA to esters before beginning the base catalyzed reaction. Otherwise, the base catalyst will react with the free fatty acids to form soap and water. The soap formation reaction is very fast and goes to completion before any transesterification begins. The upper FFA limit in a single step, base catalyzed reaction system is about 5-6 percent by weight. Above this amount of FFA, there is too much chemical water produced in the neutralization step for the transesterification step to complete.

Essentially all of the current commercial biodiesel producers use base catalyzed reactions for transesterification. Base catalyzed reactions are relatively fast, with residence times from about 5 minutes to about 1 hour, depending on temperature, concentration, mixing and the alcohol-triglyceride ratio. Most use sodium or potassium hydroxide or sodium or potassium methoxide solutions as catalysts. Some glycerol refiners prefer sodium catalysts products. However, potassium base has a higher cost but the potassium can be precipitated as potassium phosphate, a fertilizer co-product, when the products are neutralized using phosphoric acid.

Sodium or potassium methoxide, is usually sold as a 25 or 30 percent solution in methanol. Both are more powerful catalysts on a weight basis than the mixture of sodium or potassium hydroxide and methanol. This appears to be, in part, the result of the negative effect of the chemical water produced in situ when the hydroxide and methanol react to form the methoxide and water.

Acid catalyst systems are characterized by slow reaction rates and high alcohol:triglyceride ratios (20:1 and more). Generally, acid catalyzed reactions are used to convert FFA to soap (titration), or to convert soaps plus methanol to esters as a pretreatment step for high FFA feedstocks. Residence times from 10 minutes to about 2 hours are reported.

There is currently significant progress in the development of heterogeneous acid catalyst systems for FFA and soapstock esterification. These systems are of particular interest because they avoid the need for expensive materials of construction for the reactors – no acid corrosion. Also, the solid state acid catalysts do not have to be neutralized at the end of the esterification step.

All acid esterification systems need to have a water management strategy. Acid esterification systems produce a byproduct of water. In batch systems, the water tends to accumulate in the vessel to the point where it can shut the reaction down prematurely. Water may actually be easier to control in a continuous system.

Counter-current acid esterification systems have been used for decades to convert pure streams of fatty acids into methyl esters at yields above 99 percent. These systems tend to force yields by washing the water out of the system at the same time as the feedstock is converted to esters because the feedstock and the acid-methanol mix are moving in opposite directions. The sulfuric acid tends to migrate into the water and methanol rendering it unavailable for the reaction.

Good water management can minimize the amount of methanol required for the reaction. Excess methanol (such as the 20:1 ratio) is generally necessary in batch reactors where water accumulates. Another tactic is to approach the reaction in two stages: fresh methanol and sulfuric acid is reacted, removed and replaced with fresh reactant. Much of the water is removed in the first step, and the fresh reactant in the second step drives the reaction closer to completion. Acid-catalyzed esterification is discussed in more detail in the chapter on pretreatment of high FFA feedstocks.

Lipase catalyzed reactions have the advantage of reacting at room temperature without producing spent catalysts. The enzymes can be recycled or immobilized onto a substrate. If immobilized, the substrate will require replacement when yields begin to decline due to deactivation or leaching of the enzymes. The enzyme reactions are highly specific. Because the alcohol can be inhibitory to some enzymes, a typical strategy is to feed the alcohol into the reactor in three steps of 1:1 mole ratio each. The reactions are very slow, with a three-step sequence requiring from four to 40 hours or more. The reaction conditions are modest, from 35-45°C. Transesterification yields generally do not meet ASTM standards, but esterification yields for FFA or soapstock can occur relatively quickly, and yields are good. Excess free fatty acids after a lipase esterification can be removed as soaps in a subsequent transesterification or a caustic stripping step.

Biodiesel Production Process Options

Batch Processing

The simplest method for producing alcohol esters is to use a batch, stirred tank reactor. Alcohol to triglyceride ratios from 4:1 to 20:1 (mole:mole) have been reported, with a 6:1 ratio most common. The reactor may be sealed or equipped with a reflux condenser. The operating temperature is usually about 65°C, although temperatures from 25-85°C have been reported.

The most commonly used catalyst is sodium hydroxide, with potassium hydroxide also used. Typical catalyst loadings range from 0.3 percent to about 1.5 percent by weight of oil.

Thorough mixing is necessary at the beginning of the reaction to bring the oil, catalyst and alcohol into intimate contact. Towards the end of the reaction, less mixing can help increase the extent of reaction by allowing the inhibitory product, glycerol, to phase separate from the ester – oil phase. Completions of 85 to 94 percent are reported in a single step.

Some groups use a two-step reaction, with glycerol removal between steps, to increase the final reaction extent to 98+ percent. Higher temperatures and higher alcohol:oil ratios also can enhance the percent completion. Typical reaction times range from 20 minutes to more than one hour.

Figure 12 shows a process flow diagram for a typical batch system. The oil is first charged to the system, followed by the catalyst and methanol. The system is agitated during the reaction time, then agitation is stopped. In some processes, the reaction mixture is allowed to settle in the reactor to give an initial separation of the esters and glycerol. In other processes, the reaction mixture is pumped into a settling vessel or is separated using a centrifuge.

The alcohol is removed from both the glycerol and ester stream using an evaporator or a flash unit. The esters are washed gently using warm, slightly acid water to remove residual methanol and salts and then dried. The finished biodiesel is then transferred to storage. The glycerol stream may be neutralized, then sent to the glycerol refining section.

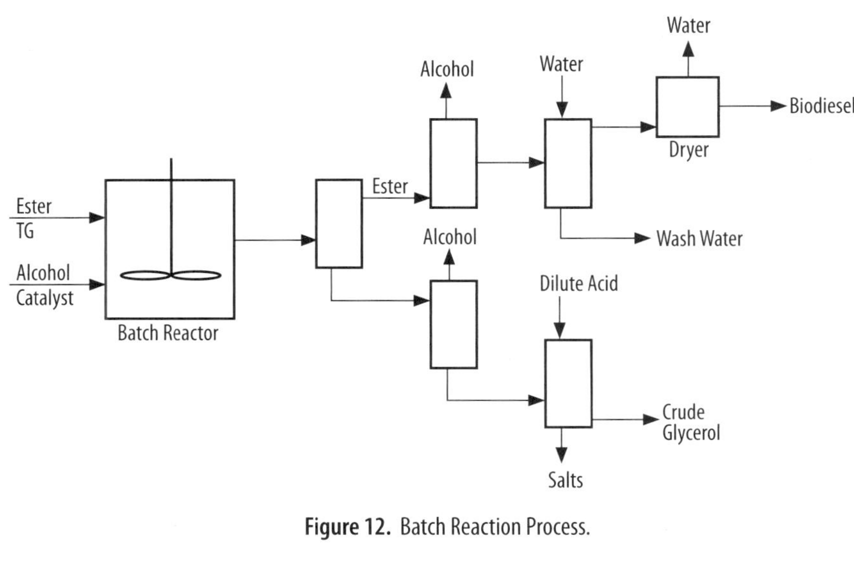

Figure 12. Batch Reaction Process.

For yellow grease and animal fats, the system is slightly modified with the addition of an acid esterification vessel and storage for the acid catalyst. The feedstock is dried to about 0.4 percent water and filtered before loading the acid esterification tank. The sulfuric acid and methanol mixture is added, and the system is agitated. Similar temperatures to transesterification are used, and sometimes the system is pressurized or a cosolvent is added. Glycerol is not produced.

If a two-step acid treatment is used, the stirring is suspended until the methanol phase separates and is removed. Fresh methanol and acid are added and the stirring resumes. Once the conversion of the fatty acids to methyl esters has reached equilibrium, the methanol/water/acid mixture is removed by settling or with a centrifuge. The remaining mixture is sent straight into transesterification where it will be neutralized using excess base catalysts. Any remaining free fatty acids will be converted into soaps by base titration in the transesterification stage.

Continuous Process Systems

A popular variation of the batch process is the use of continuous stirred tank reactors (CSTR) in series. The CSTR can be varied in volume to allow for a longer residence time in CSTR 1 to achieve a greater extent of reaction. After the initial product glycerol is decanted, the reaction in CSTR 2 is rather rapid, with 98+ percent reaction completion common.

An essential element in the design of the CSTR is sufficient mixing input to ensure that the composition throughout the reactor is essentially constant. This has the effect of increasing the dispersion of the glycerol product in the ester phase. The result is that the time required for phase separation may be extended.

There are several processes that use intense mixing, either from pumps or motionless mixers, to initiate the esterification reaction. The majority of the transesterification reaction is carried out in a pipe reactor, usually with some provision for mixing in the reactor. The reaction mixture moves through the reactor in a continuous plug, with little mixing in the axial direction. This type of reactor, called a plug-flow reactor (PFR), behaves as if it were a series of small CSTRs chained together.

The result is a continuous system that requires rather short residence times, as low as 6-10 minutes, for completion of the reaction. The PFRs can be staged to allow decanting of glycerol between stages. Often this type of reactor is operated at an elevated temperature and pressure to increase reaction rate. A PFR system is shown in **Figure 13**.

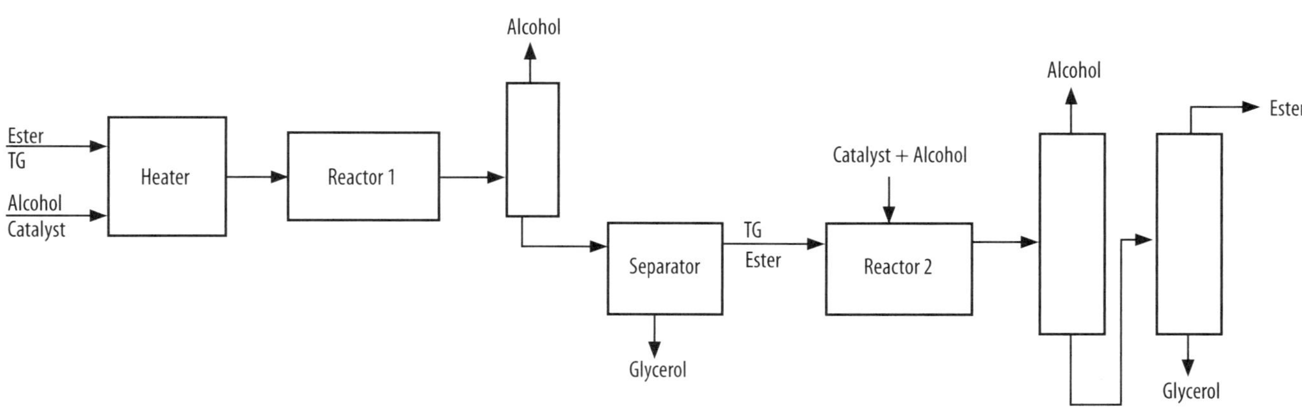

Figure 13. Plug Flow Reaction System.

High Free Fatty Acid Systems

High free fatty acid feedstocks will react with the catalyst and form soaps if they are fed to a base catalyzed system. The maximum amount of free fatty acids acceptable in a base catalyzed system is about 2 percent by weight of oil, and preferably less than 1 percent. Some approaches to using high free fatty acid feedstocks use this concept to "refine" the free fatty acids out of the feed for disposal or separate treatment in an acid esterification unit. The caustic is added to the feedstock and the resulting soaps are stripped out using a centrifuge or water extraction. This process is called caustic stripping.

Some triglycerides are lost with the soaps during caustic stripping. The soap mixture can be acidulated to recover the fatty acids and lost oils in a separate reaction tank. The refined oils are dried and sent to the transesterification unit for further processing. Rather than waste the free fatty acids removed in this manner, they can be transformed into methyl esters using an acid esterification process. As described earlier, acid catalyzed processes can be used for the direct esterification of free fatty acids in a high FFA feedstock. Less expensive feedstocks, such as tallow or yellow grease, are characteristically high in FFAs. The standard for tallow and yellow grease is < 15 percent FFA but some lots may exceed this standard.

A variation of the base-catalyzed system that avoids the problem of high FFAs forming soaps is the use of a fixed bed, insoluble base catalyst. An example of this system, using calcium carbonate as the catalyst, has been demonstrated at the bench-scale. This process is depicted in **Figure 14**. The most important issue in expanding the use of base solid state catalysts is development of catalysts that are not subject to leaching. The strongest bases result from the use of alkalai metals, such as sodium and potassium, but these metals are easily removed from the solid catalyst support. Some investigators have attempted to use barium as a transesterification catalyst with unsatisfactory results. There is, however, promising work underway using alkylguanidine organic bases.

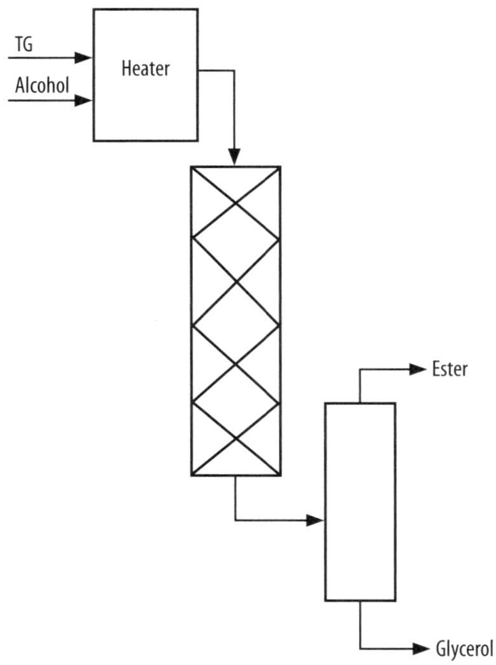

Figure 14. Fixed-Bed, Base Catalyzed Reactor System.

Direct acid esterification of a high free fatty acid feed requires water removal during the reaction or the reaction will be quenched prematurely. Also, a high alcohol to FFA ratio is required, usually between 20:1 and 40:1. Direct esterification may also require rather large amounts of the acid catalyst, depending on the process used.

The esterification reaction of FFAs with methanol produces byproduct water that must be removed, but, after drying, the resulting mixture of esters and triglyceride can be used directly in a conventional base catalyzed system. The water can be removed by vaporization, settling or centrifugation as a methanol-water mixture. Counter-current continuous flow systems will wash out the water along with the exiting stream of acidic methanol.

One approach to the acid catalyst system has been to use phosphoric acid as the initial catalyst, neutralize it with an excess of potassium hydroxide for the base step and then neutralize the potassium hydroxide with phosphoric acid upon completion. The insoluble potassium phosphate is recovered from the glycerol phase, washed and dried for use as a fertilizer. **Figure 15** shows a typical acid catalyzed direct esterification process.

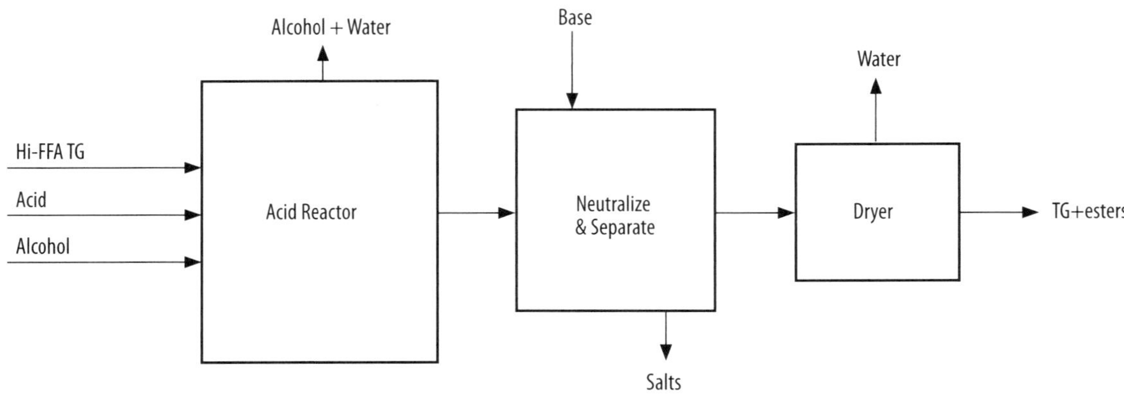

Figure 15. Acid Catalyzed Direct Esterification Process.

Another approach to utilization of high FFA feedstocks is to use a base catalyst to deliberately form soap from the FFA. The soap is recovered, the oil dried and then used in a conventional base catalyzed system. This strategy can lead to a false sense of economy. If the soapstock is discarded, the effective price of the feedstock is increased in inverse proportion to the percentage of original FFA. The soapstock can, however, be converted into esters by using an acid catalyzed reaction. The problem with this strategy is that the soapstock system contains a large amount of water that must be removed before the product esters can meet the biodiesel standard. The soapstock process is shown in **Figure 16.**

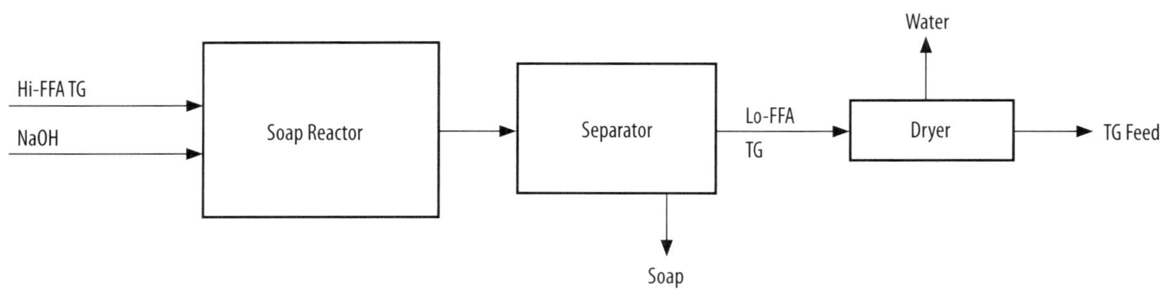

Figure 16. Preparation of Soapstock from a High FFA Feed.

Yet another procedure for processing high FFA feeds is to hydrolyze the feedstock into pure FFA and glycerine. Typically this is done in a counter current reactor using sulfuric/sulfonic acids and steam. The output is pure free fatty acids and glycerin. Any contaminates in the feedstock partition mostly into the glycerin and some may leave with the steam/water effluent. Contaminants that continue with the FFA can be removed or retained, depending on the processes and product specifications. The pure FFA are then acid-esterified in another counter current reactor to transform them into methyl esters. The methyl esters are then neutralized and dried with yields of about 99 percent. The process equipment must be acid-resistant, but generally feedstock costs are extremely low.

Co-solvent Systems – Biox Process

Co-solvent options are designed to overcome slow reaction times caused by the extremely low solubility of the alcohol in the triglyceride phase. One approach that is now offered commercially is the Biox Process. This process uses tetrahydrofuran (TFH) as a co-solvent to solubilize the methanol. The THF co-solvent is chosen, in part, because it has a boiling point very close to that of methanol. After the reaction is complete, the excess methanol and the THF co-solvent are recovered in a single vaporization step. This system uses a rather low operating temperature, 30°C. If the FFA content in the feedstock is low, the process is completed in a single step using a base catalyst. For feeds with higher FFA, the system can be operated as a two-step reaction by first using acid catalysis to convert the FFA to esters, followed by a base catalyzed transesterification. Other co-solvents, such as MTBE, have been investigated.

The equipment volume has to be larger for the same quantity of final product capacity in a co-solvent process because of the additional volume of the co-solvent. Co-solvents that are subject to the hazardous and/or air toxic EPA list for air pollutants require special "leak proof" equipment as does the entire system including methanol/cosolvent recovery and recycling. Fugitive emissions are tightly controled. The co-solvent must be completely removed from the glycerin, as well as the biodiesel. The Biox Process is depicted in **Figure 17**.

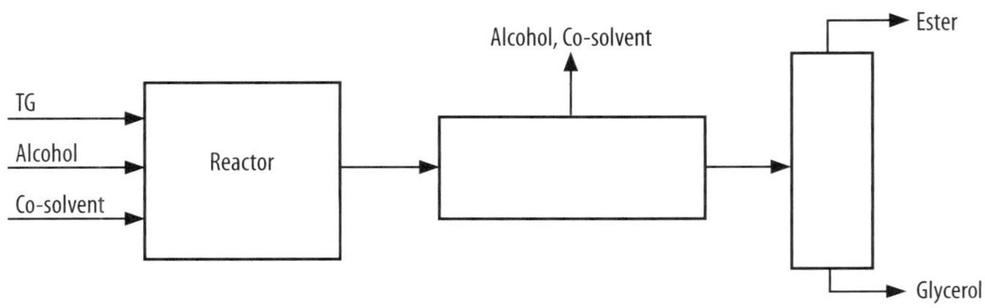

Figure 17. Biox Co-solvent Process.

Non-catalyzed Systems – Supercritical Process

When a fluid or gas is subjected to temperatures and pressures in excess of its critical point, there are a number of unusual properties exhibited. There no longer is a distinct liquid and vapor phase, but a single, fluid phase present. Solvents containing a hydroxyl (OH) group, such as water or primary alcohols, take on the properties of super-acids.

A non-catalytic approach to esters production is the use of a high (42:1) alcohol to oil ratio. Under supercritical conditions (350° to 400°C and > 80 atm or 1200 psi) the reaction is complete in about 4 minutes. Capital and operating costs can be greater and energy consumption higher than other process options.

An intriguing example of this process has been demonstrated in Japan, where oils in a very large excess of methanol have been subjected to very high temperatures and pressures for a short period of time. The result is a very fast (3-5 minutes) reaction to form esters and glycerol. The reaction must be quenched very rapidly so that the products do not de-compose. Clearly, while the results are interesting, the scale-up to a useful process may be quite difficult. **Figure 18** depicts one conception of a configuration for a supercritical esterification process.

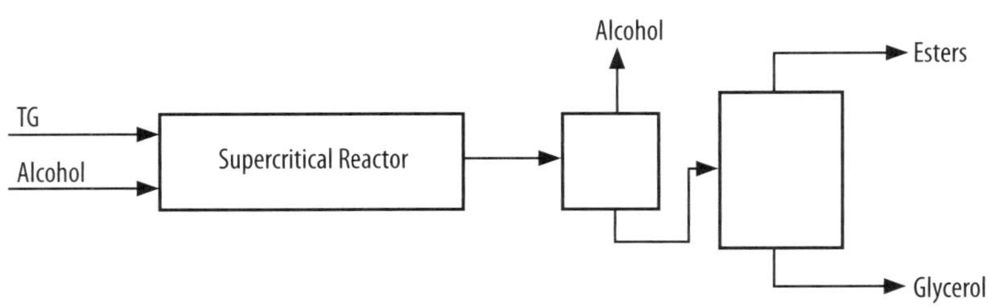

Figure 18. Supercritical Esterification Process.

Summary

Many technologies can be combined under various conditions and feedstocks in an infinite number of ways. The technology choice is a function of desired capacity, feedstock type and quality, alcohol recovery and catalyst recovery. The dominant factor in biodiesel production is the feedstock cost, with capital cost contributing only about 7 percent of the final product cost. However, some reaction systems are capable of handling a variety of feedstocks and qualities while others are not. Also, the different approaches to the esterification process result in quite different operating requirements, different water use requirements and different operating modes.

In general, smaller capacity plants and variable feedstock quality suggest use of batch systems. Continuous systems generally lead the operation on a 24/7 basis, requiring larger capacities to justify larger staffing requirements and a more uniform feedstock quality.

CHAPTER 9 | Effects of Process Parameters

There are a number of process and operating parameters that affect the transesterification reaction used to produce biodiesel. The parameters may be categorized as chemical effects, processing conditions and process structure.

In this chapter we will:

1. Review the chemistry of the transesterification reaction from the viewpoint of the quality requirements for the reactants and how the quality affects the extent of reaction and the yield.
2. Review typical specifications for reactants and the ranges of reactant concentration commonly used.
3. Describe the categories of process conditions important to the biodiesel reaction and common ranges of application.
4. Discuss the choices of batch vs. continuous reactions and the use of single-step vs. multi-step reactions.

This information will assist the designer in providing specifications for the process that best fits a specific application and for the operator to troubleshoot reasons for off-specification products.

Chemical Parameters in Transesterification

The chemical parameters that affect biodiesel production include:

- Reaction path characteristics of the esterification and transesterification mechanisms.
- Choice of primary reactants.
- Quality requirements for the reactants.
- Choice of catalysts.
- Effects of reaction products on the overall reaction.

Esterification, whether direct esterification of a free fatty acid or the transesterification of a triglyceride, is a three-step reaction. The initial step involves a solubility limited phase where the reaction rate is determined by the rate at which the alcohol reactant is incorporated into the oil or ester phase. The duration of the solubility limited phase can be reduced through the use of a solubilizing catalyst (base for transesterification and acid for direct esterification), intense mixing and/or co-solvents to make the alcohol miscible with the oil phase. In addition, longer chain alcohols are more soluble in the oil phase than shorter chain alcohols.

The quantity of free fatty acids (FFAs) present in the base-catalyzed reaction is very important. FFAs react very rapidly with bases to form insoluble soaps. With low (< 2 percent wt/wt) FFA oils it is common to add additional catalyst to allow for the "titration" of the FFAs before the esterification reaction. The soap product is then removed in the final ester water-wash step in post-processing. FFA concentrations greater than 2 percent wt/wt require pretreatment to either remove the excess FFA as soapstock or to convert the FFA to esters by direct esterification before the base-catalyzed reaction.

As the ester concentration increases, the alcohol reactant becomes soluble in the mixed oil-ester phase. The reaction becomes limited only by the intrinsic rate of reaction to form the ester. This rate depends upon the reaction temperature and the relative concentrations of the alcohol and oil, but it is generally rather fast; this stage of reaction is termed the "rate limited phase."

As the reaction progresses towards completion, the products of the reaction become inhibitory to further reaction. In the case of the base-catalyzed transesterification, the inhibitory product is glycerol. As discussed elsewhere, the overall transesterification reaction is a set of equilibrium reactions. The ultimate products, glycerol and esters, can react in reverse to reform mono-, di- and triglycerides. Further, the glycerol and oil-ester phases are immiscible, and the alcohol and catalyst distribute preferentially into the glycerol phase. This effect lowers the alcohol concentration in the reacting phase, further inhibiting the reaction.

As noted earlier, base catalysts are used primarily for fats/oils with < 2.0 percent FFA (< 0.5 percent is preferred). Some processes use base catalysts to convert high FFA feedstocks to soapstock as a preparation step before the conventional transesterification reaction.

Water is inhibitory to the transesterification reaction in quantities in excess of 0.5 wt percent. In the case of the base-FFA reaction, water is a product. Water is also a product in the acid catalyzed FFA-alcohol direct esterification reaction.

The primary strategy for water management is carefully controlling the amount of water entering the reaction system. Some reaction schemes also use stepwise reactions or purging as a way of controlling water build-up. Another means of controlling water entering the system is to use sodium or potassium methoxide as a base catalyst.

Glycerol is inhibitory to the transesterification reaction. Strategies for reducing this effect include multi-step reactions, with glycerol decantation, and use of elevated temperatures (shift the equilibrium) and excess alcohol (shift the equilibrium).

Sulfuric and phosphoric acids are used as neutralizers for the base catalysts and as acid catalysts for direct esterification. Either material must be managed carefully, since there are very tight specifications on amount of sulfur and phosphorus allowed in the final biodiesel product.

Typical amounts used for direct esterification with sulfuric acid are 0.5 to 25 wt percent, based on the FFA content. Some operators prefer to use phosphoric acid and potassium hydroxide in their process. The insoluble neutralization product, potassium phosphate, has value as a fertilizer co-product.

<u>Under no circumstances</u> is the use of hydrochloric acid recommended in a biodiesel operation. Typically, the construction material of choice for biodiesel process equipment is stainless steel, which is incompatible with HCl. A typical material of construction for an acid catalyzed system is monel, while the base-catalyzed and downstream components can be 304 or 316 stainless steel.

Operating Parameters

There are several operating degrees of freedom used to control the reactions that produce biodiesel. The final ester product must meet stringent specifications under ASTM 6751 to merit the designation "biodiesel." Several controllable parameters are used to adjust the performance of the reactions within a given system configuration. The parameters include:

- Temperature
- Pressure
- Triglyceride-to-alcohol ratio
- Reactor residence time
- Mixing intensity

In the case of base-catalyzed reactions, the typical alcohol to triglyceride ratio ranges from 6:1 to 10:1 on a mole basis. The 6:1 mole ratio is most commonly used. This ratio provides an alcohol concentration that is twice the stoichiometric requirement for the total transesterification of the triglyceride. Higher mole ratios are used to push the reaction to higher rates and greater extent of reaction in the rate-controlled regimen.

The acid catalyzed esterification reactions are much slower than the base catalyzed case. Alcohol to triglyceride ratios for acid catalyzed reactions range from 9:1 for direct esterification to as high as 40:1 for transesterification reactions. The difference in intrinsic reaction rates is also seen in the residence times used for the conversion reactions. Base catalyzed reactions are completed in residence times ranging from 5 minutes to 4 hours, depending on the other reaction conditions used. By contrast, the acid catalyzed reactions require residence times of 15 minutes to more than 8 hours.

In general, an increase in reaction temperature of 10°C will double the reaction rate. In the case of an alcohol transesterification reaction, an increase in temperature will also increase the solubility of the alcohol in the oil phase, thus shortening the duration of the solubility limited phase. Higher temperatures affect the glycerol-monoglyceride equilibrium by shifting the reaction towards the mono-glyceride with the equilibrium temperature about 70°C.

Since methanol boils at about 65°C, many reactor systems are designed to operate near that temperature. This can be accomplished using a closed reactor, with or without a reflux condenser. Operation at temperatures above 65°C requires the use of pressurized vessels. As a rule of thumb, the pressure vessel should be operated at a pressure about 10 to 15 psig (1 bar) above the vapor pressure of methanol at the operating temperature.

The critical time for mixing in transesterification reactions is the initial solubility-limited phase. Mixing is a mechanical means of distributing the alcohol reactant throughout the oil phase. This initial distribution also carries the catalyst into intimate contact with the oil phase reactants. The initial mixing can be done using stirred tanks, motionless mixers, liquid injection or high-shear pumps. Of these options, stirred tanks are the least intense.

Most reaction schemes continue the mixing throughout the reaction, but this is not truly necessary. Indeed, mixing to-wards the end of the reaction period can be inhibitory, because the glycerol product does not have an opportunity to separate from the reacting oil-ester phase.

Operating Strategy

The choice of operating strategy, whether to use a batch or continuous system and/or to use a single-step or multi-step operation, is made in the early design phase of the facility. However, this choice is determined in large part by the chemistry and operating conditions selected for the process.

Batch reactions are favored for smaller capacity plants that may operate only one or two shifts per day. Similarly, batch operations are favored when the reaction residence time is long. Continuous operations are favored for larger capacity plants, where 24/7 operating reduces the overall size of the operating equipment. If somewhat long residence times are required to meet product quality specifications, a design using sequential continuous stirred tank reactors (CSTR) is indicated.

The choice of a single-step or multiple-step reaction is commonly determined by the rate of accumulation of reaction inhibitors. Generally, the base catalyzed reactions can be done in a single step if the inlet FFA and water concentrations are very low. It is common, however, to use two or more reaction steps for base catalyzed transesterification to ensure meeting the bound glycerol specification of ASTM D 6751. Acid catalyzed reactions generally require multi-step reactions to allow for the removal of the chemical water product.

Summary

Biodiesel production is deceptively simple. The basic transesterification reaction is a multiple equilibrium reaction that can be completed reliably to a fairly high extent of reaction. However, the production of a continually high quality product that meets the exacting requirements of ASTM 6751-03a is not simple. It requires control of the input reactant quality, continuous monitoring and adjustment of operating parameters, and appropriate up-stream and down-stream processing.

CHAPTER 10 Post Reaction Processing

The objective of this chapter is to describe in more detail the steps in processing the ester phase resulting from the transesterification. We will look at the recovery of the esters from the reaction mixture and at refining them to meet the requirements of ASTM D 6751.

Ester/Glycerol Separation

The ester/glycerol separation is typically the first step of product recovery in most biodiesel processes. The separation process is based on the facts that fatty acid alcohol esters and glycerol are sparingly mutually soluble and that there is a significant difference in density between the ester and glycerol phases. The presence of methanol in one or both phases affects the solubility of ester in glycerol and glycerol in ester.

Fatty acid alcohol esters have a density of about 0.88 gm/cc, while the glycerol phase has a density on the order of 1.05 gm/cc, or more. The glycerol density depends on the amount of methanol, water and catalyst in the glycerol. This density difference is sufficient for the use of simple gravity separation techniques for the two phases.

However, the rate of separation is affected by several factors. Most biodiesel processes use relatively intense mixing, at least at the beginning of the reaction, to incorporate the sparingly soluble alcohol into the oil phase. If this mixing continues for the entire reaction, the glycerol can be dispersed in very fine droplets throughout the mixture. This dispersion requires from an hour to as much as 24 hours to allow the droplets to coalesce into a distinct glycerol phase. For this reason, mixing is generally slowed as the reaction progresses to reduce the time required for phase separation.

The more nearly neutral the pH of the reaction mixture, the quicker the glycerol phase will coalesce. This is one reason to minimize the total catalyst use. In some batch systems, the reaction mixture is neutralized at the beginning of the glycerol/ester phase separation step.

The presence of significant quantities of mono-, di- and triglycerides in the final mixture can lead to the formation of an emulsion layer at the ester – glycerol interface. At best, this layer represents a net loss of product, unless it is recovered and separated. At worst, the ester phase will not meet the biodiesel specification and will have to be re-run. If problems with mono-, di- and triglycerides occur, you should reevaluate the entire reaction to see where changes can be made to improve the extent of reaction in the preceding steps.

The esterification process is run with an excess of alcohol to ensure complete reaction and attain higher reaction rates. The residual alcohol is distributed between the ester and glycerol phases. The alcohol can act as a dispersant for the ester into the glycerol phase and for the glycerol into the ester phase. The result can be a need for additional processing of the products to meet specifications. Some claim that methanol aids in phase separation, which is one reason that product is generally phase separated before methanol recovery.

Process Equipment for the Ester/Glycerol Separation

Decanter systems: There are three categories of equipment used to separate the ester and glycerol phases. Decanter systems rely solely on the density difference and residence time to achieve the separation. For relatively small throughput or batch processes, the 1 to 8 hours required for complete separation of the phases may be acceptable. However, a separation that requires a residence time of 1 hour requires a decanter with a volume of at least 700 gallons to affect separation for a 5,000,000 gallon per year continuous plant. For lower extent of reaction, the separation is slower, and the decanter will have to be much larger.

The primary determinant for designing a decanter for biodiesel production is the desired residence time. This, plus the product mixture flow rate determines the size of the unit. Decanter units should be rather tall and narrow to allow physical separation between the ester and the glycerol withdrawal points. Length to diameter (L/D) ratios of 5-10 can work well.

The temperature in the decanter affects the solubility of the alcohol in both phases and the viscosity of the two liquids. Too high a temperature in the decanter can cause residual alcohol to flash, potentially restricting the flow of the ester phase out of the tank. On the other hand, too low a temperature increases the viscosity in both phases. The increased viscosity will slow the coalescence rate in the system.

The presence of an emulsion layer is indicative of mono- and diglycerides. The emulsion layer will form between the phases. In a continuous operation, there must be a provision for removing the emulsion, so it does not fill the decanter.

Centrifuge system: Many of the continuous plants use a centrifuge for the phase separation. The centrifuge creates an artificial, high gravity field by spinning at very high speeds. The separation can be completed rapidly and effectively. The disadvantage of the centrifuge is its initial cost, and the need for considerable and careful maintenance. However, centrifuges are used extensively in the food processing, rendering and biodiesel industries. At smaller capacities, either a batch or a continuous centrifuge can be used. The use of a batch centrifuge in a continuous process requires a surge tank to match the batch cycle time with the continuous processing rate.

Hydrocyclone: An intriguing, relatively new device is the hydrocyclone. A liquid – liquid hydrocyclone uses an inverted conical shape and the incompressibility of the liquids to accelerate the liquid entering the cyclone. The effect is similar to a centrifuge, with the heavier material being forced towards the wall and downward, and the lighter material forced to the center and upward. The result is a density-based separation. Although they are now used for oil – water separations, hydrocyclones are at the experimental stage for application in biodiesel production.

A hydrocyclone operates on the Bernoulli principle of trading pressure for speed in an incompressible flowing system. The relative density of the fluids determines the separating force applied, while the relative viscosity determines the resistance to separation. The liquid mixture enters the hydrocyclone at a moderately high pressure (about 125 psig). The pressure decreases and the velocity increases as the liquid passes from the wider to the narrower part of the inverted cone. The more dense liquid is accelerated towards the outside wall, while the less dense liquid concentrates in the center. The result is a separation based upon induced g-forces.

The presence of volatiles creates a problem for centrifuges and hydrocyclones. The rapid reduction of pressure in the device will induce flashing of the volatile liquid (alcohol), disrupting or stopping the separation process. Excess methanol should be removed from the system before introducing the reaction mixture to a hydrocyclone or a centrifuge.

Ester Washing

The primary purpose of the ester washing step is the removal of any soaps formed during the transesterification reaction. In addition, the water provides a medium for addition of acid to neutralize the remaining catalyst to remove the product salts and as a means for removal of free glycerol. Any residual methanol should be removed before the wash step. This prevents the addition of methanol to the wastewater effluent. However, some processes remove the methanol with the wash water then later remove it from the wash water. This strategy is not recommended.

The use of warm, soft water (120-140°F) for the wash step prevents precipitation of hard water soaps and retards the formation of emulsions. It is imperative that the wash water be introduced using a gentle washing action. There are a number of strategies, including bubbling air through the wash tank filled with the esters and water, misting the surface of the esters and the use of ester-water contactors.

Softened water (slightly acidic) eliminates calcium and magnesium contamination and neutralizes remaining base catalysts. Similarly, removal of iron and copper ions eliminates a source of catalysts that decrease fuel stability. Gentle washing prevents the formation of emulsions and results in a rapid and complete phase separation.

The phase separation between esters and water is typically very clean and complete after all of the soaps are removed. However, the equilibrium solubility of water in esters is higher than the specified water content for B100. After the washing step, there will be more than the equilibrium amount of water present. Ester drying is required to meet the stringent limits on the amount of water present in the final biodiesel product. In addition, there may be other treatments used to reduce color bodies in the fuel, remove sulfur and/or phosphorus from the fuel or remove glycerides.

Ester Drying

The simplest approach to ester drying is the use of a single stage, isothermal flash vaporization unit. For large capacity operations it is common to use vacuum driers to save heat energy. Vacuum driers can either be batch or continuous devices for removing water. The system is operated at a highly reduced pressure, which allows the water to evaporate at a much lower temperature. A variation that also allows for rather high heating and evaporation rates is the falling film evaporator. This device operates at reduced pressure. As the esters pour down the inside wall of the evaporator, the direct contact with the heated wall evaporates the water rapidly. Care should be taken with high temperature evaporators to avoid darkening the fuel, which is a sign that the polyunsaturated methyl esters are polymerizing.

Because the total water burden in the esters is low, molecular sieves, silica gels, etc. can also be used to remove the water. An advantage of these systems is that they are passive. However, disadvantages include the fact that the units are very large, and they must be periodically regenerated.

Other Ester Treatments

Filtering is an essential part of all biodiesel production. While feedstocks entering the plant should be filtered to at least 25 μm, biodiesel leaving the plant should be filtered to at least 5 μm to ensure no contaminants are carried with the fuel that could damage the engine. It has been suggested that the fuel could be cooled before filtering to capture some of the saturated esters as they crystallize, thereby lowering the cloud point of the fuel. The crystallized esters can then be melted and used within the plant as boiler fuel.

By 2006, all U.S. biodiesel must meet new sulfur standards of 15 ppm or less. Therefore, biodiesel producers need to be aware of the sulfur content of their feedstock and must incorporate sulfur reduction technology as needed. Vacuum distillation is an option for sulfur removal that has the added benefit of deodorization and the removal of other minor contaminants, which may provide a benefit to those firms that use highly degraded feedstocks like trap grease.

There are absorbents on the market that selectively absorb hydrophyllic materials, such as glycerol and mono- and diglycerides (e.g. Magnesol from the Dallas Group). This treatment, followed by an appropriate filter, has been shown to be effective in lowering glycerides and total glycerol levels.

Some vegetable oils and many yellow greases and brown greases have an objectionable color and/or odor that is transferred to the biodiesel. There is no color specification in ASTM D 6751, but an activated carbon bed is an effective way to remove excessive color and odor. The fats and oil industry literature has other bleaching and deodorization technologies that also may be explored.

Additization of Esters

Petroleum-based diesel fuels are treated with a wide range of additives to improve lubricity, detergency, oxidative stability, corrosion resistance, conductivity and many other properties. Additive technology for biodiesel is less advanced, so fewer additives are available to enhance performance.

Biodiesel producers need to give serious consideration to oxidative stability. Biodiesel, because it contains large numbers of molecules with double bonds, is much less oxidatively stable than petroleum-based diesel fuel. Fortunately, stability-enhancing additive technology is well-developed in the food industry, and many of these additives can be carried over to stabilizing biodiesel.

Figure 19 shows typical oxidation behavior for vegetable oils and animal fats. In this test, the oil and fat were exposed to heat and aeration to accelerate the oxidation process. The progress of the oxidation is monitored by measuring the peroxide value, or the fraction of the oil that has been converted to a peroxide molecule, the first step in the oxidation process. Vegetable oil and animal fat have no conversion for a period of time (called the induction period), but then oxidize quickly. In this case, the lard actually oxidizes earlier than the vegetable oil, probably due to the presence of natural antioxidants in the vegetable oil. Lipid oxidation is an autocatalytic process. When it starts, it progresses at an ever-increasing rate.

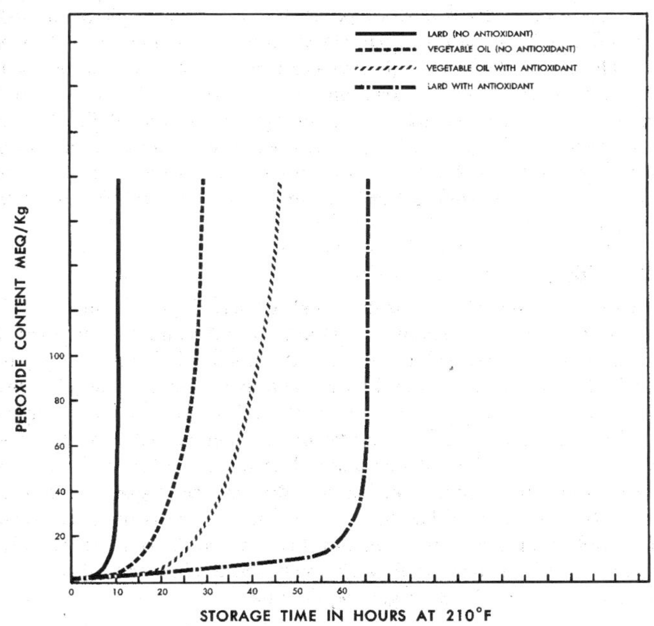

Fig. 7. Peroxide content versus storage time of some typical fats.

Figure 19. Oxidation Behavior of Vegetable Oils and Animal Fats [1].

Figure 20 shows some of the most common antioxidant additives used in food products: BHA, BHT, PG and TBHQ. The properties of these additives are well known and they are considered safe.

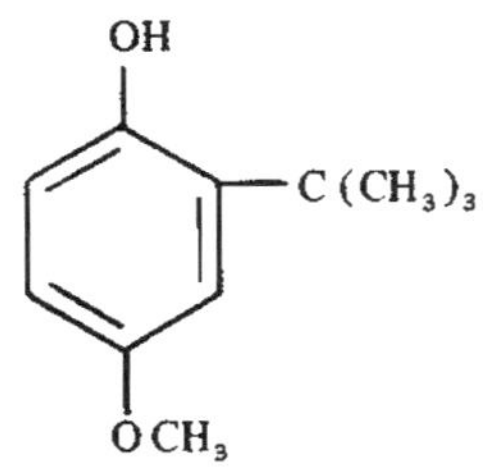

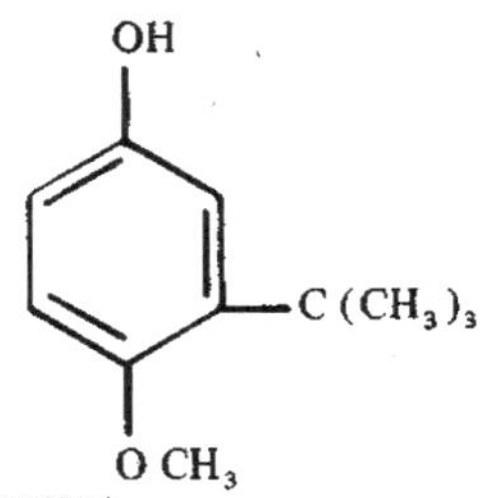

Butylated Hydroxyanisole (BHA)

3-BHA (3-*tert*-butyl-4-hydroxyanisole; 2-*tert*-butyl-4-methoxyphenol)

2-BHA (2-*tert*-butyl-4-hydroxyanisole 3-*tert*-butyl-4-methoxyphenol)

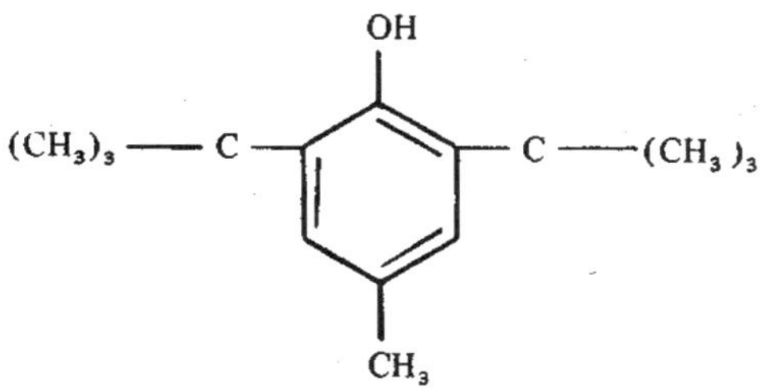

Butylated Hydroxytoluene (BHT)
(2,6-di-*tert*-butyl-*o*-cresol; 4-methyl-2,6-di-*tert*-butylphenol)

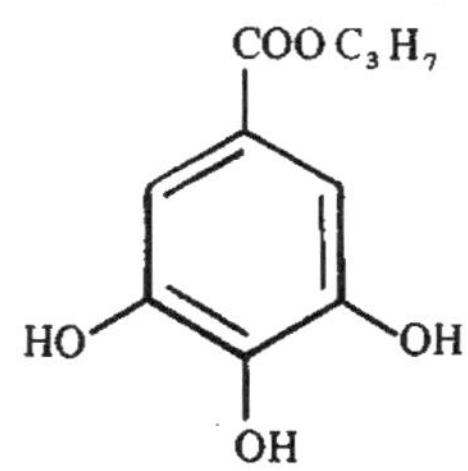

Propyl Gallate (PG)
Propyl gallate (*n*-propyl-3,4,5-trihydroxybenzoate)

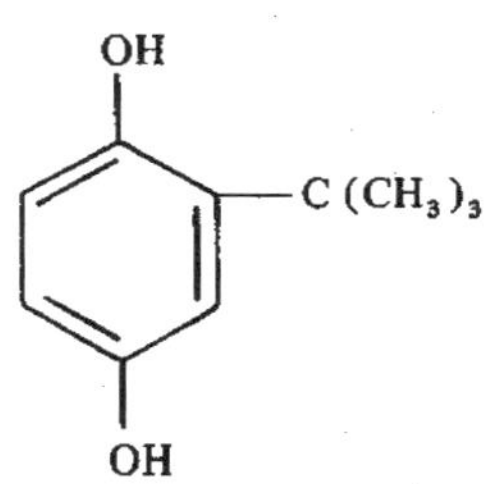

Mono-*tert*-butylhydroquinone (TBHQ)

Figure 20. Antioxidants for Oils and Fats [2].

Figure 21 shows a comparison of the effectiveness of the different antioxidants in three different oils. This data was obtained from oils, not biodiesel; however, since the oxidation processes are similar for the two classes of compounds, the relative effectiveness of the additives should be similar. **Figure 21** clearly shows that TBHQ is the most effective additive for equivalent treat rates.

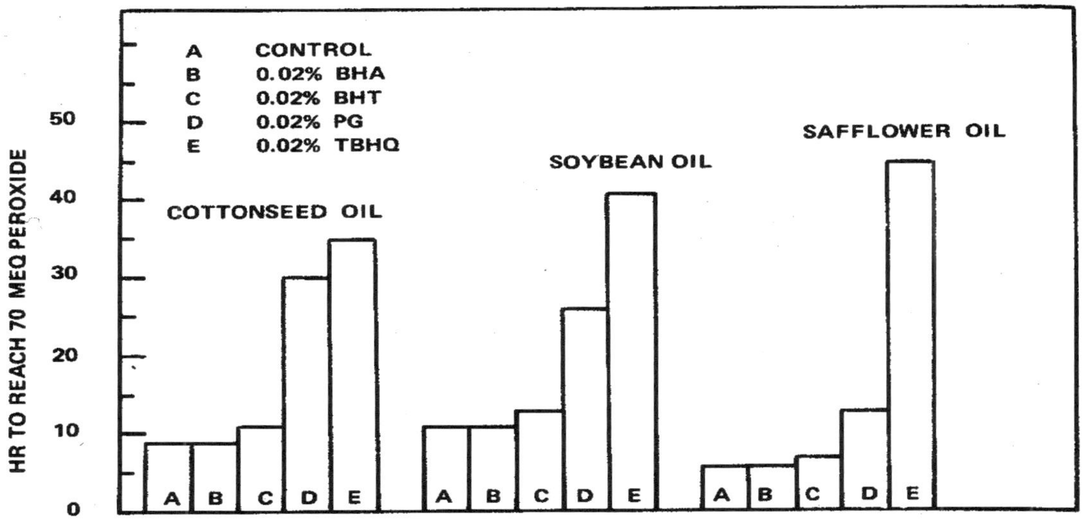

Figure 21. Comparison of Tert-butylhydroquinone (TBHQ) with other Anioxidants by the Active Oxygen Method. (AOM). From Sherwin, E. R. and Thompson, J. W., Food Technol. (Chicago), 21(6) 106, 1967.

Additional comparisons between the additives for variable treat rates are shown in **Table 11**. Again, TBHQ is almost twice as effective as the nearest other additive.

COMPARISON OF TBHQ WITH BHA, BHT, AND PG IN VEGETABLE OILS BY AOM AND OVEN STORAGE TESTS[a]

Oil treatment (wt % conc)	AOM stability (hr to reach 70 meq peroxide content)			Storage life at 115°F of potato chips prepared with oils (days)		
	Cottonseed	Soybean	Safflower	Cottonseed	Soybean	Safflower
Control (no antioxidant)	9	11	6	11	8	4
0.010 TBHQ	24	29	29	17	—[b]	—[b]
0.020 TBHQ	34	41	40	23	25	23+
0.030 TBHQ	42	53	77	33	—[b]	—[b]
0.010 BHA	9	12	8	9	—[b]	—[b]
0.020 BHA	9	10	6	10	9	7
0.030 BHA	9	10	7	7	—[b]	—[b]
0.010 BHT	10	12	10	12	—[b]	—[b]
0.020 BHT	11	13	7	7	10	9
0.030 BHT	13	15	8	15	—[b]	—[b]
0.010 PG	19	21	13	12	—[b]	—[b]
0.020 PG	30	26	10	14	20	7
0.030 PG	37	31	12	17	—[b]	—[b]

[a] TBHQ = *tert*-butylhydroquinone, BHA = butylated hydroxyanisole, BHT = butylated hydroxytoluene, PG = propyl gallate, and AOM = active oxygen method.
[b] Dash indicates test not conducted.

Reprinted from Sherwin, E. R. and Thompson, J. W., *Food Technol.* (Chicago), 21, 914, 1967.

Table 11. Comparison of TBHQ with BHA, BHT, and PG in Vegetable Oils by AOM and Oven Storage Tests.

Since metals, especially copper, can act as catalysts for oxidation, compounds called chelators are frequently added to the oils and fats along with the antioxidants. Chelators bind up metals into stable molecules that do not participate in promoting oxidation. Citric acid is a common chelator used in the food industry [1]. **Figure 22** shows the effect of adding citric acid to antioxidants. It extends the oxidative induction period and improves the stability of the oil.

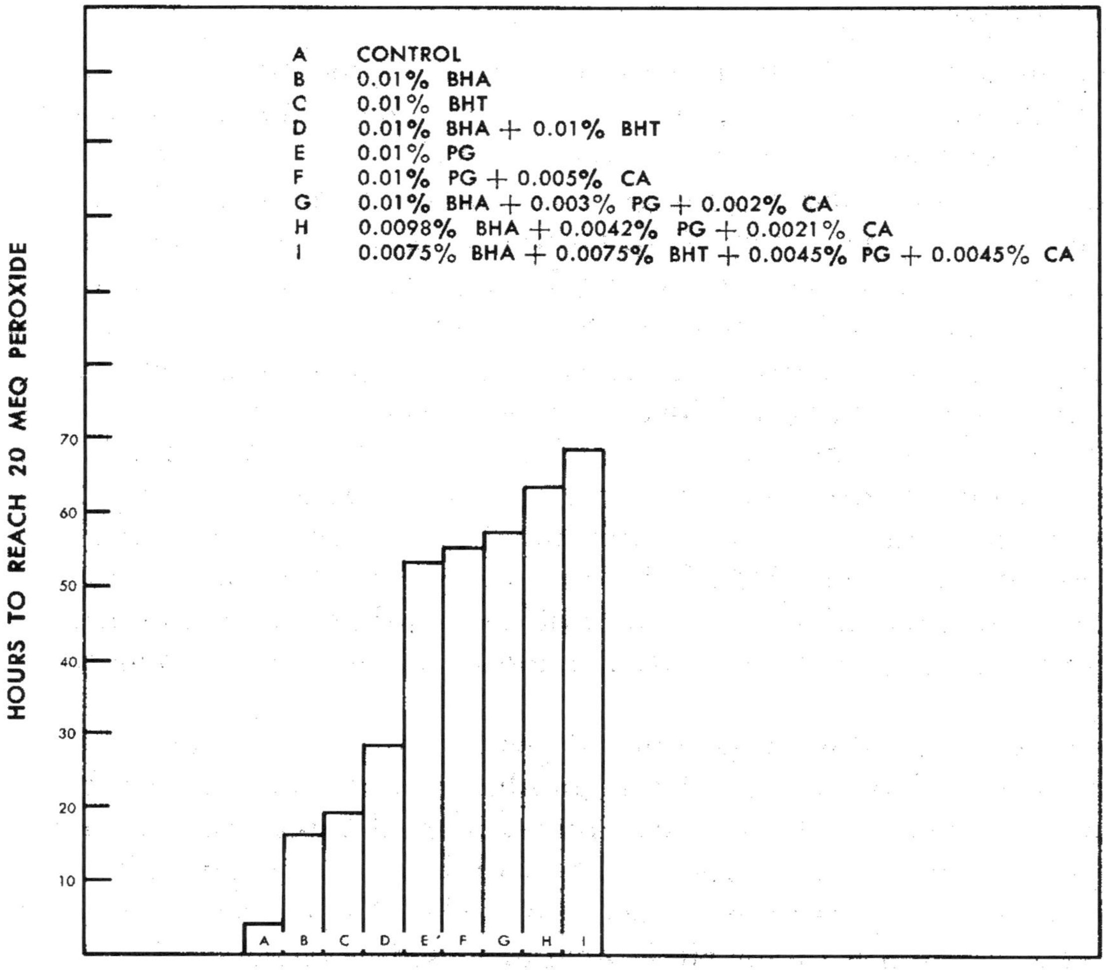

Figure 22. Effect of Citric Acid on Antioxidant Effectiveness.

References

1. Furia, T.E., Editor, Handbook of Food Additives, CRC Press, Cleveland, OH, 2nd Edition, 1972.
2. Furia, T.E., Editor, Handbook of Food Additives, Vol. 2, CRC Press, Cleveland, OH, 2nd Edition, 1980.

CHAPTER 11 Treatment and Recovery of Side Streams

There are three non-ester side streams that must be treated as a part of the overall biodiesel process.

1) The excess alcohol that is recycled within the process.
2) The glycerol co-product.
3) The wastewater stream from the process.

Here, it is assumed that methanol is the alcohol used in the process. Similar comments would apply to other alcohols.

Methanol recycling is necessary because an excess of methanol is required for an effective esterification and trans-esterification reaction system. The recovery of the unused methanol saves input costs for the process and essentially eliminates the emissions of methanol to the surroundings. The emissions reduction is mandatory because methanol is highly flammable, toxic and highly controlled under air and water quality permitting requirements.

Glycerol is recovered and partially refined as a co-product from biodiesel production. About 10 percent by weight of the input reactants are converted to glycerol in the transesterification reaction. The glycerol co-product is useful in a wide variety of applications.

Wastewater constitutes an operating cost for the plant, both because of the water consumption and because of the water treatment costs to the plant. The amount and quality of wastewater and its means of disposal, are subject to local and national permitting standards.

Methanol Management

Physical parameters are important to the recovery and recycling of methanol. Methanol's relatively low boiling point, 64.7°C, means that it is fairly volatile and can largely be removed from the oil, ester and aqueous streams by flash evaporation and re-condensation. The low boiling point, along with a low flashpoint of, 8°C, also means that methanol is a Class I-b liquid, highly flammable.

Methanol is fully miscible with water and with glycerol. However, it has a low solubility in fats and oils (approx. 10 percent wt/wt at 65°C in tallow). Methanol is more soluble in esters, but it is not fully miscible. The methanol's complete solubility in glycerol and water means that methanol will preferentially dissolve in these phases when there is a two-phase system present. The low solubility in fats and oils is the reason for the solubility-limited phase of the overall transesterification reaction.

When the two phases present are esters and glycerol, the methanol will distribute between the phases. At 90:10 percent wt/wt ester and glycerol, the methanol distributes approximately 60:40 wt percent between the ester and glycerol phases, respectively.. This is important, since the reaction is complete at 90:10 wt percent ester:glycerol.

If the methanol is allowed to remain in the system during the initial ester – glycerol phase separation, the methanol acts as a phase stabilizer, retarding the rate of gravity separation. It is advantageous to remove the methanol before this phase separation.

Methanol can be recovered using distillation, either conventional or vacuum, or in a single stage flash unit. An alternative to distillation is a falling-film evaporator. Small amounts of residual methanol in the ester phase can be removed in the water wash step in ester post-processing. Product esters are typically washed with warm (140°F), softened water to remove soaps and residual methanol.

Glycerol Refining

Generally, the quality of the glycerol produced and its value as a co-product is an important economic variable. The typical glycerol produced by a biodiesel plant is 50 percent glycerol or less and contains water, salts, methanol, methyl esters, and unreacted glycerides, color and odor bodies, dimers, timers and other minor compounds. This is commonly referred to as "biodiesel crude" and is generally worth less than 5 cents per pound. Removing the water and methanol and bringing the glycerol content up to 88 percent can generate a credit based on the value of crude glycerol. Salts, and particularly sulfur salts or potassium salts, reduce the value of the crude glycerol produced.

The recovered glycerol from the transesterification reaction contains residual alcohol, catalyst residue, carry-over fat/oil and some esters. The glycerol from rendered feedstocks may also contain phosphatides, sulfur compounds, proteins, aldehydes and ketones, and insolubles (dirt, minerals, bone, or fibers).

Chemical refining: There are several factors that are important in the chemical refining of glycerol. First, the catalyst tends to concentrate in the glycerol phase where it must be neutralized. The neutralization step leads to the precipitation of salts. Also, some of the soaps produced in the esterification will be in the glycerol. They must be removed by coagulation and precipitation with aluminum sulfate or ferric chloride if the goal is to produce refined glycerol. The removal may be supplemented by centrifuge separation. The control of the pH is very important for glycerol because low pH leads to dehydration of the glycerol and high pH leads to polymerization of the glycerol. The processed glycerol may then be bleached using activated carbon or clay.

Physical refining: The first step in physical refining of glycerol is to remove fatty, insoluble or precipitated solids by filtration and/or centrifugation. This removal may require pH adjustment. Then, the water is removed by evaporation. All physical processing is typically conducted at 150-200°F, where glycerol is less viscous, but still stable. Glycerol decomposes at about 290°C.

Glycerol purification: The final purification of glycerol is completed using vacuum distillation with steam injection, followed by activated carbon bleaching. The advantage of this approach is that this is a well-established technology. The primary disadvantage is that the process is capital and energy intensive. Vacuum distillation of glycerol is best suited to operations greater than 25 tons per day glycerol capacity.

Ion exchange purification of glycerol is an attractive alternative to vacuum distillation for smaller capacity plants. The ion exchange system uses cation, anion and mixed bed contactors to remove catalyst and other impurities. The glycerol is first diluted with soft water to a 15 to 35 percent glycerol-water solution. The ion exchange is followed by vacuum distillation or flash drying for water removal, often to an 85 percent partially refined glycerol.

The advantage of the ion exchange process is the fact that all purification takes place in the resin vessels, so the system is suited to smaller capacity operations. The disadvantage is the system is subject to fouling by fatty acids, oils and soaps. The system also requires regeneration of the beds, producing large quantities of wastewater. Regeneration requires parallel systems to operate and regenerate simultaneously.

Wastewater Considerations

Ester washing produces about 1 gallon of water per gallon of ester per wash. All process water must be softened to eliminate calcium and magnesium salts and treated to remove iron and copper ions. The ester wash water will have a fairly high BOD (~ 25,000 ppm), mostly from the soaps, but there may also be residual fat/oil, ester and glycerol in the water.

The glycerol ion exchange systems can produce large quantities of low-salt waters as a result of the regeneration process. In addition, water softening, ion exchange and cooling water blow-down will contribute a moderate dissolved salts burden.

The aggregate process wastewaters must meet local municipal waste treatment plant disposal requirements. Methanol must be fully recovered in the plant. In many areas, internal treatment and recycling of the process water may lead to cost savings and easier permitting of the process facility.

CHAPTER 12

Fatty Acid Composition and Total and Free Glycerol

The most common fatty acids (and their methyl esters) are listed in **Tables 12 and 13**. There are numerous other fatty acids, but the ones given here comprise the vast majority of those contained in biodiesel.

Fatty acid (trivial name/rational name)	Structure	Common acronym	Methyl ester (trivial name/rational name)
Palmitic acid / Hexadecanoic acid;	$R\text{-}(CH_2)_{14}\text{-}CH_3$	C16:0	Methyl palmitate / Methyl hexadecanoate
Stearic acid / Octadecanoic acid	$R\text{-}(CH_2)_{16}\text{-}CH_3$	C18:0	Methyl stearate / Methyl octadecanoate
Oleic acid / 9(Z)-octadecenoic acid	$R\text{-}(CH_2)_7\text{-}CH{=}CH\text{-}(CH_2)_7\text{-}CH_3$	C18:1	Methyl oleate / Methyl 9(Z)-octadecenoate
Linoleic acid / 9(Z),12(Z)-octadecadienoic acid	$R\text{-}(CH_2)_7\text{-}CH{=}CH\text{-}CH_2\text{-}CH{=}CH\text{-}(CH_2)_4\text{-}CH_3$	C18:2	Methyl linoleate / Methyl 9(Z),12(Z)-octadecadienoate
Linolenic acid / 9(Z),12(Z),15(Z)-octadecatrienoic acid;	$R\text{-}(CH_2)_7\text{-}(CH{=}CH\text{-}CH_2)_3\text{-}CH_3$	C18:3	Methyl linolenate / Methyl 9(Z),12(Z),15(Z)-octadecadienoate

Table 12. Chemical Structure of Fatty Acids.

Fatty acid Methyl ester	Formula	Acronym	Molecular weight	Melting point (°C)	Cetane number
Palmitic acid Methyl palmitate	$C_{16}H_{32}O_2$ $C_{17}H_{34}O_2$	C16:0	256.428 270.457	63-64 30.5	— 74.5
Stearic acid Methyl stearate	$C_{18}H_{36}O_2$ $C_{19}H_{38}O_2$	C18:0	284.481 298.511	70 39	— 86.9
Oleic acid Methyl oleate	$C_{18}H_{34}O_2$ $C_{19}H_{36}O_2$	C18:1	282.465 296.495	16 -20	— 47.2; 55
Linoleic acid Methyl linoleate	$C_{18}H_{32}O_2$ $C_{19}H_{34}O_2$	C18:2	280.450 294.479	-5°C -35°C	— 28.5; 42.2
Linolenic acid Methyl linolenate	$C_{18}H_{30}O_2$ $C_{19}H_{32}O_2$	C18:3	278.434 292.463	-11 -52 / -57	— 20.6; 22.7

Table 13. Formula, Molecular Weight and Properties of Fatty Acids and their Methyl Esters.

The data in **Table 13** indicates that two important properties of fatty compounds (melting point and cetane number) for fuel use vary with structure of the fatty acid or ester. For example, both cetane numbers and melting points increase with increasing chain length and increasing saturation. The fatty acid profile (composition) is the major factor influencing the fuel properties of a biodiesel fuel. Analytical methods for its determination are of great significance.

The methyl esters are derived from the parent oil or fat by the *transesterification* reaction shown in **Figure 23**. Note that triacylglycerols (the glycerol esters of fatty acids) are the major components of vegetable oils.

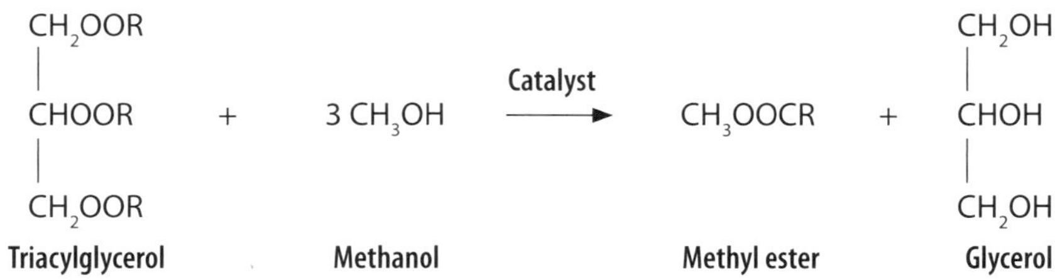

Figure 23. Transesterification Reaction.

The main reason transesterification is conducted to produce biodiesel is that the viscosity of the parent vegetable oil or animal fat is an order of magnitude greater than that of the corresponding methyl esters (biodiesel). The high viscosity of triacylglycerols leads to operational problems, such as engine deposits, etc. Other alcohols, such as ethanol, propanol or *iso*-propanol, could be used for producing biodiesel; however, methanol is the least expensive alcohol and is therefore most commonly used. Here, we will assume, unless mentioned otherwise, that methanol is the alcohol used. Some analytical methods may require modification when using an alcohol other than methanol.

As in most chemical reactions, intermediates are formed during the course of the transesterification reaction. In case of vegetable oils or animal fats (triacylglycerols), it is probably not very surprising that *mono*- and *di*-acylglycerols shown in **Figure 24** are formed, i.e., partially reacted triacylglycerols.

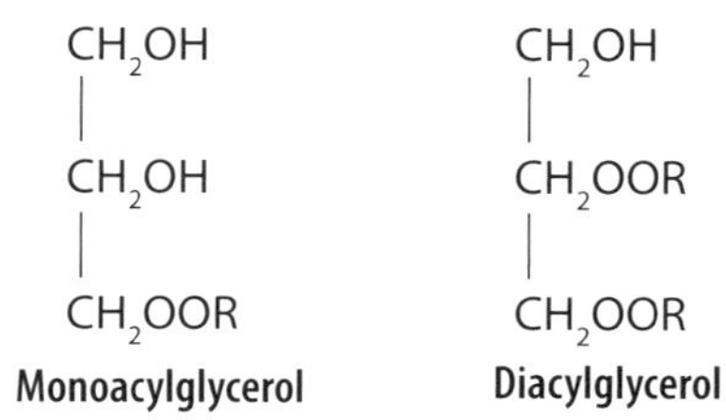

Figure 24. Intermediate Products of Transesterification.

Besides the acylglycerols, residual alcohol (methanol) and catalyst (usually KOH or NaOH) from the transesterification reaction can contaminate the final biodiesel product. Contaminants not resulting from the transesterification reaction, like non-fatty materials occurring naturally in fats and oils, may also contaminate the final product.

This chapter describes the analysis of biodiesel (fatty acid methyl esters) contaminated with glycerol and the various acylglycerols as free and total glycerin (the terms glycerin and glycerol can be used interchangeably). The limiting values given in the ASTM standard for biodiesel, D 6751, are 0.02 percent (by wt.) for free glycerol, and 0.24 percent for total glycerol. Obviously, glycerol remaining in the biodiesel after purification is the free glycerin parameter. The three acylglycerols in combination with the free glycerol are the total glycerin parameter.

Generally, the analytical methods used here can be divided into three categories: chromatographic methods, spectroscopic methods and physical property based methods. Chromatographic and spectroscopic methods can be combined (with a spectroscopic method automatically analyzing the effluent of chromatographic methods) to give what are often termed *hyphenated* methods, because the two methods are then combined through a hyphen.

Chromatographic Methods

Chromatographic methods are used to separate a mixture of compounds based on their physical properties. The major chromatographic methods are *gas chromatography* (GC) and *liquid chromatography* (LC), often termed *high-performance liquid chromatography* (HPLC).

In GC, the mixture is separated mainly by the boiling point and the structure (imparting a feature termed *polarity*) of the individual compounds. To carry out a GC analysis, the sample is usually dissolved in low concentrations in an organic solvent and then injected into the gas chromatograph. In some cases, a sample needs to be *derivatized* with a specific reagent in order to obtain a useful gas chromatogram. This is the case for biodiesel. Glycerol and the mono- and diacylglycerols contain free hydroxyl groups, causing these materials not to perform well in GC. Derivatization (in the case of biodiesel with a *silylating* reagent) improves their performance considerably. Derivatization can provide better resolution between compounds with similar properties.

After injection into the gas chromatograph, the sample is separated on a *column*, a long, thin path (*capillary*) tube that contains a material with which the sample components interact more or less strongly depending on their structure (polarity) when heated. Usually, the solvent used and, if applicable, residual derivatizing agent used for preparing the sample are the first materials to *elute* (come out) from the GC and will be registered by means of a *detector*. When the detector detects a material eluting from the column at a certain *retention time* (the time after which a compounds comes out from the column), a *peak* in the *chromatogram* (the record of the chromatographic analysis) will show. Generally, the integrated value of the peaks' amplitude over time is proportional to the amount of material causing them. This constitutes the usefulness of GC in quantifying the amounts of components in a mixture.

As indicated, the structure (polarity) of the individual components and their boiling points are the major factors determining the retention time. Frequently, larger molecules will have longer retention times in GC.

Often, *standards*, known compounds which will indicate when compounds of a specific nature can be expected to elute, are used in GC. They are useful in establishing the nature of the compounds in a mixture.

For biodiesel, GC forms the basis of the standard ASTM D 6584, which is the prescribed method for measuring free and total glycerol. It determines the amount of glycerol (in derivatized form), mono- and diacylglycerols (both also in derivatized form), triacylglycerols and methyl esters in a biodiesel sample. The derivatized glycerol is the first material to elute, followed sequentially by the methyl esters and the derivatized monoacylglycerols, diacylglycerols and triacylglycerols. While many individual compounds can be separated by GC, some overlap in elution time occurs in complex mixtures (see the chromatogram in ASTM D6584), especially when they are major components with similar properties. On the other hand, to determine the amount of contaminants in biodiesel, you actually do not need to know the complete fatty acid profile in the form of the methyl esters. All that is required is the total amount of the specified contaminants relative to the methyl esters.

Details on the preparation of the samples and conducting the GC runs are given in the ASTM D 6584 standard. Manuals provided by the manufacturers provide information on how to operate a GC.

LC generally separates a mixture based on the solubility of its components in a solvent (or mixture of solvents) while passing the mixture through a column (similar to GC). It is usually conducted at room temperature. Since similar compounds have similar solubilities and interactions with column material, complex mixtures containing many different compounds will usually be separated by classes of compounds and not necessarily by individual compounds. There are modifications to LC to improve on this aspect but they will not concern us here. In the case of biodiesel, for example, methyl palmitate and methyl stearate (both being saturated long-chain methyl esters) will elute simultaneously in most systems. Many terms such as *detector*, *retention time*, etc. are also used in LC and have the same meaning as with GC.

LC can be as suitable as GC for determining the contaminants in biodiesel relative to the total amount of methyl esters. Free glycerol can be detected in fatty acid methyl esters by AOCS (American Oil Chemists' Society) Recommended Practice Ca 14b-96 entitled "Quantification of Free Glycerine in Selected Glycerides and Fatty Acid Methyl Esters by HPLC with Laser Light-scattering Detection." However, it is not included in the standard ASTM D6751.

Spectroscopic Methods

In contrast to chromatographic methods, spectroscopic methods analyze the intact sample at the same time, i.e., impurities contribute to the results and must be taken into consideration. However, when coupling spectroscopic methods to chromatographic methods, uncertainties in the use of the latter are eliminated. The reason is that the (pure) compounds have obtained unique spectra. This information is related directly to the structure of the compound. Chromatographic methods only detect if a compound is eluting, not its identity or structure. The identity or structure needs to be established through the use of standards as far as possible. Thus, using a spectroscopic method of detection in combination with a chromatographic method yields more detailed information. However, it usually results in additional instrument costs and need for additional knowledge to interpret the data.

The most common method in combination with chromatography is *mass spectrometry* (MS). The spectra in MS record how a compound is broken up into fragments by means of energy (usually a beam of electrons). The way a compound splits into fragments is characteristic of its structure. Any detailed discussion of this is beyond the scope of this course. It may be noted, however, that GC-MS, the combination of gas chromatography with mass spectrometry, is currently the most commonly used hyphenated analytical technique in organic chemistry. Mass spectrometry can also be coupled to LC, yielding LC-MS.

Recently, stand-alone spectroscopic methods that have been used for biodiesel include *nuclear magnetic resonance* (NMR) and *near-infrared* (NIR) spectroscopy. In both cases, certain peaks characteristic for triacylglycerols and methyl esters in the spectra indicate how far the conversion of triacylglycerols to methyl esters (biodiesel) has progressed. NIR is especially easy to use and can give spectra in less than a minute. One of the advantages of these stand-alone methods is that no derivatization is needed.

Physical Property-based Methods

The difference in viscosity between the parent vegetable oil and the corresponding methyl esters can serve to monitor the progress of the transesterification reaction. Otherwise, physical property-based methods do not yield as much detailed analytical information as the other two categories of methods. It also appears that other physical properties may be suitable in a fashion similar to viscosity.

The following referenced publication, prepared by one of the authors of this book, provides an in-depth overview of biodiesel analytical methods: Knothe, G., "Analytical Methods Used in the Production and Fuel Quality Assessment of Biodiesel," ASAE Transactions, v. 44 (2), 2001, pp. 193-200.

Wet Chemical Methods

Fatty acid profile: The fatty acid profile can only be crudely determined by wet chemical methods. Wet chemical methods include the iodine and saponification values. The iodine value is based on the theoretical addition of iodine to the double bonds of fatty compounds. The iodine value is an indicator of the total amount of unsaturated fatty compounds in a sample. Otherwise, it does not give any information on the nature of the unsaturated compounds nor, of course, the saturated compounds. The saponification value is related to the average molecular weight of the sample of fatty compounds. It is probably best to use the iodine and saponification values in conjunction with each other. There are several wet chemical methods for each one, each having its advantages and limitations. Neither of these values is contained in the biodiesel standard ASTM D6751.

There are several wet-chemical AOCS methods for determining glycerol. One is AOCS Official Method Ca 14-56 entitled "Total, Free and Combined Glycerol Iodometric - Periodic Acid Method." The LC method mentioned earlier appears preferable. Other methods deal with the determination of glycerol in specific products or under specific circumstances and are not applicable.

Generally, wet chemical methods are being replaced by the more sophisticated chromatographic, spectroscopic methods or hyphenated methods because of the superior information obtained. The wet chemical methods are often more time-consuming, because they can require complex sample preparation.

CHAPTER 13

ASTM Methods for Determining the Chemical Properties of Biodiesel

The instructional goals of this chapter are to provide an overview of the ASTM standard tests for determining:

1. Acid value (or acid number).
2. Phosphorus content.
3. Sulfur content.
4. Sulfated ash content.
5. Water and sediment.

In particular, we look at tests as they apply to biodiesel. Information does not replace the need for a testing facility to have access to and explicitly follow the ASTM standard test methods for determining these properties. Individual ASTM standards can be found and purchased at www.astm.org. They are also commonly found in a technical library's reference section.

ASTM Methods for Key Biodiesel Properties

The American Society for Testing and Materials (ASTM) International provides standards and methods that are accepted and used in research and development, product testing, quality control protocols and commercial transactions. ASTM standard tests are explicit outlines of procedures and the apparatus necessary to carry out those procedures. In an effort to provide an accurate, yet concise chapter, these standards will be directly cited. The full titles of these ASTM standards are:

- D 664 Standard Test Method for Acid Number of Petroleum Products by Potentiometric Titration.
- D 4951 Standard Test Method for Determination of Additive Elements in Lubricating Oils by Inductively Coupled Plasma Atomic Emission Spectrometry.
- D 5453 Standard Test Method for Determination of Total Sulfur in Light Hydrocarbons, Motor Fuels and Oils by Ultraviolet Fluorescence.
- D 874 Standard Test Method for Sulfated Ash from Lubricating Oils and Additives.
- D 2709 Standard Test Method for Water and Sediment in Middle Distillate Fuels by Centrifuge.

All ASTM standards use a similar format. They are divided into sections that may or may not be present in all standards. Every section contained in each of these five standards is not covered, only those necessary to convey the general contents of the test method. These five standards all contain the following sections:

- Scope – discusses the materials and objects to which the test applies.
- Referenced documents – lists ASTM standards that are referred to in the standard.
- Summary of the test method – brief discussion of test.
- Significance and use – discusses points pertaining to the test method.
- Apparatus – detailed specifications for all equipment needed to perform the test.
- Reagents – detailed specifications for all chemicals needed to perform the test.
- Procedure – detailed instructions of performance of the test method.
- Precision and bias – discussion of the reliability and repeatability of the test.
- Keywords – used in database searching.

The standards also contain sections pertaining to concerns specific to each individual test.

ASTM D – 664: Determination of Acid Number by Potentiometric Titration (biodiesel limit 0.80 mg of KOH/g sample).

Scope - This test method covers procedures for the determination of acidic constituents in petroleum products and lubricants that are soluble in mixtures of toluene and isopropanol. In B100 (biodiesel), the acid number is a measure of free fatty acids. The free fatty acids can lead to corrosion and are a symptom of water in the fuel or fuel oxidation.

Discussion - This test method determines the quantity of base that is required to titrate a sample from its initial reading to a reading corresponding to a basic buffer solution or an inflection point as specified in the test method. It also provides the additional information on the quantity of base required to titrate a sample from its initial reading to a reading corresponding to an acidic buffer solution or an inflection point, as specified in the test method. This quantity is reported as the *strong acid number*. The causes and effects of strong acids and of other acids can be very different. Therefore, this test method differentiates and reports the two values. Since biodiesel generally contains only weak acids, this distinction is not relevant. An alternative method that is simpler to perform and gives similar results is the American Oil Chemists Society Official Method Cd 3a-63. This method uses phenolphthalein indicators to determine neutrality. However, since it is not included in ASTM D 6751, it should be used only for process development, not for verifying product quality.

Summary of test method - The sample is dissolved in a mixture of toluene and isopropanol that contains a small amount of water. The sample is titrated potentiometrically with alcoholic potassium hydroxide. The meter readings are plotted against the respective volumes of titrating solution and the end points are taken at well-defined inflections in the resulting curve.

Terminology - Acid number is the quantity of base, expressed as milligrams KOH per gram of sample, required to titrate the sample to a specified end point.

Apparatus - Exacting specifications are given in the standard for the potentiometer, electrodes, stirrer, buret, titration beaker and titration stand to be used. Specifications of an automatic titration apparatus accepted as an adequate substitute are also given.

Procedure for acid number and strong acid number - Small volumes of KOH are added to samples in alcohol solution while measuring the electrode potential in mV. A curve of KOH volume vs. potential is then plotted.

Detailed procedures for the manual and automatic titration methods are given, including blank titrations of solvent.

Calculations – The electrode potential curves are plotted as the volume of KOH needed to titrate the sample vs. mV or pH. The end (inflection) point is chosen as the point at which the curve is most nearly vertical. A procedure is also given for curves without a definite inflection point. **Figure 25** is an example of the titration curves that are produced using this method.

Acid number = 56.1M(A-B)/W

where M = concentration of KOH (moles/L)

A = volume of KOH used to reach end point corresponding to basic buffer (ml)

B = volume corresponding to A for blank titration (ml)

W = mass of sample (g)

Strong acid number = 56.1(CM + Dm)/W

where C = volume of KOH used to reach end point corresponding to acidic buffer (ml)

D = volume HCl used to reach end point corresponding to C (ml)

m = concentration of HCl (moles/L)

The acid number and strong acid number are to be reported as follows:

Acid number (Test Method D 664) = *result*

Strong acid number (Test Method D 664) = *result*

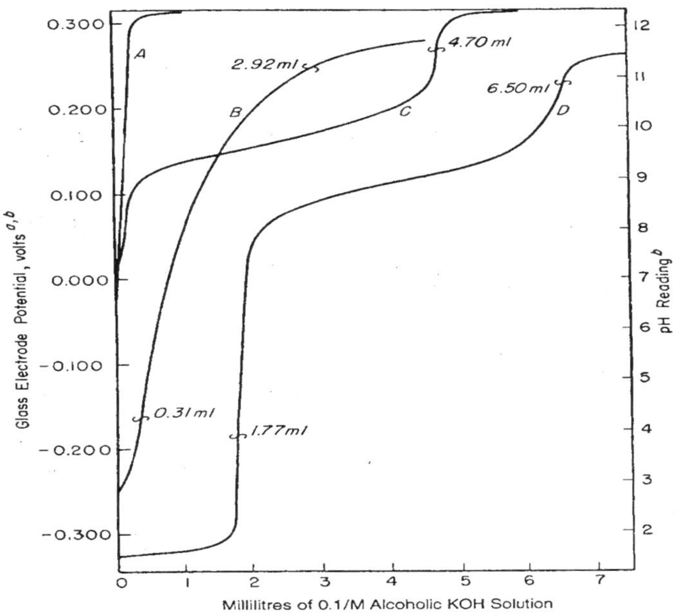

Figure 25. Sample Titration Curve for ASTM D 664.

Curve A – Blank of 125 ml of titration solvent.

Curve B – 10.00 g of used crankcase oil, plus 125 ml of titration solvent. Since no sharp inflections are apparent, the end points are chosen at the meter readings obtained with the two non-aqueous buffer solutions.

Curve C – 10.00 g of oil containing a weak acid plus 125 ml of titration solvent. The end point is chosen as the point at which the curve is most nearly vertical.

Curve D – 10.00 g of oil containing weak and strong acids, plus 125 ml of titration solvent. The end points are chosen as the points at which the curve is most nearly vertical.

STM D – 4951: Standard Test Method for Determination of Additive Elements in Lubricating Oils by Inductively Coupled Plasma Atomic Emission Spectrometry (biodiesel limit – 0.001 mass % phosphorus).

Scope - This test method covers the quantitative determination of barium, boron, calcium, copper, magnesium, phosphorus, sulfur and zinc in unused lubricating oils and additive packages.

Additive packages are blends of individual additives, which can act as detergents, antioxidants, antiwear agents, etc. Many additives contain one or more elements covered by this test, which was developed for testing oil additives to meet industry specifications. In biodiesel processing, this test is used primarily for the determination of phosphorus content. The presence of phosphorus may be due to the incomplete refining of vegetable oil or bone and proteins remaining from the rendering process in recovering animal fats.

This test can also be used to determine the presence of sulfur if the ICP instrument used can operate at a wavelength of 180 nm. However, this test is not the recommended method for sulfur determination when testing to meet the specifications of B100 biodiesel as outlined in ASTM D6751. That test method (D5453) is discussed later in this chapter.

Summary of method – A sample is diluted (1 to 5 percent by mass of oil in solvent) with mixed xylenes or kerosene. An internal standard is included in the solvent and calibration standards. The solutions are run through the ICP. By comparing emission intensities of the sample to those of the standards, concentrations of the elements present are determined.

Apparatus – Required and optional apparatus are listed in the apparatus section of the standard. The primary instrument is an atomic emission spectrometer that is equipped with a quartz inductively-coupled plasma (ICP) torch and r-f generator to form the plasma. Also needed are an analytical balance and solution containers. Optional equipment includes a peristaltic pump, solvent dispenser, vortexer and ultrasonic homogenizer. Mixed xylenes, o-xylene or kerosene can be used as solvents. Base oil (U.S.P. white oil or a lubricating base oil free of analytes) is required for the internal standard solution (several possible internal standards are listed). Calibration standard solutions containing known concentrations of the elements being determined are prepared in accordance with the ASTM Practice D 4307.

Phosphorus is the only element of interest when this test method is applied to biodiesel production specifications.

Sampling – The objective of sampling is to obtain a test specimen that is representative of the entire quantity. Lab samples are recommended to be taken in accordance to ASTM D4057, which provides procedures for sampling petroleum products from a wide variety of containers. The specific sampling technique can affect the accuracy of this test.

Preparation of apparatus – Recommendations for the operation of the atomic emission spectrometer, ICP excitation source and peristaltic pump are given in this section of the test method. There are several different adequate instruments, so the recommendation is to follow the manufacturer's instructions regarding operation of the ICP unit.

Preparation of test specimens – Procedures are given for the weighing out and dissolution of oil sample specimens.

Preparation of calibration standards – The user is given freedom to determine the identity and concentration of calibration standards. Guidelines are given for solution preparation, and instrument check standards are suggested.

Internal standardization – This section of the test method outlines the required use of internal standards.

Calibration – A comprehensive discussion of calibration curves and the application of internal standards to those curves in ICP spectrometry is provided in this section. After compensating for the internal standard, the calibration curve is a plot of the intensity of emission for the analyte vs. analyte concentration.

Analysis – Samples should be analyzed using the same procedure as the calibration standards. An instrument check standard should be run after every fifth test specimen to ensure that the instrument is still calibrated.

Quality assurance – QA/QC is required by this standard. When QA/QC protocols are already established in the testing facility, these may be followed. If no QA/QC protocols exist, the appendix to this standard (Appendix X1) can be used. Appendix X1 discusses the use of quality control charts to determine if the procedure is under statistical control.

Calculations – $C = S(W_1+W_2+W_3)/W_1$

where C = analyte concentration in sample (mass %)

S = analyte concentration in test specimen (mass %)

W1 = sample mass (g)

W2 = diluents mass (g)

W3 = base oil mass (g)

Generally, ICP software performs this calculation automatically.

Reporting – Report mass percent to three significant figures.

ASTM D – 5453: Standard Test Method for Determination of Total Sulfur in Light Hydrocarbons, Motor Fuels and Oils by Ultraviolet Fluorescence (biodiesel limit – 0.05 mass percent).

Scope – This test method covers the determination of total sulfur in liquid hydrocarbons. It is applicable to naphtha, distillates, oils, gasoline, diesel, biodiesel and jet fuel.

Some process catalysts used in petroleum and chemical refining can be poisoned when trace amounts of sulfur are present. Biodiesel feedstocks typically have very little sulfur, but this test is an indicator of contamination of protein material and/or carryover catalyst or neutralization material.

Summary of method – The sample is inserted into a high temperature combustion tube where sulfur is oxidized to sulfur dioxide (SO_2) in an oxygen-rich atmosphere. Sample combustion gases are next exposed to UV light. The SO_2 is excited, from which fluorescence is emitted as it returns to its stable state. The fluorescence is detected by a photomultiplier tube, and the resulting signal is a measure of the amount of sulfur contained in the sample.

Apparatus – Required and optional apparatus are discussed in this section of the standard. Primary pieces of equipment include a Quartz Combustion Tube and a UV Fluorescence Detector. Sample input systems can either be a boat inlet system or a direct injection design. Another necessity for this test is a furnace capable of holding a temperature (1075±25°C) sufficient to pyrolyze all of the sample and oxidize sulfur to SO_2.

Also required is a dryer tube or permeation dryer to remove water vapor that is produced in the pyrolysis tube. Water must be removed prior to the fluorescence measurement by the detector. The test method allows the option of using a self-calibrating analyzer.

Reagents – High purity, dry oxygen is used in the pyrolysis tube. Solvents such as toluene, xylene or isooctane are needed to prepare standards and dilute samples. A sulfur-containing compound such as dibenzothiophene, butyl sulfide or benzothiophene can be used for calibration standards.

Procedures for mixing appropriate standard are provided. QC samples, representative of the samples of interest, can be used to check the validity of the testing process.

Sampling – The objective of sampling is to obtain a test specimen that is representative of the entire quantity. Lab samples should be taken in accordance with ASTM D4057, which provides procedures for sampling petroleum products from a wide variety of containers. The specific sampling technique can affect the accuracy of this test.

Preparation of apparatus – This section of the standard test method discusses the typical operating conditions including sample inlet rate, furnace temperature, oxygen flow rate and carrier gas flow rate.

Calibration and standardization – This section provides the procedure for preparation of a calibration curve. Three different suggestions for curve preparation are given, based on the anticipated sulfur concentration in the sample. Again, procedures for both direct injection and inlet boat delivery into the apparatus are provided.

Procedure – The procedure allows for either gravimetric or volumetric dilution of samples and either direct injection or inlet boat delivery into the apparatus. It requires repeat instrument calibration prior to reanalysis of the sample. Also, each test specimen must be measured three times and the average response must be calculated.

Calculation – Either analyzers calibrated using a standard curve or analyzers using a self-calibration routine are accepted. Sulfur content can be calculated using either the mass or volume of the test sample for either type of analyzer used.

Using analyzer with standard curve:
(using sample mass) Sulfur (ppm) = $(I-Y)/SMK_g$
- I = integrated detector response (counts)
- Y = y-intercept of standard curve (counts)
- S = slope of standard curve (counts/μg sulfur)
- M = mass of test specimen (g)
- K_g = gravimetric dilution factor; mass sample/mass sample+solvent (g/g)

(using sample volume) Sulfur (ppm) = $(I-1000Y)/SVK_V$
- V = volume of test specimen (g)
- K_V = volumetric dilution factor; mass sample/volume sample+solvent (g/ml)

Using self-calibrating analyzer:
(using sample mass) Sulfur (ppm) = $1000G/MK_g$
- G = sulfur found in test specimen (μg)
- M = mass of sample solution injected (mg)

(using sample volume) Sulfur (ppm) = $1000G/VD$
- D = density of sample in solution (mg/μL)
- V = volume of sample solution injected (μL)

Quality control – Required confirmation of the performance of the test procedure is performed by analyzing a quality control sample. When QA/QC protocols are already established in the testing facility, these may be followed. If no QA/QC protocols exist, the appendix to this standard (Appendix X1) can be used. Appendix X1 discusses the use of quality control charts to determine if the procedure is under statistical control.
Appendices – As stated above, App. X1 discusses QA/QC techniques. App. X2 discusses direct injection system input. App. X3 discusses boat inlet system input.

ASTM D – 874: Standard Test Method for Sulfated Ash from Lubricating Oils and Additives (biodiesel limit – 0.02 mass percent).

Scope – This test method determines the amount of mineral ash remaining after a fuel is burned. For biodiesel, this test is an important indicator of the quantity of residual metals in the fuel that could come from the catalyst used in the esterification process. The lower limit of this test is 0.005 mass percent sulfated ash.

Terminology – Sulfated ash is the residue remaining after the sample has been carbonized, and the residue subsequently treated with sulfuric acid and heated to constant weight.

Summary of the test method – The sample is ignited and burned until only ash and carbon remain. After cooling, the residue is treated with sulfuric acid and heated at 775°C until oxidation of carbon is complete. The ash is then cooled, re-treated with the sulfuric acid and heated at 775°C to constant weight.

Apparatus – Necessary for the test method are a furnace capable of maintaining 775±25°C and evaporating dishes or crucibles. A 50 or 100 ml crucible is recommended for samples containing more than 0.02 mass percent sulfated ash. A 120 or 150 ml crucible is recommended for samples containing less than 0.02 mass percent sulfated ash.

Reagents – Required reagents include concentrated sulfuric acid and 50 volume% sulfuric acid in water. Isopropanol and toluene may be needed if the sample contains sufficient moisture to cause foaming and loss of material from the dish/crucible.

Sampling – It is recommended to follow the instructions in ASTM Standard Practice D4057.

Procedure – The general steps required to complete this test are:

- Prepare the sample dish by heating at 775°C for at least 10 min., cooling and weighing.
- Weigh in the sample to be determined.
- Heat the sample until the contents will sustain a flame and burn the sample until there is no further smoke or fumes.
- If the sample has sufficient moisture to foam, discard the sample, and follow procedures given to reduce the moisture content before proceeding.
- Cool the sample, add water and sulfuric acid and heat until no further fumes are evolved.
- Place the dish in a furnace and heat at 775°C until oxidation of the carbon is complete. Continue the cooling, acidification and heating process until successive weighings differ by no more than 1.0 mg.

Calculation – Sulfated ash is calculated as a percentage of the original sample.

Sulfated Ash (mass%) = w/W*100
w = mass of sulfated ash (g)
W = mass of sample used in test (g)

Report – Report the result to the nearest 0.001 mass percent for samples below 0.02 mass percent and to the nearest 0.01 mass percent for higher levels.

ASTM D 2709 Standard Test Method for Water and Sediment in Middle Distillate Fuels by Centrifuge (biodiesel limit – 0.05 volume percent).

Scope – This test method covers the determination of the volume of free water and sediment in middle distillate fuels having viscosities at 40°C in the range of 1.0 to 4.1 mm^2/s and densities in the range of 770 to 900 kg/m^3. This test is a measure of cleanliness of the fuel. For B100 biodiesel it is particularly important, because water can react with the esters to make free fatty acids and can support microbial growth in storage tanks.

Summary of test method – A 100 ml sample of undiluted fuel is centrifuged in a tube readable to 0.005 ml. After centrifugation, the volume of water and sediment which has settled into the tip of the centrifuge tube is read to the nearest 0.005 ml and recorded as the volumetric percent water and sediment by centrifuge.

Apparatus – Centrifuge which can be controlled to give a relative centrifugal force (rcf) of 800±60 at the tip of the tubes. A table is provided that relates centrifuge diameter and rpm to rcf. Specifications and suggested suppliers for centrifuge tubes are listed.

Sampling - It is recommended to follow the sampling instructions in ASTM Standard Practice D4057. The sample for a laboratory test will normally be an aliquot of a much larger sample taken for full or partial specification testing.

Procedure – 100 ml of a well-shaken sample is poured into a centrifuge tube and spun at 800±60 rcf for 10 minutes. The volume of combined water and sediment present at the bottom of the tube is recorded to the nearest 0.005 ml.

Report – Report the volume of the combined water and sediment read from the tube as a percentage of the total sample, since 100 ml of sample was used. Results lower than 0.005 percent may be reported as either 0 or 0.005 volume percent.

Apparatus – Centrifuge which can be controlled to give a relative centrifugal force (rcf) of 800±60 at the tip of the tubes. A table is provided that relates centrifuge diameter and rpm to rcf. Specifications and suggested suppliers for centrifuge tubes are listed.

Sampling - It is recommended to follow the sampling instructions in ASTM Standard Practice D4057. The sample for a laboratory test will normally be an aliquot of a much larger sample taken for full or partial specification testing.

Procedure – 100 ml of a well-shaken sample is poured into a centrifuge tube and spun at 800±60 rcf for 10 minutes. The volume of combined water and sediment present at the bottom of the tube is recorded to the nearest 0.005 ml.

Report – Report the volume of the combined water and sediment read from the tube as a percentage of the total sample, since 100 ml of sample was used. Results lower than 0.005 percent may be reported as either 0 or 0.005 volume percent.

CHAPTER 14

Soap and Catalyst Measurement

Production of biodiesel using an alkali catalyst always produces some amount of soap. There will be more soap with recycled restaurant waste and animal fats and less with refined vegetable oils.

After the transesterification reaction is complete, the leftover catalyst and soap tend to concentrate in the glycerol phase. However, some soap and a small amount of catalyst may be left in the biodiesel phase. During process development, it can be useful to know the amount of soap formed, where the catalyst resides and how effective the washing process is in removing these two compounds.

A simple titration procedure can be used to measure the amount of soap and catalyst. The basic procedure consists of titrating a sample of the phase to be analyzed, such as glycerol, ester, wash water, etc., with a 0.1 N solution of HCl to the phenolphthalein end point. This gives an estimate of the amount of catalyst. Then, a few drops of bromophenol blue indicator are added and the titration continued to the color change for that indicator. This gives an indication of the amount of soap.

In the first titration, the HCl neutralizes the alkali catalyst, so when the phenolphthalein indicates that the solution has become neutral, then all of the catalyst has been counted. Then, if the titration is continued, the HCl, as a strong acid, begins to split the soap molecules to free fatty acids and salt. When the pH reaches about 4.6, where the bromophenol blue changes color, then this indicates that the HCl has split all of the soap. It is now lowering the pH, so it has protons to donate since the soap has all been split.

The following procedure is a modified version of AOCS method Cc 17-79, soap in oil:

1. Dissolve sample in 100 ml of acetone containing 2 percent distilled water. The amount of sample will depend on the expected level of catalyst and soap. Use 0.5 g of sample for crude glycerol, 5 g of sample for unwashed methyl esters, and 100 g of sample for washed methyl esters. When testing glycerol, it can be helpful to add 10 ml of distilled water to enhance the solubility of the glycerol in the acetone.
2. Add 6 to 8 drops of 1 percent phenolphthalein indicator (in isopropyl alcohol).
3. Titrate with 0.01 N hydrochloric acid until you get the phenolphthalein color change (red to clear). This indicates that you have neutralized the free catalyst in the sample. Designate this amount of solution as "A."
4. Add 1 ml of bromophenol blue indicator (0.4 percent in water). This indicator changes color at about pH = 4.5.
5. Titrate until the bromophenol changes from blue to yellow. This indicates that the pH is low enough that all of the soap should have been split into FFA and salt. The strong acid that we are adding (hydrochloric) is more strongly attracted to the metal ion on the soap than the fatty acid chain. So, the metal ion combines with the Cl from the hydrochloric acid to produce NaCl, and the hydrogen freed from the hydrochloric acid converts the fatty acid chain to a free fatty acid. Designate this quantity of solution as "B."

Now, the amount of HCl added during the first titration tells us how much free catalyst is in the sample and the amount added during the second titration tells us the amount of soap.

Take the ml of HCl added for the first titration and do the following calculation:

["A" ml of 0.1 N HCl] x [1 liter/1000 ml] x [0.1 moles of HCl/liter] x [1 mole of KOH/mole of HCl] x [56.1 g/mole KOH] / [grams of sample] = grams of KOH/gram of sample

or

$$\frac{A \times 0.1 \times 56.1}{1000 \times W} = \text{grams of KOH catalyst/gram of sample}$$

This gives the amount of free catalyst in the sample, done here assuming the catalyst was KOH. You can substitute the appropriate molecular weight for other catalysts (KOH = 56.1, NaOH = 40.0, $NaOCH_3$ = 54.0).

If you take the ml of HCl for the second titration and do the following calculation:

["B" ml of additional 0.1N HCl added] x [1 liter/1000ml] x [0.1 moles of HCl/liter] x [1 mole of soap/mole of HCl] x [320.56 g/mole of soap] / [grams of sample] = grams of soap (as sodium oleate) per gram of sample.

or

$$\frac{B \times 0.1 \times 320.56}{1000 \times W} = \text{grams of soap (as potassium oleate)/gram of sample}$$

This often is expressed as ppm, so this number should be multiplied by one million. This calculation assumed the soap was potassium oleate. When using sodium catalysts, the molecular weight of sodium oleate is 304.4 g/mole. Oleic acid is a pretty good average compound for soybean oil, but you might want to change to a higher molecular weight for rapeseed oil, which contains significant amounts of heavier fatty acids.

When the level of catalyst and soap are small, it is a good idea to use blank samples. Just prepare a second beaker of the acetone-water solution but don't add any sample. Titrate the blank at the same time as the sample and subtract the quantities for the blank from the numbers for the sample.

CHAPTER 15

Pretreatment of High Free Fatty Acid Feedstocks

Many low-cost feedstocks are available for biodiesel production. Unfortunately, many of these feedstocks contain large amounts of free fatty acids (FFAs). As discussed elsewhere, these free fatty acids will react with alkali catalysts to produce soaps that inhibit the reaction.

Table 14 shows the ranges of FFAs commonly found in biodiesel feedstocks.

Refined vegetable oils	< 0.05 %
Crude vegetable oil	0.3 – 0.7%
Restaurant waste grease	2 – 7%
Animal fat	5 – 30%
Trap grease	40 – 100%

Table 14. FFA Ranges in Feedstocks.

Generally, when the FFA level is less than 1 percent, and certainly if it is less than 0.5 percent, the FFAs can be ignored. Common catalyst amounts are shown in **Table 15**.

Sodium hydroxide	1% of triglyceride weight
Potassium hydroxide	1% of triglyceride weight
Sodium methoxide	0.5% of triglycerides weight

Table 15. Common Catalyst Amounts.

Soaps may allow emulsification that causes the separation of the glycerol and ester phases to be less sharp. Soap formation also produces water that can hydrolyze the triglycerides and contribute to the formation of more soap. Further, catalyst that has been converted to soap is no longer available to accelerate the reaction.

When FFA levels are above 1 percent, it is possible to add extra alkali catalyst. This allows a portion of the catalyst to be devoted to neutralizing the FFAs by forming soap, while still leaving enough to act as the reaction catalyst.

Since it takes one mole of catalyst to neutralize one mole of FFA, the amounts of additional catalyst can be calculated by the following formulas shown in **Table 16** (which include the amounts shown above that are needed to catalyze the reaction).

Sodium hydroxide	[%FFA](0.144) + 1%
Potassium hydroxide	[%FFA](0.197)/0.86 + 1%
Sodium methoxide	[%FFA](0.190) + 0.5%

Table 16. Calculations for Additional Catalyst Amounts.

For example, when adding sodium methoxide to a feedstock with 1.5 percent FFA, the amount of catalyst would be:

$$(1.5)(0.190) + 0.5\ \% = 0.79\ \%\ \text{of the triglyceride weight}$$

Note that a factor of 0.86 has been included with the potassium hydroxide calculation to reflect that reagent grade KOH is only 86 percent pure. If other grades of catalyst are used, this factor should be adjusted to their actual purity.

This approach to neutralizing the FFAs will sometimes work with FFA levels as high as 5 – 6 percent. The actual limit depends on whether other types of emulsifiers are present. It is especially important to make sure that the feedstock contains no water. FFAs of 2-3 percent may be the limit if traces of water are present.

For feedstocks with higher amounts of FFA, the addition of extra catalyst may create more problems than it solves. The large amount of soap created can gel. It can also prevent the separation of the glycerol from the ester. Moreover, this technique converts the FFAs to a waste product when they could be converted to biodiesel.

When working with feedstocks that contain 5-30 percent FFA or even higher, it is important to convert the FFAs to biodiesel or the process yield will be low. There are at least four techniques for converting the FFAs to biodiesel:

1. **Enzymatic methods**: These methods require expensive enzymes but seem to be less effected by water. At the present time, no one is using these methods on a commercial scale.
2. **Glycerolysis**: This technique involves adding glycerol to the feedstock and heating it to high temperature (200°C), usually with a catalyst such as zinc chloride. The glycerol reacts with the FFAs to form mono- and diglycerides as shown in the following reaction.

   $$\text{FFA + glycerol} \rightarrow \text{monoglyceride + water}$$

   Figure 26 shows the rate of decrease of the fatty acid level in a batch of animal fat. This technique produces a low FFA feed that can be further processed using traditional alkali-catalyzed techniques.

   The drawbacks of glycerolysis are the high temperature that is required and relative slowness of the reaction. An advantage is that no methanol is added during the pretreatment, so that as water is formed by the reaction, the water immediately vaporizes and can be vented from the mixture.

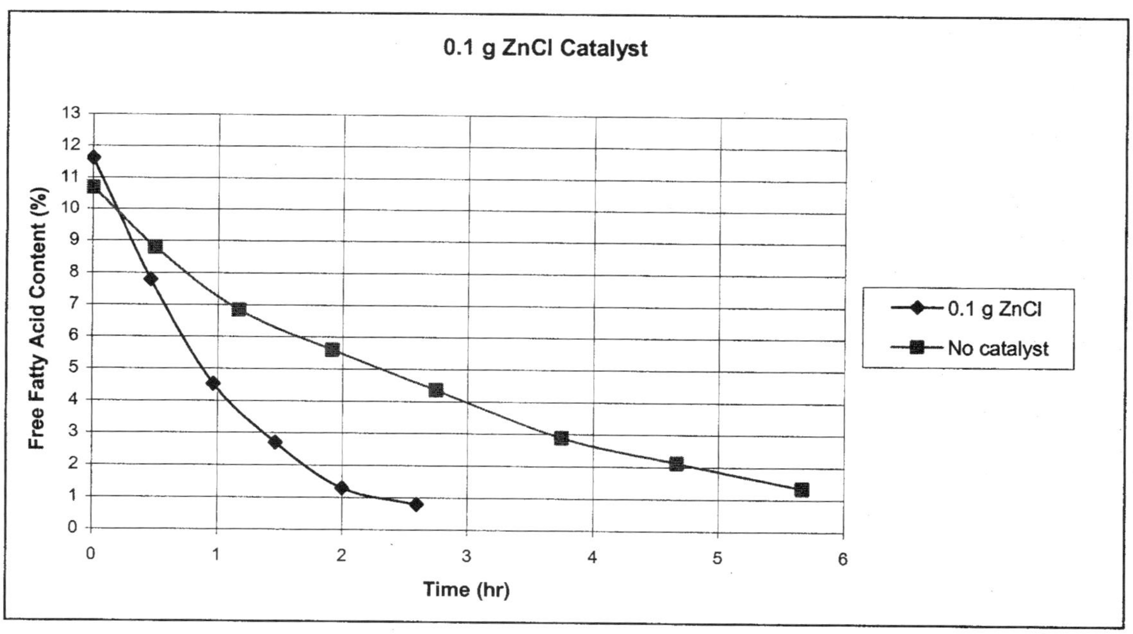

(100 g animal fat, 13 g glycerol, 0.1 g ZnCl, 200°C, 11 psi vacuum)

Figure 26. Reduction of Fatty Acids by Glycerolysis.

3. **Acid catalysis**: This technique uses a strong acid such as sulfuric acid to catalyze the esterification of the FFAs and the transesterification of the triglycerides. The reaction does not produce soaps because no alkali metals are present. The esterification reaction of the FFAs to alcohol esters is relatively fast, proceeding substantially to completion in 1 hour at 60°C. However, the transesterification of the triglycerides is very slow, taking several days to complete. Heating to 130°C can greatly accelerate the reaction, but reaction times will still be 30-45 minutes. Another problem with acid catalysis is that the water production from the following reaction stays in the reaction mixture and ultimately stops the reaction, usually well before reaching completion.

$$\text{FFA} + \text{methanol} \rightarrow \text{methyl ester} + \text{water}$$

4. **Acid catalysis followed by alkali catalysis**: This approach solves the reaction rate problem by using each technique to accomplish the process for which it is best suited. Since acid catalysis is relatively fast for converting the FFAs to methyl esters, it is used as a pretreatment for the high FFA feedstocks. Then, when the FFA level has been reduced to 0.5 percent, or lower, an alkali catalyst is added to convert the triglycerides to methyl esters. This process can convert high FFA feedstocks quickly and effectively. Water formation is still a problem during the pretreatment phase. One approach is to simply add so much excess methanol during the pretreatment that the water produced is diluted to the level where it does not limit the reaction. Molar ratios of alcohol to FFA as high as 40:1 may be needed. The disadvantage of this approach is that more energy is required to recover the excess methanol. Another approach would be to let the acid-catalyzed esterification proceed as far as it will go until it is stopped by water formation. Then, neutralize the acid and boil off the alcohol and water. If the FFA level is still too high, then additional methanol and, if necessary, acid catalyst can be added to continue the reaction. This process can be continued for multiple steps and will potentially use less methanol than the previous approach. Again, the disadvantage is the large amount of energy required by the distillation process.

   A less energy intensive approach is to let the acid-catalyzed reaction mixture settle. After a few hours, a methanol-water mixture will rise to the top and can be removed. Then, additional methanol and acid can be added to continue the reaction. (Hammond and Wang, U.S. Patent 6,965,044. 2005.) It is also possible to use fluids such as glycerol and ethylene glycol to wash the water from the mixture.

Procedure for High FFA Feedstocks

1. Measure FFA level.
2. Add 2.25 g methanol and 0.05 g sulfuric acid for each gram of free fatty acid in the oil or fat. Sulfuric acid and methanol should be mixed first and then added slowly to the oil.
3. Agitate for 1 hour at 60-65°C.
4. Let mixture settle. Methanol-water mixture will rise to the top. Decant the methanol, water and sulfuric acid layer.
5. Take bottom fraction and measure new FFA level.
6. If FFA is > 0.5 percent, return to step 2 with new FFA level. If FFA is < 0.5 percent, proceed to step 7.
7. Add an amount of methanol equal to 0.217 x (grams of unreacted triglycerides) and an amount of sodium methoxide equal to (0.5 + [%FFA]0.190)/100 x (grams of unreacted triglycerides). Mix the sodium methoxide with the methanol and then add to the oil. This corresponds to a 6:1 molar ratio of methanol to oil for the unreacted triglycerides. It ignores any methanol that may have carried over from the pretreatment.
8. Agitate for 1 hour at 60°C.

Example: 100 g of 12% FFA animal fat

Pretreatment: 2.25 x 12 g = 27.0 g methanol

0.05 x 12 g = 0.6 g H_2SO_4 (sulfuric acid)

Mix acid with methanol, then add mixture to fat. Agitate for one hour at 60°C. Let settle and separate bottom phase. Acid value should decrease substantially, to at least 5-6 mg KOH/g. Therefore, FFA = 2.5%.

Second step of pretreatment:

2.25 x 2.5 g = 5.6 g of methanol

0.05 x 2.5 g = 0.13 g H_2SO_4

Mix with oil, agitate at 60°C for 1 hour. FFA should be < 0.5%.Removal of upper phase is usually optional at this point.

Then add: 0.217 x (88) = 19.1 g methanol

(0.5 + [0.5][0.190])/100 x 88 = 0.52 g sodium methoxide

Agitate at 60°C for one hour. If glycerol and ester do not separate, add 50 g of warm distilled water to encourage separation. Wash 3-4 times.

CHAPTER 16 | Alternative Feedstocks

This chapter describes several categories of biodiesel feedstocks from the point of view of their chemical and physical properties. The key properties reviewed are the fatty acid profiles for vegetable oil resources, free fatty acid content, non-triglyceride content and grade specifications for tallows and greases.

The fatty acid profile of the feedstock determines the physical and performance properties of the biodiesel product. This is further described elsewhere in the book Other physical and chemical properties of the raw materials, such as the amount of free fatty acids and the MIU (moisture, insolubles and unsaponifiables) content, affect the steps necessary to prepare the feedstocks for the esterification reaction used to produce biodiesel.

In this chapter, you will:

1. Become familiar with the similarities and differences in composition among the various feedstock materials for biodiesel production.
2. Learn the terminology used to describe different sources of fats and oils and their properties and understand the differences in specifications applied to these sources.
3. Compare the different feedstocks on the basis of their physical properties and chemical compositions.

Feedstock Categories

Vegetable oils are recovered from plant seeds by a series of physical processing steps. The typical vegetable oil recovery process begins with crushing the seed to break the cell membranes retaining the oil and to partially expel the oil from the seed meal. The seed meal may be further extracted using a solvent, such as hexane, to recover additional oil from the meal. The recovered oils may then be treated to remove non-oil components, reduce color and odor and to remove free fatty acids. The treatment process is termed "refining." The degree of refining employed is determined by the ultimate use for the oil.

Tallows and fish oils are recovered from animal sources by a process called "rendering." Essentially, the rendering process is a controlled cooking of animal tissues and bones to separate the fats, proteins, water and bones. The fat (lipid) products are not water soluble and rise to the surface where they are skimmed off. The skimmed fats are further refined, deodorized, bleached and dried to produce tallow (meltable solids), fats (semi-solid) or oils (liquid). Beef and pork lipids form tallows, chicken lipids are fats and fish lipids are typically oils. The differences in physical state arise because of the fatty acid profiles (compositions) characteristic of each source.

Greases are recovered using a similar rendering process, but they may come from a number of sources. Fats, greases and oils (FOG) have somewhat different measures of quality, compared to vegetable oil standards. They include the percent free fatty acids, MIU (moisture, insolubles and unsaponifiables) and the titer (melting point) of the material.

There are a number of waste materials associated with fats and oils processing that contain significant quantities of fatty acid (lipid) material. These materials are expensive to dispose of because of their high BOD (biological oxygen demand). They can, however, provide a low quality feedstock for the production of biodiesel if pretreated before the esterification process. Three of these materials are described in this chapter.

Key Physical and Chemical Properties of Biodiesel Feedstocks

Fats and oils are composed of a specific category of organic compounds called "triacylglycerols" or "triglycerides." A triglyceride (TG) molecule is made up of three fatty acid units attached to a three carbon backbone. The biodiesel reaction uses an alcohol, such as methanol, to split the TG molecule into a single molecule of glycerol (glycerine) and three molecules of fatty acid esters. For example, if the alcohol is methanol, the esters will be called fatty acid methyl esters (FAME).

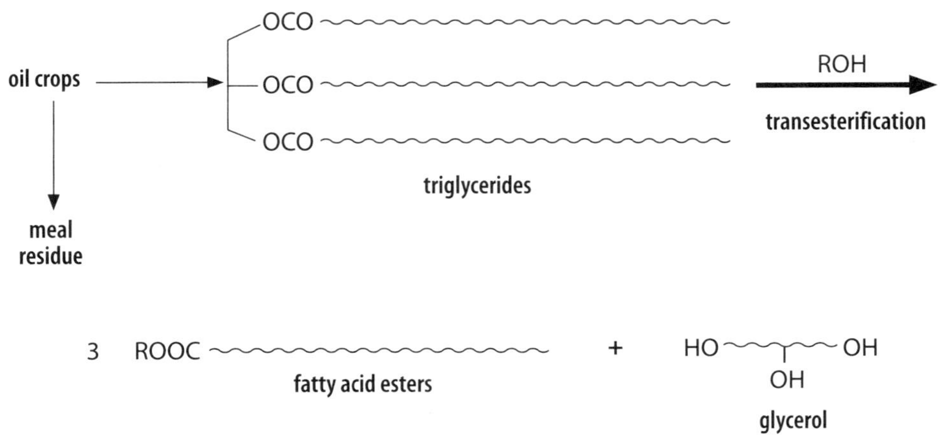

The fatty acids present and their relative amounts depend upon the source of the fat or oil. There is a relatively small variation in composition in fats and oils from the same source material. The fatty acid profile of any source can be determined using gas or liquid chromatography. More importantly, the fatty acid profile can be used to predict both physical and performance properties of the biodiesel product.

Free fatty acids (FFA) can be formed during the recovery process for the fats and oils or when there is water present to break the glyceride–fatty acid bond. FFAs require different processes than triglycerides, but can be made into biodiesel. Hence, the FFA content is a significant measure of feedstock quality, because it indicates the degree of processing required. FFA content is reported as percent FFA.

The free fatty acid percentage of a fat or oil is measured by the acid number. The acid number is the number of milligrams of KOH required to saponify the free fatty acids in one gram of fat or oil. The assumption is made that the FFAs present are oleic acid. This results in the relation that one unit of acid number = 0.503 percent FFA.

Specific gravity is a measure of the density (grams per milliliter) of the fat or oil. The specific gravity (sp. gr.) is reported as a ratio of the density of the sample at a specified temperature to the density of water at a specified temperature. For example, sp. gr. 25/25 means that the temperature of the sample and the water is 25°C. There are a number of different "standard" temperatures used in the industry, so it is very important to note the temperatures associated with any data reported or standards used. In the U.S., the typical temperature standard is 25/25°C.

The refractive index of a liquid is the ratio of the velocity of light in a vacuum to the velocity of light in the liquid. In spite of the esoteric definition, the refractive index is easily measured. It is useful in identifying substances and in determining the purity of a material. In the case of fats and oils, the refractive index is a quick measure of the degree of unsaturation in the fatty acids present. For biodiesel, this measure is important because the degree of unsaturation affects the viscosity, pour point, cloud point and cetane number of the product.

The iodine number is a direct chemical measure of the degree of unsaturation present in a triglyceride or the derivative esters or fatty acids. The iodine number is defined as the number of grams of iodine reacted with 100 grams of trigylceride. The reactant source for the iodine is iodine monochloride or iodine monobromide. The iodine number and refractive index for a given material are related by the general expression:

$n_D = a + b\,I$, where a and b are empirical constants that depend on the triglyceride.

Saponification is the reaction of a fat or oil with an alkali hydroxide (NaOH, KOH, etc.) to form glycerol and fatty acid salts. The saponification number is the number of milligrams of potassium hydroxide (KOH) required to saponify one gram of fat or oil. The saponification number is a crude indicator of the average molecular weight of the oil.

Titer is a measure of the solidification temperature of a tallow or grease. The titer is used as a measure of quality and to distinguish between inedible tallow and inedible grease. A material with a titer less than 40°C is classified as a grease, and a material with a titer above 40°C is a tallow.

MIU is a combined measure of the percentage of moisture (M), insoluble impurities (I) and unsaponifiable matter (U) present in the tallow or grease. The I and U components indicate the presence of a variety of possible non-triglyceride materials that can serve as contaminants for the biodiesel product and can reduce product yields. The moisture value is particularly important, because moisture can lead to the production of free fatty acids instead of esters in the reaction system.

Color is measured by comparison with a set of permanent color standards. The sample must be liquid when making the comparison. The color blocks and comparator are purchased items (The Tintometer Company).

Vegetable Oil Sources for Biodiesel Production

Palm oil is the largest volume triglyceride resource in the world. As the name indicates, palm oil is a seed oil derived from the oil palm tree. Palm kernel oil is derived from the seed kernel, while palm oil is recovered from the seed pulp. Palm oil is a material of interest for biodiesel production. It is characterized by high (32-40 percent) palmitic acid and high (38-52 percent) oleic acid contents. The oil is semi-solid at room temperatures. The ester product has to be winterized to meet pour point and cloud point standards in temperate regions.

Soybean oil is recovered as a co-product from the production of soybean meal, a high protein food and feed material. This temperate seed oil has received the greatest attention as a source for biodiesel in the U.S. The fatty acid profile of soybean oil yields an ester product that requires minimal winterization. There is an extensive infrastructure in the U.S. for the recovery and refining of soybean oils.

There are two types of soybean oil available. The majority of the soybean oil is extracted oil, which means that the oil was recovered from the meal using a solvent extraction process. There are a number of smaller facilities that specialize in expeller pressed oils and meals. Both types of oil can be used to make biodiesel.

Rapeseed oil is the primary feedstock used in Europe for the production of biodiesel. The European variety of rape is a low erucic acid, edible oil. Industrial rapeseed is high in erucic acid (C 22:1), which is used in a variety of high performance lubricants, corrosion inhibitors and other products, as well as biodiesel. The primary source of industrial rapeseed oil in the U.S. is the Northwest, where it has been tested extensively as a source for biodiesel. Canola oil is a low erucic acid oil derived from the rape family that can be used for biodiesel, but it commands a premium price as an edible oil.

Corn oil is recovered as a co-product from a corn wet milling operation. The properties of corn oil are very similar to those of soybean oil, but the oil is typically priced $0.02 to $0.03 per pound higher than soybean oil.

Other oils that can be used for biodiesel production include cottonseed oil, sunflower oil and peanut oil. The U.S. Department of Energy (DOE) is sponsoring the development of an oilseed mustard variety that has very high oil yields per acre as a new source for biodiesel.

Table 17 shows typical values for the fatty acid compositions of common vegetable oils that have been used for biodiesel production. The numbers describing each acid indicate the number of carbon atoms in the acid chain, followed by the number of unsaturated carbon-carbon bonds in the chain. For example 16:0 is palmitic acid, with 16 carbon atoms and no unsaturated bonds. Oleic acid (18:1) has 18 carbon atoms and one unsaturated bond.

Fatty acid	Palm	Soybean	Industrial rapeseed	Corn
Palmitic 16:0	32 - 45	7 - 11	1 - 3	8 - 12
Stearic 18:0	2 - 7	2 - 6	0.4 – 3.5	2 - 5
Oleic 18:1	38 - 52	15 - 33	12 - 24	19 - 49
Linoleic 18:2	5 - 11	43 - 56	12 -16	34 - 62
Linolenic 18:3	-	5 - 11	7 - 10	-
Erucic 22:1	-	-	40 - 55	-

Table 17. Fatty Acid Percentages in Commodity Vegetable Oils.

Rendered Products as Sources for Biodiesel Production

Tallows and greases are triglyceride materials that are recovered using a rendering process. The rendering process is a large-scale cooking of animal processing residues or used cooking oils in water. The fats and oils rise to the surface and are recovered.

Because of the fatty acid profiles of most animal fats, the rendering products are generally semi-solid or solid at room temperature. Beef and pork tallows are always solid at room temperature. Chicken fats are semi-solid. The rendered cooking oils may be semi-solid, depending on what other materials were processed with them. **Table 18** shows typical fatty acid profiles for the three primary sources of animal fats in the U.S.

Acid	Beef	Pork	Chicken
Myristic 14:0	1.4 – 6.3	0.5 – 2.5	1
Palmitic 16:0	20 - 37	20 - 32	25
Palmitoleic 16:1	0.7 – 8.8	1.7 - 5	8
Stearic 18:0	6 - 40	5 - 24	6
Oleic 18:1	26 - 50	35 - 62	41
Linoleic 18:2	0.5 - 5	3 - 16	18

Table 18. Fatty Acid Percentages in Selected Animal Fats and Oils.

The nature of the rendering operation does not lend itself to making highly consistent products. The input stream is too variable. The specifications used to characterize tallows and greases reflect the inherent variability of the materials. The specifications are in terms of limits, rather than ranges, as shown in **Table 19**.

Analysis	Yellow grease	Brown grease	Inedible tallow
Min. titer	36.0°C	38.0°C	40.0 °C
Max. FFA	15%	50%	15%
Max. color	37	none	33
Max. MIU	2 %	2 %	2 %
Sp. Gr.	-	-	0.86 – 0.87
Iodine #	-	-	35 - 48

Table 19. Grade Specifications for Tallows and Greases.

The decision to use tallows and greases in a biodiesel facility is a commitment to careful monitoring of each new lot of feed material. Also, these materials will require the addition of pre-processing equipment to ensure that the feedstock quality to the esterification process is suitable for the technology in use in the facility.

Low Quality Sources for Biodiesel Production

Trap grease materials are typically recovered from the bottoms of commercial frying systems and from grease traps installed as a part of the water discharge system. These materials are highly variable in composition, but will have significant quantities of water and non-lipid materials associated with them. They require dewatering, filtering and grit removal and may also require deodorization and bleaching prior to use.

Soapstock is a large-volume, low-value material produced as a part of the oil refining process. The oil is treated with an alkali solution to convert the free fatty acids into a soap suspension that is decanted and discarded. A typical composition of soapstock is shown in **Table 20**.

Glycerides	12%
Phosphoglycerides	8%
Free fatty acids	10%
Water	45%
Other	25%

Table 20. Typical Composition of Soapstock.

Spent bleaching clay also is a waste product of the oil refining process. Activated clays are contacted with the oil as an adsorbent to remove coloring agents from the oil. The clays can contain from 20-50 percent by weight oil that can be recoverable. A danger in working with spent clays is that the materials are pyrophoric and must be handled appropriately.

Summary

The two most important factors in feedstock selection for any biodiesel operation are the cost of the material(s) to be used and the quality (particularly the free fatty acid content) of the materials. The cost of the TG material used to make biodiesel is typically 70 to 85 percent of the total production cost of the product. Hence, there is a significant pressure to use the lowest cost TG source possible.

In general, rendered products are consistently lower in cost than vegetable oils. Because of its massive production and refining infrastructure, soybean oil tends to be the lowest cost vegetable oil available in the U.S. **Figure 27** illustrates the relative pricing of soybean oil, inedible tallow and inedible yellow grease over the period from 1990 to 2001. Typically, soybean oil costs about \$0.05 to \$0.07 per pound more than tallow, and tallow is \$0.03 to \$0.05 per pound more than yellow grease.

The choice of feedstock also affects biodiesel's properties and the overall process scheme used to produce the product. Key properties that are affected by the fatty acid profile of the raw material include the pour point and cloud point (measure of cold weather performance), the cetane number, viscosity and storage stability. In general, the greater the fraction of saturated fatty acids, the poorer the cold weather performance. Also, typically, the greater the number of polyunsaturates, the lower the cetane number and the poorer the storage stability.

Free fatty acids present in a feedstock at > 0.5 percent require additional pretreatment before the material can be used in a typical base-catalyzed biodiesel process. This means that use of all rendered products requires pretreatment to either remove the free fatty acids or to convert them into fatty acid esters before entering the standard esterification process step. Because of the inherent variability of the rendered products, there must be careful monitoring of the raw material as well as the final product.

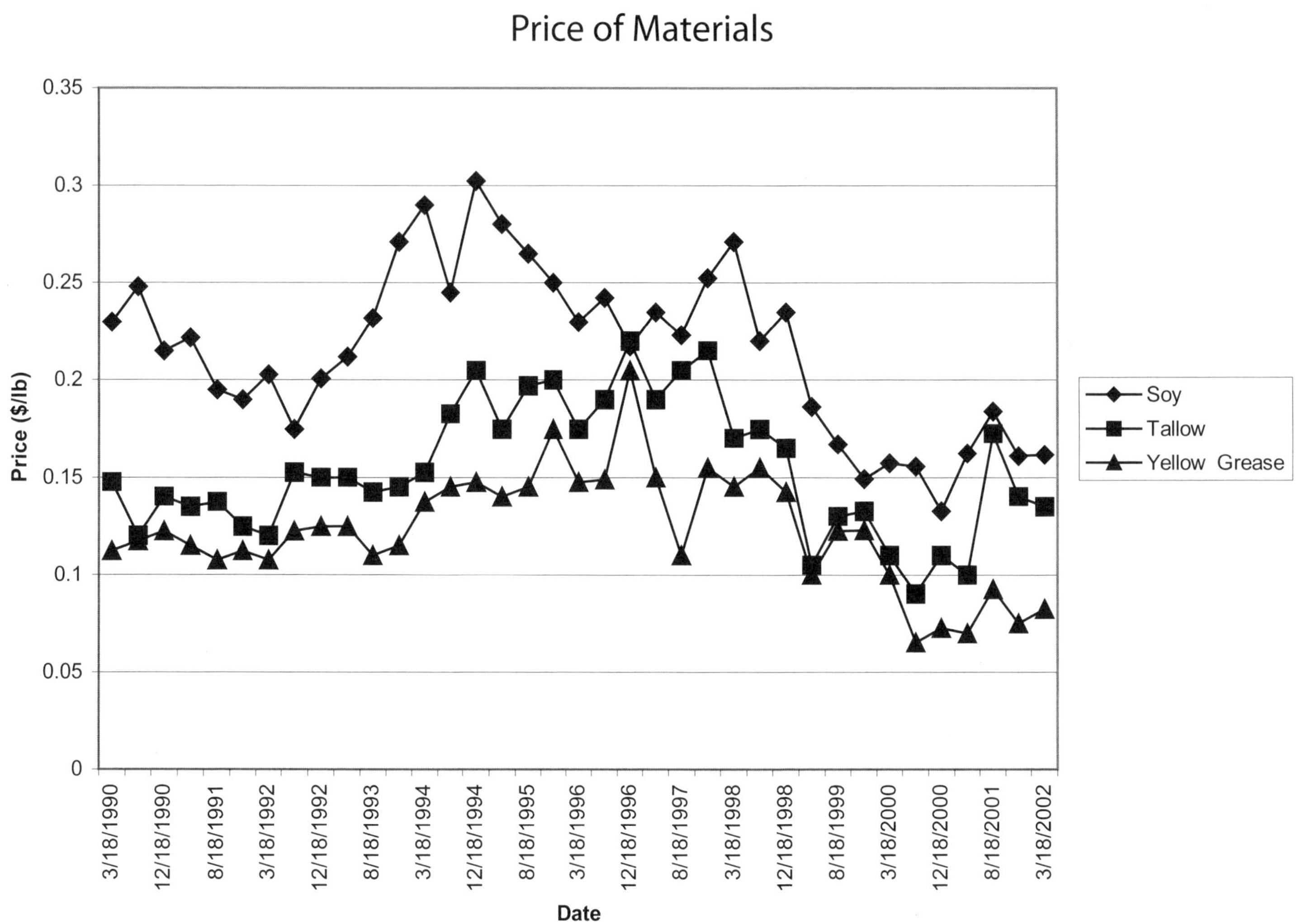

Figure 27. Comparison of Soybean Oil, Tallow and Yellow Grease Prices,

CHAPTER 17

Feedstock Acquisition

This chapter will review the categories of feedstocks available for the production of biodiesel and a number of key feedstock decisions that affect the business structure, technology options and operations of a biodiesel plant.

In it, you will:

1. Understand the advantages and disadvantages of combining a biodiesel plant with a commodity processing facility such as an oilseed crush facility or a rendering operation.
2. Review the categories of feedstocks available.
3. Examine the advantages and disadvantages of single feedstock versus multiple feedstock operations.

The decision to couple a biodiesel plant with a new or existing processing facility defines the primary feedstock early in the business planning process. Operating a biodiesel facility in conjunction with a processing facility offers a measure of stability for the sourcing of raw materials and may lead to opportunities to take advantage of a number of financing and/or tax incentives for producers/processors. Also, the processor may be able to justify an internal transfer price for the feedstock that is lower than the posted market price, in essence, deferring profits to the biodiesel product or transferring costs to another product in the overall operation.

The categories of feedstocks chosen and the decision to use single or multiple feedstocks affect the expected pricing of the raw materials and also impact the choices of technologies to be used in the plant. Election of multiple or lower quality feedstocks will lead to additional process steps to prepare some of the inputs for the biodiesel production or to manage process co-products.

Many of the potential biodiesel feedstocks are traded commodities. This may allow the operation to benefit from trading on the futures and spot markets for securing raw materials at below-market prices. Since the raw materials costs are as much as 70 to 85 percent of the finished product cost, advantageous pricing of the raw materials is a key to the ultimate profitability of the operation.

A raw material producer can gain a number of advantages by the addition of a biodiesel plant to a new or existing facility. The incremental investment for a biodiesel operation is relatively small, compared with the cost of an oilseed refinery or a rendering plant.

A key person in a feedstock purchasing facility is the commodity trader for the organization. A significant factor in the profitability for the plant will be the skills exercised in developing both long-term feedstock contracts and utilization of the fats and oils spot markets and futures markets.

The decision to purchase the majority of the feedstock will affect the plant infrastructure. As noted, a single processor of a seasonal feedstock will have to either provide storage for year-round operation or plan for purchase of supplies in the off-season. The purchaser option will lessen total storage requirements, but will increase transportation expenses. Determining the best balance between these factors requires analysis of the available suppliers for the fats and oils needed, transport costs and storage requirements.

The use of multiple suppliers and multiple feedstocks will require additional care in the characterization of each new batch of raw material and in monitoring of the products. There can be considerable variation among suppliers and from batch to batch from the same supplier. This is particularly true of tallows and greases during the summer months.

Vegetable oils: Vegetable oils suitable for producing biodiesel include the commodity oils: palm, soybean, corn, canola, sunflower and rapeseed. Some oils, such as olive and peanut oils, can be used for biodiesel, but their prices are generally too high. Coconut oil can be used for biodiesel, but the fatty acid composition is such that the pour point and cloud point properties are very poor. Cottonseed oil can be used also, but it is a primary feedstock for the margarine industry. Mustard oil from some very high yield varieties is under consideration by DOE.

A key factor in the decision to use vegetable oils for the biodiesel feedstock is the fact that these oils are premium food products. This means that the pricing of all vegetable oils is somewhat higher than other triglyceride (TG) sources.

Vegetable oils have the advantage that whether they are partially or fully refined, they have a low free fatty acid content and typically have very low concentrations of most process contaminants, including water. The fatty acid profiles of most temperate climate vegetable oils allow the esters to meet cloud point and pour point requirements without need for winterization. The polyunsaturate content desirable for most premium food oils can lower the cetane number of the biodiesel and may contribute to a lower storage stability for the product.

Futures trading for the plant's oil supply can offer opportunities to lower the average cost of the vegetable oil feedstock. Also, off-specification oils from regional processors may be suitable for use in biodiesel production, although not available in large quantities.

Rendered products: The primary animal fats considered for biodiesel are beef tallow (inedible grade) and inedible grade pork lard. Because of the palmitic and stearic acid content, biodiesel from these sources may need to be winterized or blended to meet pour point and cloud point specifications.

Poultry fat has not been explored extensively for biodiesel, but it has the potential to be a significant feedstock. Fish oils could be used for biodiesel, but they tend to be priced at a premium. In addition, fish oils are highly unsaturated, which may lead to stability problems unless these fatty acids are removed for the specialty health markets. Animal fats tend to have higher free fatty acid contents than vegetable oils because they have been cooked (rendered) once in their preparation.

Rendered products from the meat processing facility (integrated meat packers and renderers) will almost always be from a single source: beef, pork, chicken, mutton, etc. There may be some variation in quality of the product, although the product will meet the standards for which it is labeled. Most fats produced by beef or pork meat packers are considered edible fats, and command slightly higher prices in the marketplace. Other products produced by renderers include protein, bone, blood and feather meal.

Independent renderers are not integrated with a meat packing operation and often accept a wider variety of animal products from different sources. Their fats and greases are considered inedible for humans and are largely used as enhancements in animal feed for their calorie content. Independent renderers typically accept recycled fryer greases and process these (occasionally with other greases from their rendering operation) into a product sold as "yellow grease."

Because of its lower price, yellow grease is frequently considered for biodiesel production. Yellow grease has a rather high (maximum of 15 percent) free fatty acid content. The original materials are a mix of vegetable oils and animal fats and oils. The compositions will vary over time and among sources.

The fatty acid profile is one of the most important elements in determining biodiesel properties, so both edible and in-edible sources can be used. The price, however, guides biodiesel producers towards the inedible materials – the inedible grades are generally about $0.03 per pound cheaper.

The free fatty acid contents tend to be the highest in the summer and the lowest in the winter. The feedstocks used to make the grease products can vary greatly, even from a single plant. The result is variability in the product grease.

There are several low-value crude sources of fats and oils that require pre-processing before use in any esterification process. These include trap grease, brown grease and soapstock. These sources are commonly associated with water, proteins, polysaccharides and mineral matter. The addition of appropriate pre-processing capabilities may be economical, since these materials commonly constitute a disposal cost, rather than a product value to the processors.

Single or multiple feed stocks: There are advantages to using a single feedstock. The processing is inherently simpler because feedstock variability is limited. There does not have to be multiple tankages for feedstock or product storage. If the biodiesel producer is taking advantage of some of the tax or subsidy incentives, accounting for raw materials and products is simpler. Some process technologies are optimized for use with a single feedstock, typically soybean oil.

Multiple feedstocks allow more flexibility in minimizing feedstock costs by "playing the market" to acquire the lowest cost materials available. However, multiple sources will require strict quality control for the raw materials as well as the final product. The process will have to be more complicated, with provision for one or more pretreatment steps available to ensure suitable feed to the primary esterification unit.

Multiple feedstocks impact a business in these ways: feedstock cost flexibility, higher capital and operating costs, product performance, public perception (including color and odor acceptance), public acceptance (some customers prefer some feedstocks over others), sulfur removal and glycerin quality.

Summary

Feedstock choice relies on these key factors:

- The average raw material cost.
- The availability of raw materials (sources and timing).
- The process technology most suitable for the plant.
- Access to certain, feedstock specific, tax and/or subsidy advantages.

CHAPTER 18

Feedstock Preparation

It has been mentioned earlier that biodiesel can be produced from a wide variety of feedstocks including vegetable oil, animal fats, restaurant waste oils and trap grease. Another way to categorize these feedstocks is to identify those coming from oilseeds and those that originate from rendering plants. This section will discuss the processes used to produce these two categories of feedstocks and the implications for biodiesel processing.

Production of Oil from Oilseeds

Soybean oil is the largest source of vegetable oil in the United States and rapeseed oil (canola) is the largest source in Europe. Most of these oils are used as food products but they require considerable processing before they are considered edible. **Table 21** shows the composition of crude soybean oil compared with a fully refined oil. Fully refined oils are also known as RBD oils where RBD strands for refined, bleached and deodorized. The table shows how a crude soybean oil typically contains 1.5 to 2.5 percent phosphatides, 1.6 percent unsaponifiable matter including 0.15 to 0.21 percent tocopherol (Vitamin E) and 0.3 to 0.7 percent free fatty acids. These compounds contribute color, odor and strong flavors to the oil that are usually undesirable for cooking. Some of the compounds also contribute to poor storage stability.

It has been shown that high quality biodiesel can be produced even from crude soybean oil [8]. However, use of partially or fully refined oil simplifies the biodiesel production process so some producers have chosen to use degummed oils or even RBD oils. The following discussion describes current practice for processing soybean oil in the U.S. Similar processes are used for rapeseed oil in Europe.

	Crude oil	Refined oil
Triglycerides, %	95 – 97	>99
Phosphatides, % [a,b]	1.5 – 2.5	0.003 – 0.045[c]
Unsaponifiable matter, %	1.6	0.3
Plant sterols, % [d]	0.33	0.13
Tocopherols, % [e]	0.15 – 0.21	0.11 – 0.18
Hydrocarbons, % [f]	0.014	0.01
Free fatty acids, %	0.3 – 0.7	< 0.05
Trace metals[a]		
Iron, ppm	1 – 3	0.1 – 0.3
Copper, ppm	0.03 – 0.05	0.02 – 0.06

[a] Evans, C.D., G.R. List, R.E. Beal, L.T. Black. 1974. JAOCS. 51:444-448.

[b] List, G.R., A.J. Heakin, C.D. Evans, L.T. Black, T.L. Mounts. 1978. JAOCS. 55:521-522.

[c] Corresponding to 1-15 ppm of phosphorus.

[d] Weihrauch, J.L., J.M. Gardiner. 1978. J. Am. Diet. Assoc. 73:39-47.

[e] Bauernfeind, J.C. 1977. Crit. Rev. Food Sci. 8:337-382.

[f] Gutfinger, T., A. Letan. 1974. J. Sci. Food Agric. 25:1143-1147.

Table 21. Average Composition for Crude and Refined Soybean Oil [7, p. 14].

Oil is removed from soybeans by two techniques: mechanical crushing and solvent extraction. Mechanical crushing is prefered for smaller plants because it requires a smaller investment. Typically plants that process less than 100,000 kg/day use mechanical crushing and plants that process more than 300,000 kg/day use solvent extraction.

Before the oil can be extracted, the seeds must be prepared. This involves removing stems, leaves, stones, sand, dirt and weed seeds. The stems and stones can be screened out, because they are usually larger than the seeds. Weed seeds, sand and dirt are usually smaller and can be removed by finer screens. Magnets can remove iron particles.

After cleaning, the seeds are often dehulled. The hulls are abrasive and contain very little oil (less than 1 percent). Removing the hulls reduces the wear on the screw press. It can also increase the extraction of oil because the hulls absorb oil. On oilseeds with thick shells, dehulling typically involves cracking the shells and then separating the shells by screening or aspiration. With soybeans, dehulling generally includes removal of residual fines.

Oilseeds that are to be processed by solvent extraction are usually flaked to increase the exposure of the oil to the solvent. Hexane in a percolation extractor is the most common technology used today. This type of extractor drips the solvent down on the flaked soybeans, so it can dissolve the oil in a manner similar to a coffee percolator. The oil-laden solvent, or *miscella*, is then filtered and removed from the extractor. To remove the solvent, the miscella is heated to vaporize the solvent. This brings the the miscella to about 5 percent solvent [4]. Then the remaining solvent is stripped by injecting steam into the oil. The steam and hexane vapors are condensed and, since the two fluids are insoluble, they can be separated in a settling tank. **Figure 28** shows a schematic of a typical solvent extraction plant that includes several stages of solvent recovery.

The alternative to solvent extraction is mechanical extraction or crushing. The beans are generally preheated to destroy enzymes that cause problems with using the meal as animal feed. One popular way of cooking the bean is to use an expanding extruder. This device compresses the beans to very high pressure so that the temperature rises to 150°- 160°C, which is high enough to deactivate the enzymes. The extrusion also frees the oil and produces a frothy liquid that quickly solidifies as the water flashes to vapor and reduces the moisture level from 10-14 percent to 6-7 percent. After cooking or extrusion, the oil seeds are usually fed to a screw press where they are compressed to high pressure. Drainage slots are provided for the oil to leave the press. Moisture control is important to maintain proper temperature. If the moisture level is too high, then the beans will be soft and the oil will contain excessive solids. If the moisture level is too low, then the oilseeds will overheat. Ideally, the moisture should be 2-3 percent for full pressing and 4-6 percent for pre-pressing [4]. If the beans have not been extruded, then pre-pressing may be used to break the cell walls and release the oil from the seed. It can preceed both mechanical and solvent extraction.

After extrusion from the seed, crude soybean oil contains impurities that can be classified into two categories: oil insoluble and oil soluble [1]. The oil-insoluble impurities consist of seed fragments and meal fines, free water and long-chain hydrocarbons or waxes that cause cloudiness when the oil is refrigerated. Most of this material can be removed by filtration. The oil-soluble materials include free fatty acids, phosphatides, gummy or mucilaginous substances, color bodies, protein, tocopherols, sterols, hydrocarbons, ketones and aldehydes. Some of these compounds, particularly the ketones and aldehydes, can cause the oil to have an unpleasant taste and smell. Other compounds are actually desirable, such as the tocopherols which act as antioxidants. The sterols are relatively inert and can remain in the oil. The food-grade oil typically contains 0.5 - 1.5 percent non-oil compounds that are known collectively as *unsaponifiable matter*.

Processing of crude soybean oil to food-grade oil generally consists of three steps: refining (degumming and caustic refining), bleaching and deodorization. As mentioned earlier, these oils are known as RBD oils (refined, bleanched, and deodorized).

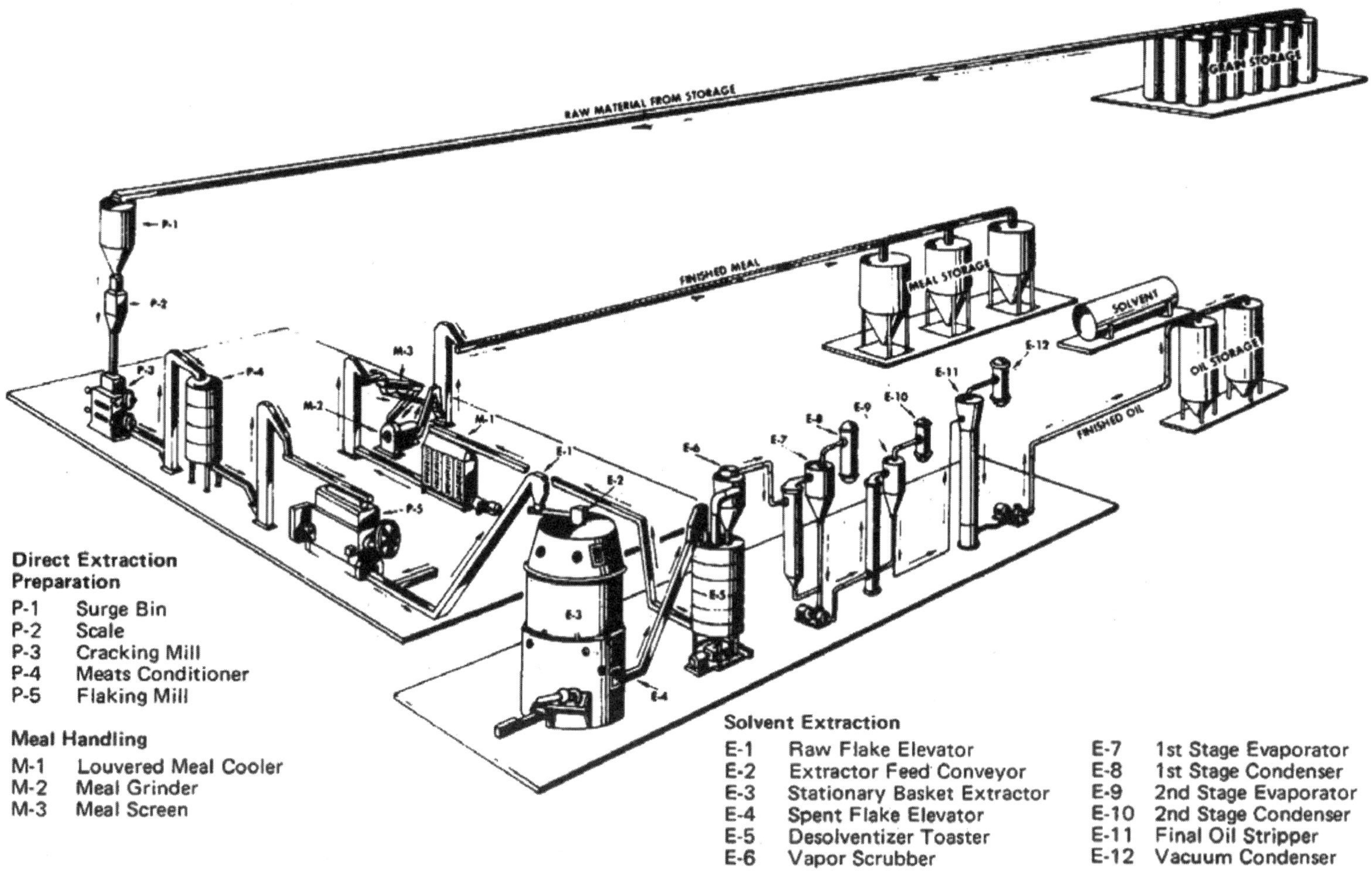

Figure 28. Soybean Oil Extraction Plant [7].

Refining

The refining step is designed to remove the phospholipids and free fatty acids from the crude oil. Crude soybean oil typically contains 500-700 ppm of phosphorus, which corresponds to 1.5 to 2.1 percent phospholipids [1]. The main reason to remove these phospholipids is that some of the compounds, particularly the calcium and magnesium salts of phosphatidic and lysophatidic acids, are strong emulsifiers. If these compounds are still present during the later alkali neutralization step, they will inhibit the separation of the soaps and lower the yield of neutral oil. Phospholipids (also called phosphotides) will also react with water to form insoluble sediments that are not desirable when cooking with the oil.

Refining may consist of two steps. The first step is degumming. In this step, crude soybean oil is mixed with 1- 3 percent of water and the mixture is agitated mechanically for 30-60 minutes at 70°C (158°F). This hydrates the phospholipids and gums, and these hydrates are insoluble in the oil. They can be separated by settling, filtering or centrifuging. The phosphorus content can be lowered to 12-170 ppm by this procedure. The byproduct of water degumming has value as a feedstock for lecithin production.

Table 22 shows the profile of phospholipids from lecithins produced from various oilseeds. It shows that there are significant differences in the distribution of compounds between the different feedstocks.

Phospholipid	Soy	Rapeseed	Sunflower seed
Phosphatidylcholine	21.9	24.6	25.4
Phosphatidylethanolamine	13.6	22.1	11
Phosphatidylserine			0.8
Phosphatidylinositol	12	14.7	19.4
PG + DPG*	2.3		1.2
Phosphatidic acid	5.8		3.3
N-A cyl-phosphatidylethanolamine	2.8		1
Lyso-phospholipids	2.9	19.4	
Others	3.6	19.2	

* PG + DPG = Phosphatidylglycerol and Diphosphatidylglycerol. Note: some commercial lecithins contain glycolipids (phosphorus-free polar lipids) that are not included in the totals given in the table.

Table 22. Phospholipid Composition [wt%] of Commercial Lecithins [2, p. 52].

A portion of the phosphatides are not hydratable by contact with water alone. The addition of citric or phosphoric acid will hydrate those remaining. Citric acid is preferred if the byproduct is to be used for lecithin. The acid hydration is accomplished by adding 0.05-0.2 wt percent of concentrated (85 percent) phosphoric acid to crude oil, which has been heated to 60-85°C. The residence time varies from a few seconds to 1-2 minutes depending on the type and quality of the oil. The extent of gum removal by these techniques is related to the strength of the acid treatment. **Figure 29** shows a schematic diagram of the processes involved to degum soybean oil and produce edible lecithin.

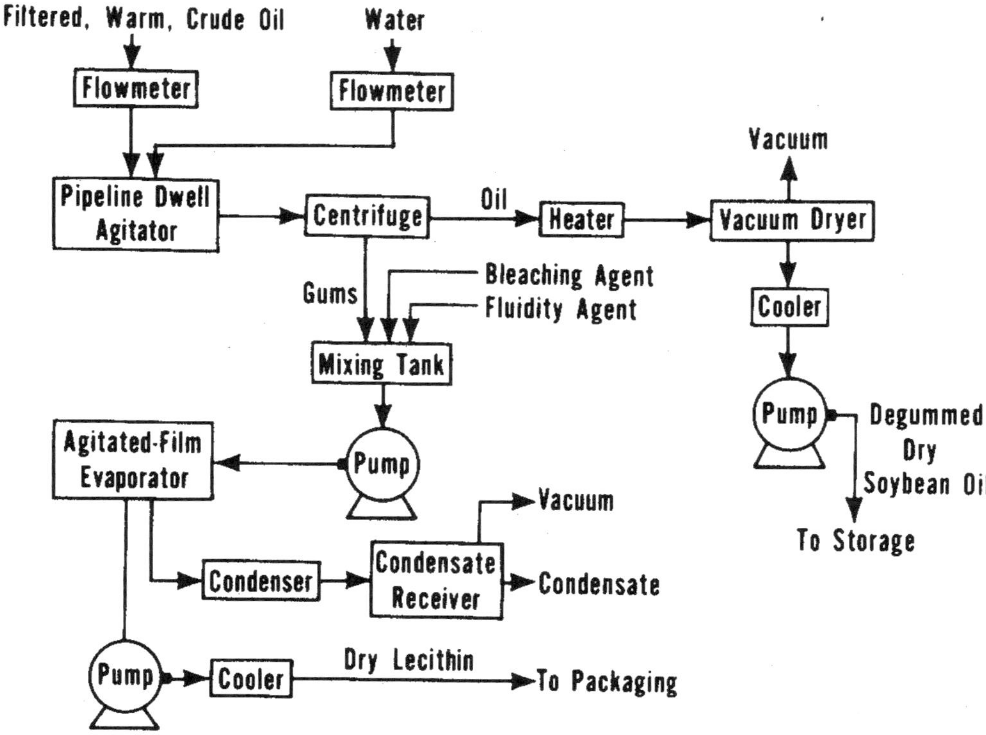

Figure 29. Flow Diagram for Soybean Oil Degumming [7].

The second phase of refining is neutralization or *caustic refining*. This process is to remove the free fatty acids present in the crude oil. An alkali solution, usually sodium hydroxide, is added that reacts with the free fatty acids to produce soaps. The soaps are insoluble in the oil and easily separated by water washing. The alkali solution also neutralizes any acid remaining from the degumming stage.

The alkali will react with the triglycerides in the oil also, so the neutralization parameters (type of alkali, solution strength, temperature, agitation and time) must be optimized to minimize the yield loss. There can be additional losses from emulsification and suspension of oil droplets in the soap solution. A byproduct of caustic refining is a mixture of soap, water and oil known as soapstock. This has been considered as a feedstock for biodiesel as its cost is low, but its high water content and converting the soaps to methyl esters are significant obstacles to cost-effective utilization. **Figure 30** shows a diagram of the caustic refining process.

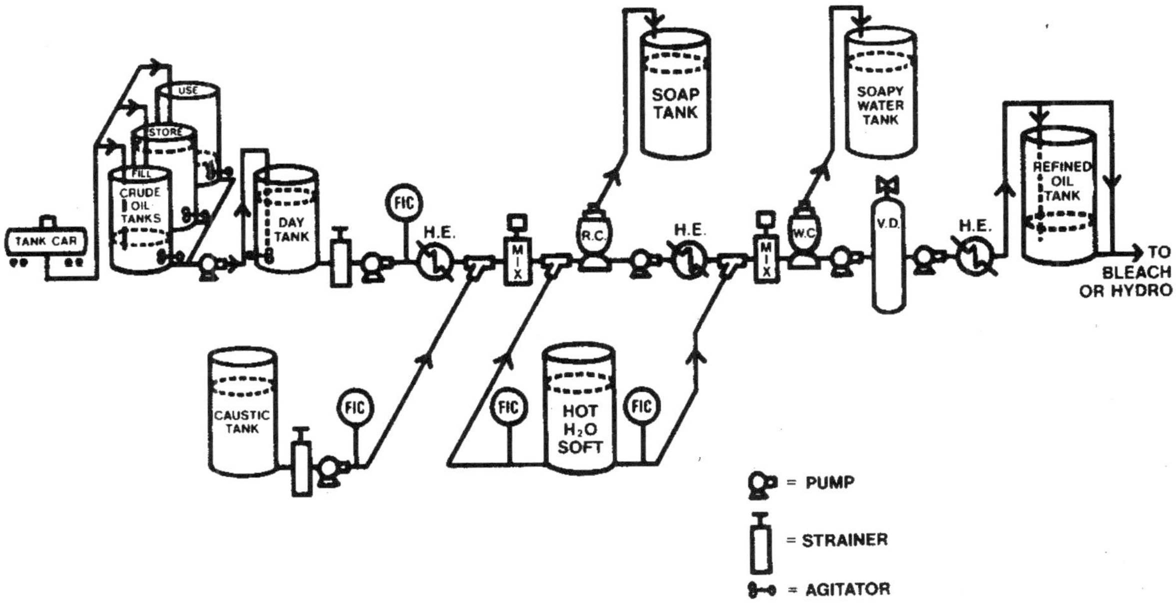

Figure 30. Flow Diagram for Caustic Refining of Soybean Oil [7].

Bleaching

The primary purpose of bleaching is to remove the color pigments from the oil. It also helps to remove remaining soap, trace metals, phosphatides and sulfur compounds. Hydroperoxides, the initial products of oxidation are broken down during the bleaching process and some of the final oxidation products, the carbonyl compounds, are removed.

Bleaching involves mixing bleaching clays, consisting of natually occurring bentonite and montmorillonite clays, with the oil and agitating for 10-30 minutes. The oil is usually heated to 90-120°C, and the process occurs under a slight vacuum to exclude oxygen. A consequence of the vacuum and high temperature is that residual water from the neutralization and water washing is removed. This is not always desirable because the effectiveness of the clays is usually enhanced by the presence of a small amount of water. The clays are usually activated before use with a mineral acid such as sulfuric acid. This removes some of the minerals from the clay and produces a larger volume of micropores and smaller clay particles.

The bleaching clays are mixed with the oil, either directly or by premixing with a small amount of oil to produce a slurry and then adding the slurry to the oil. After the adsorption processes have reached equilibrium, the clay is filtered out by self-cleaning filters.

The peroxide value of the oil is very low after bleaching because the hydroperoxides have been eliminated, but the free fatty acid level may be slightly higher. This may be due to some of the sulfuric acid leaching out of the clay or to hydrolysis of the oil. Hydrolysis is a chemical reaction between the oil and water that produces free fatty acids and glycerin.

Deodorization

Deodorization is the final processing stage before the oil is ready for use in food products. The deodorization removes the trace components that give the oil an unpleasant taste and odor. Deodorization is essentially a distillation process that occurs at high temperature (200-260°C) and low pressure (2 - 7 torr, 2.5-9.2 mbar). The odor causing compounds are more volatile than the oil and will be removed by extended heating at high temperature and low pressure.

The first step of deodorization is deaeration. This step removes all the dissolved oxygen so that the oil does not oxidize at the high temperatures used in deodorization. In batch deodorizers, the deaeration occurs naturally while the oil is heated from the typical inlet temperature of 40-80°C to the deodorization temperature range of 200-260°C. If adequate vacuum is available, the oil should be deaerated by the time the oil reaches 100-120°C.

Preheating is sometimes identified as the second step of deodorization. In continuous flow deodorization processes, a portion of the heat required for heating the oil comes from cooling the deodorized oil. As much heat as possible is transfered from the final product to incoming oil. After preheating, the oil is heated to its final temperature by steam or another heating fluid such as Dowtherm A.

After the oil has been held at high temperature for sufficient time (2 - 5 hours in a batch deodorizer), it is cooled to 120°C where it is common practice to add 50 - 100 mg/kg of citric acid. The citric acid is added to chelate trace metals in the oil so that they can be removed by filtration. Chelation is a chemical process where a compound, known as a chelating agent, binds to a metal compound and makes it unavailable for further reaction. Then the oil is cooled further to its storage or bottling temperature. One final stage of filtration, known as polish filtration, occurs immediately before the oil leaves the plant.

As mentioned earlier, biodiesel can be produced from oil at any stage in the various processes. If biodiesel is produced from crude oil, the compounds that would be removed during the refining, bleaching and deodorization steps of oil refining are mostly removed by water washing or collected in the glycerol phase. Biodiesel produced from crude soybean oil retains somewhat more color than biodiesel from refined oil but phosphorus and free fatty acid levels seem to be equivalent [8].

Rendering

Rendering is the process whereby animal by-products, including fat, bone, offal, hides, feathers and blood are recycled into usable products. In the case of rendering operations that may be attached to slaughtering facilities, the products may be edible fats and proteins that are suitable for human consumption. These edible fats may be refined, bleached and deodorized in a manner similar to the one described earlier for vegetable oils. This provides a product that is suitable for food, cosmetics and other high value uses. **Table 23** shows the specifications for the wide variety of edible and inedible tallows and greases produced from rendered fats.

Grades	Titre, min. (°C)	FFA, max.	FAC, color, max.	R&B, max.	MIU
Edible tallow	41.0	0.75	3	None	****
Lard (edible)	38.0	0.50	**	None	****
Top white tallow	41.0	2	5	0.5	1
All beef packer tallow	42.0	2	None	0.5	1
Extra fancy tallow	41.0	3	5	None	1
Fancy tallow	40.5	4	7	None	1
Bleachable fancy tallow	40.5	4	None	1.5	1
Prime tallow	40.5	6	13-11B	None	1
Special tallow	40.0	10	21	None	1
No. 2 tallow	40.0	35	None	None	2
A tallow	36.0	15	39	None	2
Choice white grease	36.0	4	13-11B	None	1
Yellow grease	***	15	39	None	2

* from the National Renderers Association

** Lovibond color 5-1/4 inch cell – max 1.5 red. Lard peroxide value 4.0 ME/K max.

*** Titre minimum, when required, to be negotiated between buyer and seller on a contract basis.

**** Moisture maximum 0.20%. Insoluble impurities maximum 0.05%.

Table 23. American Fats and Oils Association Specifications for Tallows and Greases*.

Larger volumes of material are processed in plants that produce so-called "inedible" products, although the products are suitable for use in animal feeds. These inedible rendering plants may process fat and bone trimmings, meat scraps, restaurant grease, fallen animals, and other sources of fat and protein. **Table 24** shows the composition and tallow and grease yield of various raw materials for inedible rendered products.

Source	Tallow/grease wt%	Protein solids wt%	Moisture wt%
Packing house offal and bone			
Steers	30-35	15-20	45-55
Cows	10-20	20-30	50-70
Calves	10-15	15-20	65-75
Sheep	25-30	20-25	45-55
Hogs	25-30	10-15	55-65
Dead stock (whole animals)			
Cattle	12	25	63
Calves	10	22	68
Sheep	22	25	53
Hogs	30	28	42
Butcher shop fat and bone	31	32	37
Blood	None	16-18	82-84
Restaurant grease	65	10	25
Poultry offal	10	25	65
Poultry feathers	none	33	67

* from National Renderers Association

Table 24. Composition of Raw Materials for Inedible Rendering Products*.

Figure 31 shows a typical rendering operation. Dead stock carcasses and other raw materials are received by the plant and fed in a continuous flow to a crusher that grinds the material to a uniform paste-like consistency. This material is then fed to a cooker that heats the material to 121-135°C (250-275°F). The high temperature encourages the removal of water and separates a portion of the fat from the solids. The fat and solids are fed to a drain pan where the fat is drained off, and the solids are fed to a screw press where the remaining fat can be squeezed out. The dry solids are ground to a fine powder and sold as meat and bone meal. The fat is centrifuged or filtered to remove particulates. Since the fat may still contain some water, it is common to reheat the fat with steam injection to encourage separation of the water. After heating, the dry fat will rise to the top, the water will settle to the bottom and an interphase layer will appear that consists of emulsified water and oil and possibly protein and minerals. This material is separated and returned to the cooker for reprocessing.

Inedible fats with free fatty acid levels below 15 percent and MIU (moisture, insolubles and unsaponifiables) less than 2 percent are generally sold as *yellow grease*. Fats with higher free fatty acid levels are sold as *brown grease*. Although their products are referred to as inedible, most rendered material is reintroduced to the food chain as animal feed. Meat and bone meal are used to add protein and minerals to feed, and grease is used to increase calories or energy content. Renderers are coming under increasing scrutiny because of concerns about disease transmission. Most renderers in the U.S. will no longer accept sheep carcasses because of concerns about scrapies. The same concern also exists with deer and elk carcasses, due to chronic wasting disease, a relative of Bovine Spongiform Encephalapathy (BSE – Mad Cow Disease). Products from ruminants (cattle) are not fed to other cattle. This requirement has been reinforced by recent events in England, where BSE was spread to large herds of cattle through use of infected animal products in the feed. The risks of disease transmission are thought to be minimal in the U.S. and use of animal products across species is generally considered safe. However, concerns about BSE may cause future legislation further restricting the use of animal byproducts in animal feed. If this occurs, biodiesel may become the most significant market for animal fat.

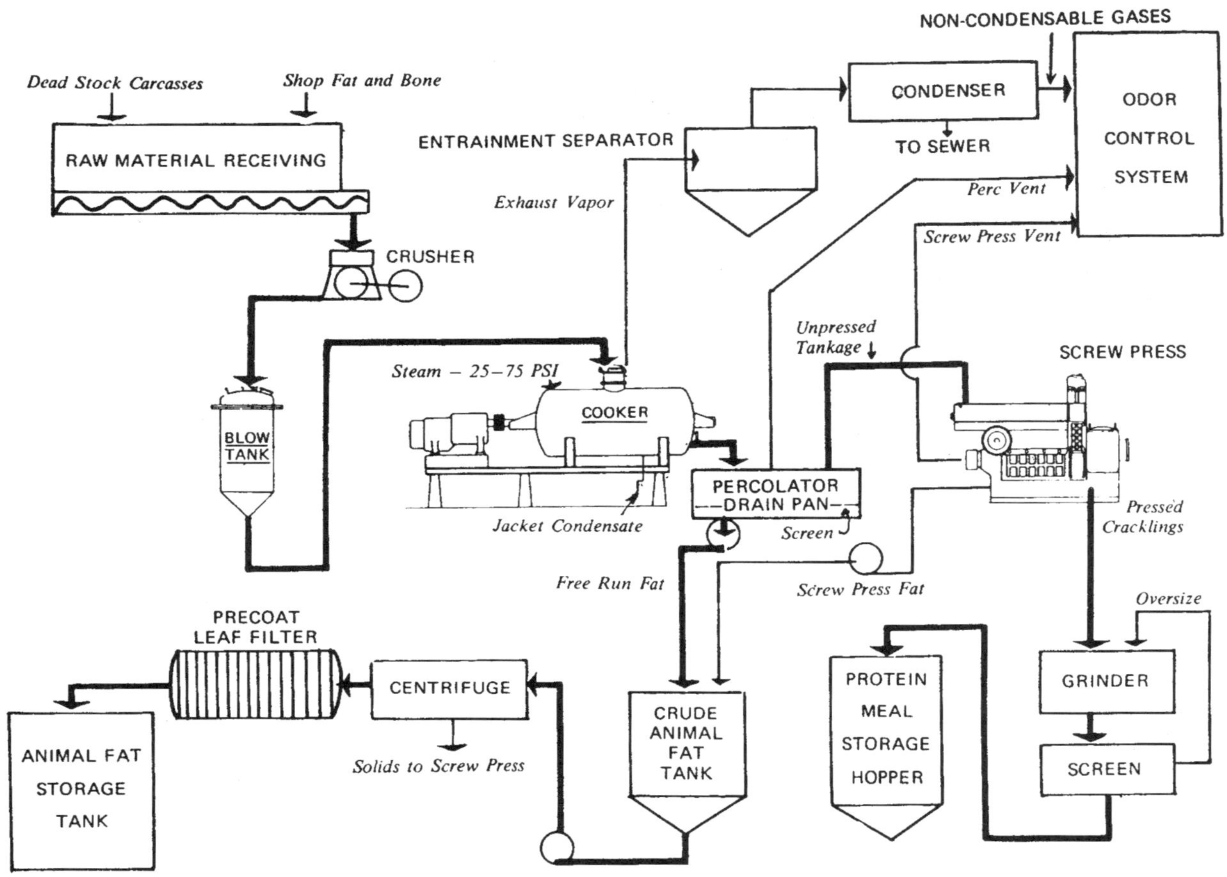

Figure 31. Rendering Plant Operations.

References

1. Hui, Y.H., Editor, Bailey's Industrial Oil and Fat Products, V. 2. Edible Oil and Fat Products: Oils and Oilseeds, Chapter 11 Soybean Oil, Fifth Edition, Wiley-Interscience, New York.
2. Schneider, M., "Phospholipids," Chapter 3 from Lipid Technologies and Applications, Edited by F.D. Gunstone and F.B. Padley, Marcel Dekker, New York. 1997.
3. Allen, D.A., "Refining," Chapter 6 from Lipid Technologies and Applications, Edited by F.D. Gunstone and F.B. Padley, Marcel Dekker, New York. 1997.
4. Williams, M.A., "Extraction of Lipids from Natural Sources," Chapter 5 from Lipid Technologies and Applications, Edited by F.D. Gunstone and F.B. Padley, Marcel Dekker, New York. 1997.
5. Peterson, C.L., Feldman, M., Korus, R., and Auld, D.L., "Batch Type Transesterification Process for Winter Rape Oil," Applied Engineering in Agriculture, Vol. 7, No. 6, pp. 711-716. Nov. 1991.
6. Peters, Richard, Process Chemical Engineer for Proctor and Gamble, Personal Communication.
7. Erickson, D.RPryde, ., E. H., Brekke, O.L., Mounts, T.L., and Falb, R.A., Editors, "Handbook of Soy Oil Processing and Utilization," published jointly by the American Soybean Association, St. Louis, MO, and the American Oil Chemists' Society, Champaign, IL. 1980.
8. Van Gerpen, J.H., and Dvorak, B., "The Effect of Phosphorus Level on the Total Glycerol and Reaction Yield of Biodiesel," presented at Bioenergy 2002, The 10th Biennieal Bioenergy Conference, Boise, ID. Sept. 22-26, 2002.
9. Franco, D.A., and Swanson, W., *The Original Recyclers*, Animal Protein Producers Industry. 1996.

CHAPTER 19

Feedstock Quality Issues

As a general rule, the closer a feedstock is to containing pure triglycerides, the easier it is to convert to biodiesel and the more it will cost. The lower the cost of the feedstock, the more difficult and more expensive it is to convert to biodiesel.

Biodiesel feedstocks will contain many possible contaminants such as water, free fatty acids, particles and phospholipids. There is potential for each of these contaminants to impact the quality of the final biodiesel product.

Water

It is important that water be kept out of the biodiesel production process. While most processes can tolerate up to 1 percent water, even this low level will increase soap production and measurably affect the completeness of the transesterification reaction. Water can be removed by several techniques. Heating to break any emulsions between the water and oil and then allowing the water to settle to the bottom can remove water. Centrifuges can also be used to separate water and oil. Heating the oil under pressure and then spraying into a vacuum chamber can bring the water content to a very low level.

Crude soybean oil typically contains about 0.3 percent water. This can be removed easily by a flash vaporization process. Yellow grease (and usually brown grease) will contain about 0.5 percent water because it has been dewatered during previous processing steps, although the sum of the water, insolubles and unsaponifiables (MIU) is allowed to be as high as 2 percent. Restaurant grease and trap grease that have not been rendered should always be checked for water content because it may be very high.

Water not only impacts the transesterification process, it can impact the effective cost of the feedstock. Analyses of two actual samples of trap grease are shown here.

	Sample A	Sample B
Water	81.5%	40.5%
Solids	6.34%	10.2%
Lipids	12.2%	49.3%

Table 25. Trap Grease Sample Analysis.

Trap grease is a very inexpensive feedstock. In most cases, restaurants will pay to have the material removed. However, even if the cost to have the material brought to the plant were as low as 2 cents/lb, the actual cost for the lipid material in Sample A would be more than eight times higher, or over 16 cents/lb. This is a pretty high price for material that is almost all free fatty acids. This price also does not include the added cost of disposing of the water after it is removed or the cost of landfilling the solids.

Solids

Feedstocks used for biodiesel should always be filtered to ensure no particles enter the processing scheme. This can be a particular problem with recycled products such as trap grease, which typically contains fine particles that may require a fine filter. A 100 μm filter will suffice to remove particles from most feedstocks although all fuel leaving the plant should be filtered to 5 μm.

Free Fatty Acids

While FFAs can impact the production process by deactivating the catalyst, creating soap and releasing water when converted to esters, they do not seem to impact the final acid value of the biodiesel. Biodiesel produced using a normal alkali-catalyzed process that meets the ASTM specification for total glycerol will usually have 0.2-0.3 percent FFAs regardless of the initial acid content of the feedstock. The alkali catalyst also serves as a caustic stripper and removes the FFAs by converting them to soap that is removed during washing.

Phosphorus

Crude soybean oil contains from 600-900 ppm of phosphorus in the form of phospholipids or gums. This material must be removed because the ASTM specification from biodiesel limits the phosphorus content to 10 ppm. Fortunately, even if the oil is not *degummed* to remove this material, the usual biodiesel production process removes the phosphorus and it is either transferred to the glycerol fraction or removed by water washing. There does not seem to be any effect of the initial phosphorus content of the feedstock on the final phosphorus content of the biodiesel.

Sulfur

One of the frequently-touted advantages of biodiesel is that it contains very low levels of sulfur. The EPA has mandated that sulfur levels for petroleum-based diesel fuel must drop to 15 ppm from 500 ppm by 2006. Most biodiesel is already below this level. However, some samples of biodiesel from yellow grease and animal fat have shown sulfur levels as high as 40-50 ppm. The source of this sulfur is still not completely understood, but it may be originating from food preservatives, proteins, hair on the animal hides and, in those cases where poultry are being processed, from chicken feet. Additional research is needed in this area.

Iodine Value

The iodine value is a very crude but commonly used indicator of the level of saturation of an oil. While saturation and fatty acid profile do not seem to have much impact on the transesterification process, they do affect the properties of the biodiesel. Saturated fats produce a biodiesel fuel with superior oxidative stability, a higher cetane number, but poor low temperature properties. Biodiesel from saturated fats is more likely to gel at ambient temperatures than biodiesel from vegetable oils, especially soybean oil.

CHAPTER 20

Glycerol

Glycerol is a byproduct of the transesterification process that produces biodiesel. It is also known as glycerin, glycerine and 1,2,3-propane-triol. Most people use glycerin and glycerol interchangeably. Occasionally, a distinction is made that identifies glycerol as the pure chemical compound and glycerin as the commercial product that is available at different levels of purity.

Glycerol will be produced by a biodiesel plant at about 10 percent of the biodiesel production level. It is often referred to as a high-value co-product but suitable markets have been hard to find.

Glycerol is a very common industrial chemical with a multitude of uses. Principle uses include: food products, cosmetics, toiletries, toothpaste, explosives, drugs, animal feed, plasticizers, tobacco and emulsifiers. Jungermann makes the following comments about uses for glycerol:

> "Glycerine is a versatile chemical. It is found in baby care products and in embalming fluids used by morticians, in glues that hold things together and in explosives to blow them apart; in throat lozenges and in suppositories." [Chapter 1 of Glycerine – A Key Cosmetic Ingredient, Edited by E. Jungermann and N.O.V. Sonntag, Marcel Dekker, Inc., New York, 1991.]

The glycerol produced by transesterification is only about 50 percent pure. It contains a significant amount of contaminants including methanol, soap, and catalyst. It is relatively easy to raise the purity level of the crude glycerol to 80-90 percent. This can be accomplished by adding hydrochloric acid to the crude glycerol until the pH is acidic (around 4.5). This splits the soaps into fatty acids and salt. The fatty acids will rise to the top of the glycerol where they can be removed. Then, the methanol can be removed by evaporation to yield 80-90 percent purity glycerol. The actual level will depend on the purity of the original oil because contaminants tend to concentrate in the glycerol.

Prices for pure glycerol have varied between $0.40 to $1.50/lb over the past several years. At a glycerol price of $0.60/lb, each gallon of biodiesel will be accompanied by an amount of glycerol worth about $0.48. Clearly, the disposition of the glycerol could be an important element of biodiesel profitability. However, it needs to be pointed out that these prices are based on glycerol that is at least 99.7 percent pure. To take the partially refined glycerol described above to this level of purity requires either vacuum distillation or ion exchange refining. Vacuum distillation is capital intensive and probably not practical for small biodiesel plant operators. Ion exchange columns involve less capital but generate large volumes of wastewater during regeneration, so they will involve additional wastewater treatment costs for large operators.

Current annual production of glycerol in the United States is about 300 million pounds per year (Jungermann and Sonntagg, op. cit.). If biodiesel production in the U.S. reaches 200 million gallons per year (a relatively modest goal that would use only 8 percent of the U.S. production of soybean oil), the amount of additional glycerol produced from this source would be equal to 50 percent of the current glycerol production. The impact of this additional glycerol on prices is unclear but it is possible that if new uses for glycerol are not found, the glycerol price will drop to a level that is consistent with its value as a burner fuel, which is about 5 cents/lb.

The selection of feedstock can affect glycerin co-product values. Color and odor bodies are not acceptable in most grades of glycerin. One of the key tradeoffs in selecting a feedstock is to consider how it will impact the quality of the crude glycerin produced.

There are color and odor standards for glycerin. Rendered feedstocks and off-spec vegetable oils càn lend undesirable color and odor bodies to the glycerin. These can be treated at the front end of the process with some feedstock pretreatment or treated at the back end of the process with more intensive glycerin refining. If nothing is done to remove these contaminants, the value of the glycerin will be discounted, and the sales price will be reduced.

Other contaminants also impact the value of the glycerin. Some glycerin buyers will not purchase glycerin with potassium salts present. There may also be limits on water and methanol.

Some glycerin is kosher, which is a desirable grade of glycerin often used for food and pharmaceutical products. It commands a higher price than other glycerin grades. Most kosher glycerin is produced from seed oils. Mixing feedstocks is not permitted.

As the amount of biodiesel production expands worldwide, the supply of glycerin will expand. There are many growing uses for glycerin, but the market is sensitive to imbalances between supply and demand. This could cause periods as long as several years in which glycerin prices become depressed. In fact, there is some concern within the industry that a major glut could occur in future years. The supply of glycerin produced from biodiesel could double or triple U.S. glycerin supplies. Biodiesel producers should consider developing a plan for making some type of marketable product from the glycerin. For example, glycerin could be produced in the first five years, and then the facility could be modified to produce a new glycerin-based product. This could reduce any risk to co-product values that low glycerin prices could cause.

Glycerol Standards

There are a wide variety of standards for glycerol. There are standards for crude glycerol, such as those shown in **Table 26**. Standards for high quality glycerol are usually based on the United States Pharmacopeia (USP) or one of the European equivalents (British Pharmacopeia, etc.). Individual companies may sell USP grade glycerol with higher quality specifications than are actually required by the USP specification.USP specifications are listed in **Table 27** and the British Standard Specifications for three grades of glycerol are provided in **Table 28**. Finally, **Table 29** shows an example of a commercial USP product from Proctor and Gamble Corp.

	Soap lye crude glycerol[a]	Hydrolyser crude glycerol[b]
Glycerol (% wt) min	80.0	88.0
Ash (%wt), max	10.0	1.0
MONG[c] (%wt), max	2.5	1.5
Water (% wt), Karl Fisher, max	10.0	--
Propane 1,3 diol (%wt), max	0.5	0.5
Arsenic as As (ppm)	2.0	2.0
Sugars (max)	Nil	Nil

[a]Soap lye crude glycerol is prepared by evaporation of the purified lyes obtained from the manufacture of soap.

[b]Hydrolyser crude glycerol is prepared by evaporation of the sweet waters obtained from the hydrolysis of fats under pressure or in the presence of catalysts.

[c] MONG = matter organic non-glycerol

Source: Woollatt, E., The Manufacture of Soaps, Other Detergents and Glyerine, Ellis Horwood, Ltd., Chichester, UK, 1985. Originally British Standard Specifications.

Table 26. Crude Glycerol Specifications.

Purity: Contains not less than 99.0 percent and not more than 101.0 percent, by assay.
Color: When viewed downward against a white surface in a 50-ml color-comparison tube, is not darker than the color of a standard made by diluting 0.40 ml of ferric chloride CS with water to 50 ml and similarly viewed in a color-comparison tube of the same diameter as that containing the glycerol.
Specific gravity: not less than 1.249
Residue on ignition: Heat 50 g in an open, shallow 100-ml porcelain dish until it ignites, and allow it to burn without further application of heart in a place free from drafts. Cool, moisten the residue with 0.5 ml of sulfuric acid, and ignite to constant weight; the weight of the residue should not exceed 5 mg (0.01%).
Water: not more than 5.0%
Chloride: A 7.0-g portion shows no more chloride than corresponds to 0.10 ml of 0.020N hydrochloric acid (0.001%).
Sulfate: A 10-g portion shows no more sulfate than corresponds to 0.20 ml of 0.020 N sulfuric acid (about 0.002%).
Heavy metals: Mix 4.0 g with 2 ml of 0.1 N hydrochloric acid, and dilute with water to 25 ml; the limit is 5 ppm.
Chlorinated Compounds: Following specified reactions with morpholine, nitric acid and silver nitrate, the turbidity should not be greater than that of a blank to which 0.20 ml of 0.020 N hydrochloric acid has been added (0.003% of Cl).
Organic volatile impurities: Use gas chromatograph to identify 5 organic volatile impurities. Levels not to exceed: benzene, 2 µg/g; chloroform, 60 µg/g; 1,4-dioxane, 380 µg/g; methylene chloride, 600 µg/g; trichloroethylene, 80 µg/g.
Fatty acids and esters: Mix 50 g with 50 ml of freshly boiled water and 5 ml 0.5 N sodium hydroxide, boil the mixture for 5 minutes, cool, add phenolphthalein, and titrate the excess alkali with 0.5 N hydrochloric acid. Perform a blank determination. No more than 1 ml of 0.5 N sodium hydroxide should be consumed.

Table 27. USP Specifications for Glycerol (USP, v. 24, 2000).

	BS 2623:1979 Technical grade glycerol	**BS 2624:1979 Dynamite grade glycerol**	**BS 2625:1979 Chemically pure glycerol**
Glycerol (% wt)	>99.0	>99.0	>99.0
Relative density (20°/20°)	1.261 < density < 1.264		
Sulfated ash, %	<0.010	< 0.010	<0.010
Acidity or alkalinity, milliequivalents/100 g.	< 0.32	< 0.32	< 0.064
Saponification equivalent, milliequivalents/100 g.	--	< 0.64	< 0.64
Heavy metals, ppm	--	--	< 5.0
Lead, ppm	--	--	< 1.0
Iron, ppm	< 2.0	--	--
Arsenic, ppm	--	--	< 2.0
Chlorides as %NaCl	< 0.010	< 0.010	No turbidity in standard test
Organic chlorides	--	--	Turbidity not greater than standard
Reducing substances (silver test)	--	Not darker than standard	Not darker than standard
Color, lovibond	< 5.0 Y and 1.2 R	As technical grade	--
Sugars	--	--	Nil
Odor and taste	Free from abnormal odor when tested as described in BS 5711.	Shall not emit an acrid odor when tested as described in BS 5711.	Free from odor in test described. Shall taste sweet and be free from any burnt flavor.

Source: Woollatt, E., *The Manufacture of Soaps, Other Detergents and Glyerine*, Ellis Horwood, Ltd., Chichester, UK, 1985. Originally British Standard Specifications.

Table 28. British Specifications for Refined Glycerols.

Chemical properties	Superol[b] glycerol
Glycerol (Bosart and Snoddy Table)	99.7% min (99.9)
Specific Gravity, 25°/25°	1.2613
Assay (percent glycerol on anhydrous basis)	99.0 – 101.0
Moisture	0.3% max
Color, APHA Pt-Co (Hazen) scale	10.0 max (6.0)
Residue on ignition	0.007% or 70 ppm max
Chlorides (as chlorine)	0.001% or 10 ppm max
Sulfates	0.002% or 20 ppm max
Heavy metals (As Pb)	0.0005% or 5 ppm max
Chlorinated compounds (as Cl)	0.003% or 30 ppm max
Fatty acids and esters	Not more than 0.3 ml of 0.5 N NaOH is reacted with 50 g of glycerol
Identification by IR	Passes test as glycerol
Identity by GC	Passes test as glycerol
Diethylene glycol (DEG) & related compounds	DEG impurity 0.1% Max Individual impurity 0.1% max Sum of all impurities 1.0% max
Organic volatile compounds	Chloroform, 60 ppm max 1,4-dioxane, 380 ppm max Methylene chloride, 600 ppm max Trichloroethylene, 80 ppm max
Readily carbonizable substances test	Passes – current FCC
CAS number	56-81-5

[a] http://www.pgchemicals.com/

[b] Trademark name for P&G refined glycerol, available as Kosher, non-Kosher, or Kosher for non-food applications.

Table 29. Proctor and Gamble Glycerol Specification[a] .

CHAPTER 21

Product Quality

Generally, the fuel quality of biodiesel can be influenced by several factors:

- The quality of the feedstock.
- The fatty acid composition of the parent vegetable oil or animal fat.
- The production process and the other materials used in this process.
- Post-production parameters.

The best current measure for biodiesel quality is the ASTM standard, ASTM D 6751: Standard Specification for Biodiesel Fuel (B100) Blend Stock for Distillate Fuels (see **Table 30**; for comparison purposes, the European standard EN 14214 is attached to this chapter). This standard specifies the properties required for a biodiesel fuel to be used in an engine without problems. Even in blends with conventional diesel fuel, most people in the industry expect that the biodiesel blending stock will meet the standard before being blended. While some properties in the standard, such as cetane number and density, reflect the properties of the chemical compounds that make up biodiesel, other properties provide indications of the quality of the production process. Generally, the parameters given in the standard ASTM D6751 are defined by other ASTM standards. However, other test methods (American Oil Chemists' Society, AOCS) may also be suitable (or even more suitable as they were developed for fats and oils and not for petroleum-derived materials addressed in the ASTM standards). This discussion will focus on the most important issues for assuring product quality for biodiesel as it is related to production as well as some post-production parameters.

Property	Test method	Limits	Units
Flash point	D 93	130.0 min	°C
Water and sediment	D 2709	0.050 max	% volume
Kinomatic viscosity, 40°C	D 445	1.9 - 6.0	mm^2/s
Sulfated ash	D 874	0.020 max	% mass
Sulfur	D 5453	0.0015 max or 0.05 max[a)]	% mass
Copper strip corrosion	D 130	No. 3 max	
Cetane number	D 613	47 min	
Cloud point	D 2500	Report	°C
Carbon residue	D 4530	0.050 max	% mass
Acid number	D 664	0.80 max	mg KOH / g
Free glycerin	D 6584	0.020	% mass
Total glycerin	D 6584	0.240	% mass
Phosphorus content	D 4951	0.001 max	% mass
Distillation temperature, atmospheric equivalent temperature, 91% recovered	D 1160	360 max	°C

[a)] The limits are for Grade S15 and Grade S500 biodiesel, respectively. S15 and S500 refer to maximum sulfur specifications (in ppm).

Table 30. Biodiesel Standard ASTM D6751.

Production Process Factors

Completion of Reaction

The most important issue during biodiesel production is the completeness of the transesterification reaction. The basic chemical process that occurs during the reaction is indicated by the following sequence of events:

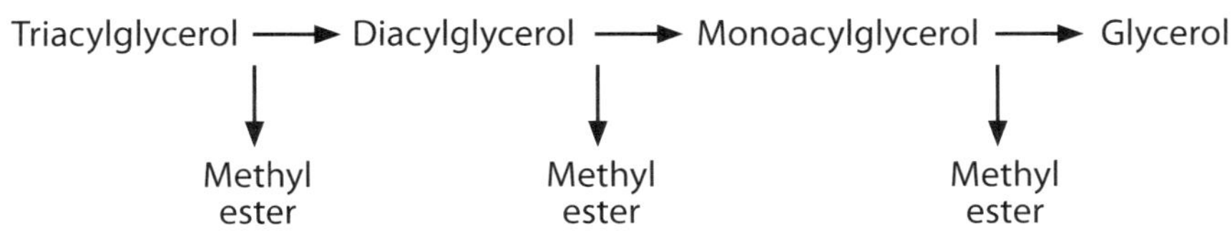

The triglycerides are converted to diglycerides, which in turn are converted to monoglycerides, and then to glycerol. Each step produces a molecule of a methyl ester of a fatty acid. If the reaction is incomplete, then there will be triglycerides, diglycerides and monoglycerides left in the reaction mixture. Each of these compounds still contains a glycerol molecule that has not been released. These compounds are referred to as *bound glycerol.* When the bound glycerol is added to the free glycerol, the *total glycerol* is obtained. The ASTM specification requires that the total glycerol be less than 0.24 percent of the final biodiesel product. The ASTM method requires that the total glycerol be measured using a gas chromatographic method described in ASTM D 6584. Gas chromatography separates the components of a complex mixture based on boiling point and compound structure. Usually, lighter compounds (within a class of compounds) will *elute* (come out of the gas chromatograph and registered by the detector of the gas chromatograph) first from the gas chromatograph. ASTM D6584 separates methyl esters, mono, di- and triacylglycerols as well as free glycerol. However, ASTM D6584 is not suitable for determining the fatty acid profile of the biodiesel sample. Other methods can be used such as high performance liquid chromatography (HPLC; for example AOCS Recommended Practice Ca 14b-96: Quantification of Free Glycerine in Selected Glycerides and Fatty Acid Methyl Esters by HPLC with Laser Light-scattering Detection) or a chemical procedure such as that described in AOCS Official Method Ca 14-56 (Total, Free and Combined Glycerol Iodometric Method). All of these techniques require a fairly high level of training and some laboratory equipment.

Free Glycerol

Free glycerol refers to the amount of glycerol that is left in the finished biodiesel. Glycerol is insoluble in biodiesel so almost all of the glycerol is easily removed by settling or centrifugation. Some glycerol may remain either as suspended droplets or a very small amount that is dissolved in the biodiesel and is known as free glycerol. Most of this glycerol should be removed during the water washing process. Water-washed fuel is generally very low in free glycerol, especially if hot water is used for washing. Distilled biodiesel tends to have a greater problem with free glycerol due to glycerol carry-over during distillation. Fuel with excessive free glycerol will usually have a problem with glycerol settling out in storage tanks, creating a very viscous mixture that can plug fuel filters and cause combustion problems in the engine.

Residual Alcohol and Residual Catalyst

Since methanol (and ethanol) and the alkaline catalysts are more soluble in the polar glycerol phase, most will be removed when the glycerol is separated from the biodiesel. However, the biodiesel may still contain 2-3 percent methanol after the separation, which may constitute as much as 40 percent of the excess methanol from the reaction. Most plants will recover this methanol by heating the biodiesel. Any methanol remaining after this heating process should be removed by the water washing process. Therefore, the residual alcohol level in the biodiesel should be very low. The allowable alcohol level is specified in European biodiesel standards, but is not included in the ASTM standard. Tests have shown that even as little as 1 percent methanol in the biodiesel can lower the flashpoint of the biodiesel from 170°C to less than 40°C. Therefore, by including a flashpoint specification of 130°C, the ASTM standard limits the amount of alcohol to a very low level (<0.1 percent). Residual alcohol left in the biodiesel will not affect its use in the engine. The amount is too small to negatively impact the fuel's performance. However, lowering the flashpoint presents a potential safety hazard as the fuel may need to be treated more like gasoline, which also has a low flashpoint, than diesel fuel.

Most of the residual catalyst is removed with the glycerol. Like the alcohol, remaining catalyst should be removed during the water washing. Although a value for residual catalyst is not included in the ASTM standard, it will be limited by the specification on sulfated ash. Excessive ash in the fuel can lead to engine deposits and high abrasive wear levels.

Equipment and Expertise Requirements for Quality Control of Production Factors

All biodiesel production facilities should be equipped with a laboratory so that the quality of the final biodiesel product can be monitored. It is also important to monitor the quality of the feedstocks.

One strategy used by many producers is to draw a sample of the oil (or alcohol) from each delivery and use that sample to produce biodiesel in the laboratory. This test can be fairly rapid (1 - 2 hours) and can indicate whether serious problems are likely in the plant.

Measuring feedstock quality can usually be limited to acid value and water content. These are not too expensive ($500 for the acid value equipment and $5,000 for the water measurement equipment) and can be operated by less experienced technicians.

To monitor the completeness of the reaction according to the total glycerol level specified in ASTM D 6751 requires the use of a gas chromatograph and a skilled operator. Large producers will find that having this equipment on-site is necessary. Commercial laboratories (i.e. Williams Laboratory Services) are available that can analyze the samples, but the cost is $100-$150/test and the time required may be several days. Smaller producers will need to use a more robust production process involving extra methanol and probably multiple reaction steps. Then the product quality can be monitored through periodic testing by an outside laboratory.

Other possibilities for monitoring the transesterification reaction and assessing fuel quality are methods based on spectroscopy (such as near-infrared spectroscopy) or physical properties (such as viscometry). These methods, although they are not yet ASTM methods, are usually faster and easier to use than gas chromatography. However, some of them require somewhat extensive calibration. They also cannot accurately quantify glycerol at the low levels called for in the ASTM standard. To circumvent this, comparison to a reaction and product known to meet ASTM standards is needed.

Post-production Factors

Water and Sediment

Water and sediment contamination are basically housekeeping issues for biodiesel. Water can be present in two forms, either as dissolved water or as suspended water droplets. While biodiesel is generally considered to be insoluble in water, it actually takes up considerably more water than diesel fuel. Biodiesel can contain as much as 1500 ppm of dissolved water, while diesel fuel usually only takes up about 50 ppm. The standards for diesel fuel (ASTM D 975) and biodiesel (ASTM D 6751) both limit the amount of water to 500 ppm. For petroleum-based diesel fuel, this actually allows a small amount of suspended water. However, biodiesel must be kept dry. This is a challenge because many diesel storage tanks have water on the bottom due to condensation. Suspended water is a problem in fuel injection equipment because it contributes to the corrosion of the close-fitting parts in the fuel injection system. Sediment may consist of suspended rust and dirt particles or it may originate from the fuel as insoluble compounds formed during

fuel oxidation. Some biodiesel users have noted that switching from petroleum-based diesel fuel to biodiesel may cause an increase in sediment that comes from deposits on the walls of fuel tanks that had previously contained diesel fuel. Because its solvent properties are different from diesel fuel, biodiesel may loosen sediments and cause fuel filter plugging during the transition period.

Storage Stability

Storage stability refers to the ability of the fuel to resist chemical changes during long- term storage. These changes usually consist of oxidation due to contact with oxygen from the air. The fatty acid composition of the biodiesel fuel is an important factor in determining stability towards air. Generally, the more unsaturated fatty acids (C18:2, linoleic acid; C18:3 linolenic acid) are more susceptible towards oxidation. As described in the section on influence of compound structure of biodiesel properties, the *allylic* and *bis-allylic* positions in the chains of unsaturated fatty acids are responsible for this susceptibility.

The changes can be catalyzed by several factors such as temperature, the presence of certain metals (including those making up the storage container) and light. If water is present, hydrolysis can also occur. Oxidation is usually denoted by an increase in the acid value and viscosity of the fuel. Often these changes are accompanied by a darkening of the biodiesel color from yellow to brown and the development of a "paint" smell. No approximate value can be given concerning how long biodiesel can be stored without deterioration of fuel quality, because it is greatly dependent on the storage conditions.

When water is present, the esters can hydrolyze to long chain free fatty acids, which also cause the acid value to increase.

There is currently no generally accepted method for measuring the stability of biodiesel, although the European standard shown in **Table 31** includes the Rancimat method for oxidative stability, which closely resembles the Oil Stability Index (American Oil Chemists' Society method). The techniques generally used for petroleum-based fuels, such as ASTM D 2274, have been shown to be incompatible with biodiesel. The Oil Stability Index or the Rancimat apparatus, which are widely used in the fats and oils industry, may be more appropriate for use with biodiesel. However, the engine industry has no experience with these tests; acceptable values are not known. Also, the validity of accelerated testing methods has not been established or correlated to actual engine problems.

Additives such as BHT and TBHQ (t-butylhydroquinone) are common in the food industry and have been found to enhance the storage stability of biodiesel. Biodiesel produced from soybean oil naturally will contain some antioxidants (tocopherols, i.e., vitamin E), providing some protection against oxidation (some tocopherol is lost during refining of the oil prior to biodiesel production). Any fuel that will be stored for more than six months, whether it is diesel fuel or biodiesel, should be treated with an antioxidant additive. However, even an antioxidant additive will not delay the onset of oxidation indefinitely.

Property	Test method	Limits		Units
		min.	max.	
Ester content	EN 14103	96.5		% (m/m)
Density; 15°C	EN ISO 3675 EN ISO 12185	860	900	kg/m3
Viscosity; 40°C	EN ISO 3104, ISO 3105	3.5	5.0	mm2/s
Flash point	pr EN ISO 3679	120		°C
Sulfur content	pr EN ISO 20846 pr EN ISO 20884		10.0	mg/kg
Carbon residue (10% dist. residue)	EN ISO 10370		0.30	% (m/m)
Cetane number	EN ISO 5165	51		
Sulfated ash	ISO 3987		0.02	% (m/m)
Water content	EN ISO 12937		500	mg/kg
Total contamination	EN 12662		24	mg/kg
Copper strip corrosion (3h, 50°C)	EN ISO 2160	1		
Oxidative stability, 110°C	EN 14112	6.0		h
Acid value	EN 14104		0.50	mg KOH/g
Iodine value	EN 14111		120	g iodine/100g
Linolenic acid content	EN 14103		12	% (m/m)
Content of FAME with ≥ 4 double bonds			1	% (m/m)
Methanol content	EN 14110		0.20	% (m/m)
Monoglyceride content	EN 14105		0.80	% (m/m)
Diglyceride content	EN 14105		0.20	% (m/m)
Triglyceride content	EN 14105		0.20	% (m/m)
Free glycerine	EN 14105 EN 14106		0.02	% (m/m)
Total glycerine	EN 14105		0.25	% (m/m)
Alkali metals (Na + K)	EN 14108, EN 14109		5.0	mg/kg
Earth alkali metals (Ca + Mg)	prEN 14538		5.0	mg/kg
Phosphorus content	EN 14107		10.0	mg/kg

Table 31. European Biodiesel Standard EN 14214.

Biodiesel Fuel Quality Management Program (BQ-9000)

In order to ensure that quality fuel is marketed and used, the National Biodiesel Board has formed a "cooperative and voluntary" program termed the BQ-9000 Quality Management Program. The website is **www.bq-9000.com**. According to the website, the "program is a unique combination of the ASTM standard for biodiesel, ASTM D 6751, and a quality systems program that includes storage, sampling, testing, blending, shipping, distribution and fuel management practices." The program is open to any biodiesel manufacturer, marketer or distributor of biodiesel and biodiesel blends in the U.S. and Canada with the companies accredited under this program listed on the website. Details of program requirements and application package are available on-line. However, a synopsis of the program is given here.

The program is run by National Biodiesel Accreditation Commission (NBAC), which is a committee of the National Biodiesel Board (NBB). The NBAC has authority to design and implement the BQ-9000 program and developed the program in its current form. The NBAC is also responsible for improvements to the program. The 10 commissioners are drawn from the range of interested and effected parties (producers, marketers, manufacturing sector, technical sector, petroleum industry, consumers, raw material producers), nominated by the president of the NBB and approved by the NBB Board of Directors.

The registration process consists of the following steps: completing the application, submitting a quality manual, paying a $750 application fee and proving EPA Registration. Thereupon, the NBAC reviews the application and application materials and accepts them for processing and audit if they are "administratively complete." Then a $2,500 Certification Audit Fee is payable. Next an independent auditor is assigned, and the auditor establishes an audit schedule. A desk audit is conducted to verify written compliance with BQ-9000 criteria and an on-site audit verifies "compliance by auditor review of certain objective evidence."

Any corrective actions necessary to comply with the program must be addressed within 30 days. The NBAC reviews the results of the audit results and votes. Accreditation is granted for two years if the audit is passed. After two years, a company needs to undergo a recertification audit in order to be accredited for another two-year term. Distributors meeting the BQ-9000 requirements receive a certification logo, marketers are awarded the Certified Marketer status, and producers receive the Certified Producer status.

For recertification, a recertification form needs to be submitted and a $1,750 surveillance audit fee must be paid. Again, the NBAC reviews the application and accepts "administratively complete" applications for further action. An on-site audit by an independent auditor is conducted to again verify compliance by "review of certain objective evidence" with any corrective action to be taken within 30 days. The NBAC reviews the audit and votes. If successful, recertification is granted for two years.

To participate in the program, a Certificate of Analysis (COA) must be generated for each production lot of fuel. Initially, full specification testing must be conducted on seven consecutive production lots. Full specification testing consists of testing according to all limits prescribed in ASTM D6751 plus visual appearance according to ASTM D4176 Procedure 2 with a limit of 2 max (outlet sample). If successful, reduced testing is allowed. The reduced testing that must be conducted on every production lot is flash point, water and sediment, cloud point, acid value, free and total glycerin, and visual appearance. Full specification testing, however, is required at least every six months.

Another requirement is that a storage tank containing a "unique lot" that has not seen activity for 30 days cannot be shipped unless a sample of the biodiesel fuel is tested for water and sediment per ASTM D4176 and acid number per ASTM D664. If any of these tests do not meet specifications, the lot must be isolated and procedures described in the requirements brochure (available at the website) for a nonconforming product apply. This brochure also describes other procedures on handling and shipping of biodiesel.

CHAPTER 22

Site Selection Criteria

The site consideration for a biodiesel plant should include the following items: land ownership and size, transportation access, feedstock supply, customer location, utility access, wastewater treatment availability, climate, governing jurisdictions, proposed plant size and future expansion plans.

Land Ownership and Size

As a general rule, the amount of land required for a plant depends on the size of the plant, but can be as small as one to three acres and is subject to city, county and state regulations. This size may be larger if waste treatment is required on site.

If the site is in conjunction with a feedstock supplier or customer, the land may be owned by them already. If so, a lease arrangement may be an option rather than an outright purchase. The influence of a partnership or joint venture can have both a financial and land ownership effect on the site selection.

If future expansion is planned at the site, then extra land may need to be secured initially.

Financial considerations are always a factor and prime sites command a high price. It is wise to look in an industrial park setting where local financial incentives might be present. A rural location may reduce the initial cost of the site, but the overall cost may be higher when infrastructure costs are considered. The cost of providing energy, natural gas, transportation access, and rail siding can greatly increase the overall cost of the site.

Transportation Access

Road, rail and water access should be considered, with road access being the most important and closely followed by rail. Road access is needed for receiving supplies as well as shipping product, especially to local and regional markets. Depending on your plant size, truck shipments of biodiesel can handle many of your customers; even if you could ship it by rail, many times your customer may not have rail access. A second reason for placing road access above rail is that if your customer is planning to deliver directly to customers and do splash blending, a truck is the primary means of accomplishing that.

Rail and water way access will most likely be used for long distance shipments and larger quantities. Rail should be considered if your future expansion plans require additional shipments for longer distances. Also, if you are to be a global player, you will need rail to transport your product to an ocean port. A biodiesel plant size of 5 million gallons per year should consider rail access, and a plant size of 10 million gallons per year or larger should have rail access. Rail access allows incoming feedstock to be delivered from long distances in large quantities. The future requirements for rail car usage need to be weighed against the cost of trucking to or from a rail spur or the cost of putting in a rail spur. If it is determined that current or future requirements do dictate the usage of rail car quantities, then it should be moved up in importance in the site selection priorities. You can also use rail cars for temporary storage, but the cost associated with demerage is high and should be used only in an emergency.

Road access should be adequate to support truck transportation and easy access to the interstate highway system. The road should be well maintained (Class A) and sufficiently wide to accommodate the passage of two semi or tank trucks. In addition, the entrance onto the access road and the highway should provide an unrestricted view in all directions.

The site should be located where truck traffic will not create a problem with local residences or towns. The city council of the town nearest your plant site should be contacted to determine if the increase in truck traffic will be a problem. Also check with the state and county departments of transportation for any regulations or permits that will be needed to handle the truck traffic. This should include a certification that the bridges to be crossed can handle the weight and height of the trucks.

Pipeline access is used for larger plants to move product to large fuel distributors or other means of transportation like barges on a water way, which could be important if you plan to use large quantities of imported feedstock like palm oil. It can also be used for the supply for fuel like natural gas used for heating or water for the process, but it is not as important a consideration here as other forms of transportation.

Feedstock Supply

Consideration should be given to the source of feedstock supply, and it is wise to locate the plant close to the source if feasible. If a partnership with a feedstock supplier is established, then this makes economic sense. There may also be synergies in other areas, such as storage, waste treatment and utilities. However, before you build your plant next to a feedstock supplier, you should make sure that your agreement covers contingencies if the partnership dissolves and it is no longer your supplier.

If you are getting feedstock from rail car, placing the plant near a railroad siding makes loading and unloading much easier. If your feedstock is going to be from multiple sources, then a central location would be best. But if your feedstock is being received by rail car, then the location close to the originating source of your feedstock becomes less critical, although transportation charges should be minimized if possible.

Customer Location

Depending on your plant size, it might make sense to locate near your major customers to make the time of delivery as short as possible. This might make more sense than locating near your feedstock supplier. Locating near your customer base will give you a marketing advantage over competitors as well.

The down side of putting too much importance on placing your plant near your customers is that your customer base may change. With the use of rail shipments, the ability to ship to your customer is less dependent on your plant location. If the majority of your business is regional, then smaller plants located throughout the region is an option to consider and can be very economical. Placing your plant near your customer can build customer loyalty, which might give you an edge if supply outpaces demand.

Utility Access

The utilities you will need for the production of biodiesel are electricity and some source of heat or fuel for heat (natural gas, fuel oil, propane, steam or electricity). Although the entire plant can use electricity, the cost for process heating is usually more than using fuels like natural gas, propane, fuel oil or your own biodiesel. A cost analysis of heating with the various sources mentioned should be conducted when the equipment is identified and the site location chosen. Keep in mind that some utility companies will charge you for the installation of a feeder line to your plant location if it is too remote or rural.

Biodiesel, unlike its sister bio-renewable fuel ethanol, takes relatively less energy to produce. Therefore, energy requirements should be reasonable and well within the means of rural supply sources.

A benefit to choosing a site location near other neighboring plants or feedstock sources is the possibility of using their excess heat, which may also save some money on utility infrastructure by sharing their energy supply.

Finally, you will need a source of water for your plant. Both potable and process water will be needed. If you choose to recycle your water, you can minimize the amount you need. If recycling your water, you will have to design a means to drop the temperature of your recycled water when it is used for process cooling.

Wastewater Treatment

Site selection should consider the means of disposing of your wastewater. The amount of wastewater depends on the process you are using and the quality of your feedstock. For example, if your feedstock has a high free fatty acid that will make soaps in your process, you will need additional water to wash those soaps out of your product. This will require additional water treatment. If your feedstock has a high amount of other impurities, they will have to be washed out.

The process and feedstock can have a definite effect on the amount of your wastewater. If treatment is not available at the site you have chosen, then you will have to install a water treatment facility, truck it to a treatment facility or design a waterless process to remove the impurities. To minimize this cost, you may have to be selective in your choice of processing techniques and feedstock quality.

Climate

Usually the climate is thought of last in site selection, but itis still an important consideration, especially for biodiesel which has a higher gel temperature. Climate can affect road access, which affects your labor, supplies and shipments to customers. It is important that orientation of the plant and location minimize impact of the weather.

Also, locating in a northern or cold climate has a bearing on the amount of energy you will need in the winter, including heating of feedstock and products from your plant. It will also affect the means you use for transportation and the ability to keep the rail car or tank truck warm.

Governing Jurisdictions

Factors like city, county, state and federal jurisdiction also have an influence on site selection. Depending on regulations set forth by governing bodies that have authority over the site you have selected, the cost of construction and equipment can be impacted. There are positives and negatives to locating outside city limits, and they have to be weighed to determine the best choice for the plant location.

Proposed Plant Size

The selection of a site is very much related to the plant size you are proposing to build. A smaller plant will need a smaller site, less energy, less tank storage and less transportation access. Small plant sizes of 1 to 5 million gallons per year can be skid mounted and require minimal space. Larger plants can require different regulations and different site considerations. Also keep in mind the option of using existing buildings and plant infrastructure.

Future Expansion Plans

Your business plan will help you decide if you should consider future expansion plans in your selection of the site for the initial plant. You need to balance finances available now and potential market growth in the future. Many times, future site expansion is covered by a right of first refusal on adjacent land.

CHAPTER 23

Solid Catalysts for Biodiesel Production

As discussed in previous chapters, the most common catalysts used in the production of alkyl esters through transesterification of oils and fats are sodium methoxide and sodium hydroxide. These catalysts, when added to the reactor, end up dissolved in the reaction media. When the catalyst used in a reaction is in the same phase as the reactants and products, the catalyst is known as a homogeneous catalyst. An advantage of homogeneous catalysts is that they allow intimate contact between the reactants and the catalysts, thereby decreasing mass transfer resistance. A disadvantage of homogeneous catalysts is that they are somewhat difficult to remove from the product stream since they exist in the same phase.

Heterogeneous catalysts are an alternative type of catalysts. As suggested by the name, these catalysts exist in a different phase than the reactants and products. The most common form for heterogeneous catalysts is as a solid. The advantage of heterogeneous catalysts is that no separation stage is required to separate the products from the catalyst. For this reason almost all large volume chemicals are produced using heterogeneous rather than homogeneous catalysts. The negative aspect of using a heterogeneous catalyst is that good mass transfer, which allows intimate contact between the reactants and the catalyst, is more difficult to achieve.

As specific reaction systems become technically more mature, catalyst and process conditions are inevitably developed to at least partially overcome the mass transfer limitation with heterogeneous catalyst so that the more desirable solid catalyst (from a separations standpoint) can be used. The progression from homogeneous catalyst use to heterogeneous catalyst use has occurred many times in the petrochemical industry and likely will eventually occur in biodiesel production as well. Due to this common progression and the fact that homogeneous catalysts (such as sodium methoxide for transesterification and sulfuric acid for esterification) have been covered elsewhere, our emphasis here will be on the status of solid catalysts for biodiesel production.

It is important to remember that while a solid catalyst significantly simplifies separation downstream of the reactor, the absolute best activity that could ever be achieved with a solid catalyst will only be equal to that with the homogeneous catalyst. Therefore, the best possible solid catalyst would operate at similar reaction conditions to that used with the current homogeneous catalyst. The more likely situation is that even the optimal solid catalyst will require a higher reaction temperature than the homogeneous catalyst, since the contact between the reactants (triglyceride and alcohol) and the catalyst is better with a homogeneous catalyst. In addition, the equilibrium reaction condition that exists at high conversion with the homogeneous catalysts will not be changed by the use of a solid catalyst. Reaction equilibrium is determined by the thermodynamic relationship of the reactants and the products. Catalysts cannot change the equilibrium relationship, but they can change the rate at which equilibrium is reached. They can speed up the reaction but not change the ultimate extent of the reaction. Therefore, the best possible solid catalyst that could ever be developed for the alkyl ester reaction system will undoubtedly be somewhat less active than the currently used homogeneous catalysts.

Two general catalyst classes, bases and acids, can convert triglycerides to alkyl esters through transesterification. The efficacy of either catalyst type in the transesterification reaction is dependent on the strength of the catalyst, i.e., a higher pKa value for a base catalyst and a lower pKa value for an acid catalyst. As the strength of the basic or acidic catalyst is increased the catalyst will become more active meaning that it will be able to perform the reaction at lower temperature. Sodium methoxide and sodium hydroxide are quite strong base catalysts, and sulfuric acid is a very strong acidic catalyst. The transesterification reaction rate with homogeneous base catalysts such as sodium methoxide and sodium hydroxide is significantly higher than homogeneous acid catalysts such as sulfuric acid. The higher activity associated with the strong base homogeneous catalyst relative to the strong acid homogeneous catalyst will also hold for solid catalysts. For example, a highly acidic solid catalyst will likely be less active than just sulfuric acid and both types of acid catalysts will be significantly less active than sodium methoxide. Therefore, the most desirable solid catalyst for transesterification would be a material that acts as a strong base.

The development of solid (heterogeneous) catalyst technology has been largely driven by petroleum refining and petrochemical processes. These processes typically involve olefin conversion, and acidic catalysts commonly play an important role. As such, significant effort has been expended on developing improved solid acid catalysts. In addition, most olefin conversion processes rely on the reactants and products being in the gas phase. Therefore, the attributes needed for a good transesterification catalyst, which include being a strong base and utilization in a liquid-phase reaction, are significantly different than those needed in petroleum refining and petrochemical processes. These differences mean that a new type of solid catalyst must be developed specifically for use in biodiesel production.

Initial studies with solid bases focused on readily available materials such as CaO, MgO, K_2CO_3, etc. [1, 2]. These catalysts were significantly less active than the homogeneous base catalysts, which was not surprising since they are also significantly less basic. Complicating the reported results was the fact that these solid, basic materials have limited solubility in methanol. Therefore, it is probable that the activity that was observed for these materials was due to the portion of the catalyst material that had been solubilized. The reaction was likely still occurring via a homogeneous catalyst reaction mechanism with the solid material only serving as a solid reservoir for the material going into solution, and little or no direct reaction was occurring with the solid material.

Suppes et al. [3] have reported the use of calcium carbonate, $CaCO_3$, as a solid catalyst in the transesterification reaction. Their studies showed no appreciable amount of the $CaCO_3$ being dissolved although it should be noted that methanol was not used. The alcohols used for the reaction studies were ethanol and ethylene glycol. The catalyst was also found to be effective in the esterification of fatty acids. $CaCO_3$ is not particularly basic, so its catalyst activity is significantly lower than that of sodium methoxide. As such, the $CaCO_3$ catalyst must be operated at a significantly higher temperature. To achieve reasonable conversion rates the catalyst had to be operated at >200°C.

A French group compared the reactivity of several solid base catalysts including a cesium-exchange zeolite, hydrotalcite, magnesium oxide and barium oxide [4]. From this study, only the barium oxide catalyst had good activity for rapeseed oil transesterification. While the barium oxide catalyst results appeared to be quite promising, very little reaction data were presented for the catalyst, so it is difficult to evaluate the performance of the catalyst.

A group of organic bases called alkylguanidines, which are a type of amine complexes, have sufficient basicity to catalyze the transesterification reaction. Unfortunately, these molecules are also homogeneous catalysts and are more expensive than sodium methoxide. An interesting paper by Sercheli et al. [5] discusses the synthesis of solid catalysts in which the alkylguanidine complexes are immobilized via tethering to a silica support with sufficiently large pores to accommodate a triglyceride molecule (>30 □). Previous work with alkylguanidine complexes in zeolite supports, which have smaller pore sizes, demonstrated that pores in the zeolite size range (<13 □) were not very active catalysts [6]. The immobilized alkylguanidine solid catalysts with larger pore diameter were not quite as active as homogeneous alkylguanidine species, but were significantly more active than any other solid base catalyst that has been reported in the open literature. While this solid catalyst system looks quite interesting, it is not currently close to being commercially viable since it would cost too much to manufacture the catalyst. A second concern with this catalyst system was leaching of the alkylguandine complex from the catalyst support.

Recently, a group from the Institut Francais Du Petrole applied for a patent on the use of a zinc aluminate, $ZnAl_2O_4$, as a solid catalyst for the transesterification of vegetable oil [7]. Since this material is less basic than sodium methoxide, it will require a higher temperature for the reaction. Few details are yet publicly available about the catalyst, so the operating conditions necessary for the catalyst are not clear. Discussions with European biodiesel producers suggest that this catalyst is beginning to be offered, so it is certainly a catalyst that bears some attention.

Several recent papers demonstrate the need to be careful in assessing the performance of solid base catalysts in the transesterification reaction. Of particular concern is that the methanol might be leaching the catalytically active component off the solid support and into solution. For these cases, the activity attributed to the solid catalyst is likely due to the leached catalyst species that is in solution. There are two recent papers in the literature that ascribe promising results for solid base catalysts [8,9] that have likely incorrectly characterized their reaction system. The results presented in these two papers are probably due to solubilization of the active catalyst component with the catalytic species then performing as a homogeneous catalyst and not as a solid catalyst.

In summary, there are no solid base catalysts reported in the literature for vegetable oil transesterification that appear to be ready for commercialization. However, several promising catalyst technologies have been presented that could lead to commercialization in the near future. If a transesterification reactor was designed for a high temperature of >200°C, which also means high pressure, there does appear to be some solid catalyst candidates.

As discussed earlier, a highly basic catalyst will be significantly more active for the transesterification reaction than a highly acidic catalyst. If feedstocks are to be processed that have high free fatty acid content, an acidic catalyst could be used as a pretreatment catalyst to esterify the fatty acids to alkyl esters [10-12]. A recent article has provided a nice review of acid catalysis in biodiesel production [13]. Several solid acid catalysts have been tested in the fatty acid esterification reaction. The performance of commercially available Amberyst acidic resin and Nafion catalysts has been determined on a model free fatty acid feed [14]. While not as active as sulfuric acid, these materials will catalyze the reaction. In addition, several novel materials consisting of sulfonic acid supported on silica have shown promise as active catalysts in the reaction [14, 15]. However, these new materials have only been demonstrated in the laboratory with a model feed, so additional information is required to assess their effectiveness with real high free fatty acid feedstocks.

1. Peterson, G.R. and W.P. Scarrah, "Rapeseed oil transesterification by heterogeneous catalysis," *J. Am. Oil Chem. Soc. 61*:1593-1596 (1984).

2. Gryglewicz, S., "Rapeseed oil methyl esters preparation using heterogeneous catalysts," *Bioresource Tech. 70*: 249-253 (1999).

3. Suppes, G.J., K. Bockwinkel, S. Lucas, J.B. Botts, M.H. Mason, and J.A. Heppert, "Calcium carbonate catalyzed alcoholysis of fats and oils," *J. Am. Oil Chem. Soc. 78*:139-145 (2001).

4. Leclercq, E., A. Finiels, and C. Moreau, "Transesterification of rapeseed oil in the presence of basic zeolites and related solid catalysts," *J. Am. Oil Chem. Soc. 78*:1161-1165 (2001).

5. Sercheli, R., R.M. Vargas, and U. Schuchardt, "Alkylguanidine-catalyzed heterogeneous transesterification of soybean oil," *J. Am. Oil Chem. Soc. 76*:1207-1210 (1999).

6. Schuchardt, U., R. Serchi, and R.M. Vargas, "Transesterification of Vegetable Oils: A Review," *J. Braz. Chem. Soc. 9*:199-210 (1998).

7. Hillion, G., S. Leporq, D. Le Pennec, B. Delfort, "Process for preparation of a catalyst based on zinc aluminate and catalyst obtained," French patent application 2003-3317 (2004).

8. Suppes, G.J., M.A. Dasari, E.J. Doskocil, P.J. Mankidy, and M.J. Goff, "Tranesterification of soybean oil with zeolite and metal catalysts," *Appl. Catal. A: Gen. 257*:213-223 (2004).

9. Kim, H.-J., B.-S. Kang, M.-J. Kim, Y.M. Park, D.-K. Kim, J.-S. Lee, and K.-Y. Lee, "Transesterification of vegetable oil to biodiesel using heterogeneous base catalyst," *Catal. Today 93-95*:315-320 (2004).

10. Srivastava, A. and R. Prasad, "Triglycerides-based diesel fuels," *Renewable & Sustainable Energy Reviews 4*:111-133 (2000).

11. Canacki, M. and J. Van Gerpen, "Biodiesel production from oils and fats with high free fatty acids," *T. ASAE 44*:1429-1436 (2001).

12. Canacki, M. and J. Van Gerpen, "A pilot plant to produce biodiesel from high free fatty acid feedstocks," *T. ASAE 46*:945-954 (2003).

13. Lotero, E., Y. Liu, D.E. Lopez, K. Suwannakarn, D.A. Bruce, and J.G. Goodwin, "Synthesis of biodiesel via acid catalysis," *Ind. Eng. Chem. Res.* 44:5353-5363 (2005).

14. Mbaraka, I.K., D.R. Radu V. S.-Y. Lin, and B.H. Shanks, "Organosulfonic acid functionalized mesoporous silicas for the esterification of fatty acid," *J. Catal. 219*:329-336 (2003).

15. Mbaraka, I.K. and B.H. Shanks, "Design of multi-functionalized mesoporous silicas for fatty acid esterification," *J. Catal., 229*:365-373 (2005).

CHAPTER 24

Influence of Compound Structure on Biodiesel Properties

The properties that will be discussed here are cetane number with some aspects of exhaust emissions, viscosity, cold flow, oxidative stability and lubricity. As the discussion of the individual properties shows, a structural feature that imparts some advantageous properties to biodiesel usually also causes some disadvantageous properties.

Cetane Number, Exhaust Emissions

As discussed elsewhere in this book, the cetane number is a dimensionless indicator of ignition quality of a diesel fuel. The reference compounds on the cetane scale are hexadecane (cetane number = 100; high-quality reference compound) and 2,2,4,4,6,8,8-heptamethylnonane (cetane number = 15; low-quality reference compound). Generally, higher cetane numbers are desirable. The cetane number increases with increasing saturation and increasing chain length. For example, methyl palmitate and methyl stearate have cetane numbers in the 85-100 range; that of methyl oleate is around 55-60; that of methyl linoleate around 40. Generally, when determining cetane numbers, the deviations in repeated tests are greater than for other properties. The cetane numbers of soybean oil-derived biodiesel are typically in the 48-52 range. The cetane numbers of biodiesel derived from more saturated feedstocks, such as used frying oils or animal fats, are usually higher due to the greater content of saturates.

Effects of compound structure are also observed on exhaust emissions. This is of particular interest for NO_x exhaust emissions, since these species are the only regulated emissions typically showing a slight increase when using biodiesel compared to petrodiesel. It has been observed that NO_x exhaust emissions are reduced when using saturated fatty esters as fuel while they increase with unsaturation. The use of biodiesel generally decreases the other regulated exhaust emissions species – particulate matter (PM), hydrocarbons and carbon monoxide (CO) – compared to petrodiesel. The effect of compound structure on exhaust emissions is complicated by the observation that there can be a strong dependence on the technology level of the engine, with more modern engines generally generating less regulated exhaust emissions. The effect of unsaturation on NO_x exhaust emissions, however, appears to be of more general nature.

Viscosity

Reduction in viscosity is the major reason why vegetable oils are used as fuel in form of biodiesel and not as neat oil.

Viscosity increases with increasing chain length. For example, methyl palmitate has a kinematic viscosity of 4.38 mm^2/sec (at 40°C, the temperature prescribed in biodiesel standards; all other values given here were also obtained at that temperature) while that of methyl stearate is 5.85 cm^2/sec. A similar effect is observed for methyl esters vs. ethyl esters, etc. A *cis* double bond, as found in the common unsaturated fatty acid chains, lowers viscosity. For example, the kinematic viscosity of methyl oleate is 4.51 mm^2/sec. A second *cis* double bond lowers it even more. Thus the kinematic viscosity of methyl linoleate is 3.65 mm^2/sec. On the other hand, compounds with *trans* double bonds exhibit higher viscosity. Methyl elaidate, the *trans* isomer of methyl oleate, has a kinematic viscosity of 5.86 mm^2/sec, almost identical to that of methyl stearate. Polar groups, such as the carboxylic acid moiety, in the chain also increase viscosity. For example, the value for oleic acid is 19.46 mm^2/sec. A typical value for the kinematic viscosity of biodiesel (methyl esters) derived from soybean oil is around 4.10 mm^2/sec.

These viscosity effects are important because of the limits on kinematic viscosity in biodiesel standards and the different viscosity of biodiesel depending on the feedstock. Thus biodiesel made from shorter-chain fatty acids will have lower viscosity. On the other hand, biodiesel obtained from used frying oils will have higher viscosity because of their greater content of saturated fatty acids as well as some content of *trans* fatty acids (during heating, *cis* double bonds can isomerize to *trans* double bonds).

In the scientific literature, viscosity is often reported as dynamic viscosity. It must be kept in mind that kinematic viscosity is the property specified in biodiesel standards. Dynamic and kinematic viscosity are connected with density as the factor: Kinematic viscosity X density = dynamic viscosity.

Cold Flow

The cold flow problem of biodiesel is highlighted by the high melting points of some of its components. Methyl palmitate has a melting point of 30°C and methyl stearate melts at 39°C, showing that saturated compounds have the highest melting points and that it increases with chain length. Introducing a *cis* double bond lowers the melting point considerably. Thus, the melting point of methyl oleate is -20°C. A second *cis* double bond in methyl linoleate lowers the melting point to -35°C. Fatty acids have higher melting points than the esters. For example, the melting point of stearic acid is 71°C and that of oleic acid 15°C. Fatty compounds with *trans* double bonds have higher melting points than their *cis* counterparts. Thus, the melting point of elaidic acid (*trans* isomer of oleic acid) is 45°C.

The high melting points of the saturated fatty acid chains are reflected in the relatively high cloud and pour points of biodiesel. Thus soybean oil-derived biodiesel has a cloud point of about 0°C and a pour point only a few °C beneath the cloud point, around -3°C. Not surprisingly, the cloud and pour points of biodiesel derived from more saturated feedstocks, such as used frying oils or animal fats, are higher.

This cold flow behavior and the properties of the individual components have led to efforts to improve this behavior. Approaches are: A) Winterization, which is removal of saturated components by cooling the fuel and filtering the formed solids from the liquid part of the fuel. This approach leads to loss of material and reduces the cetane number. B) Use of longer-chain or branched esters. For example, ethyl palmitate melts at about 19°C and *iso*-propyl palmitate melts at 13-14°C, translating into lower cloud and pour points for these esters; these esters have cetane numbers comparable to methyl esters. C) Use of additives, which often are of polymeric nature and usually affect the pour point more than the cloud point, although the latter is the more significant parameter and is used in the ASTM D6751 standard.

Oxidative Stability

Oxidative stability is one of the most complex technical issues facing biodiesel. It is complicated by the fact that it depends not only on structure of the fatty esters comprising biodiesel but also a variety of external factors. The use of antioxidant additives is a common approach to improve the oxidative stability of biodiesel. Some aspects of oxidation as related to biodiesel storage are discussed in the product quality section.

Increasing unsaturation increases susceptibility to oxidation. While methyl oleate is more susceptible to oxidation than of saturated fatty esters, methyl linoleate and methyl linoleate are the most sensitive. Values given in the literature for relative rates of oxidations are, when giving methyl oleate the value 1, about 40 for methyl linoleate and 98 for methyl linolenate. The reason is the presence of allylic and, especially, *bis*-allylic CH_2 positions (found in linoleic and linolenic acids) in the fatty acid chain.

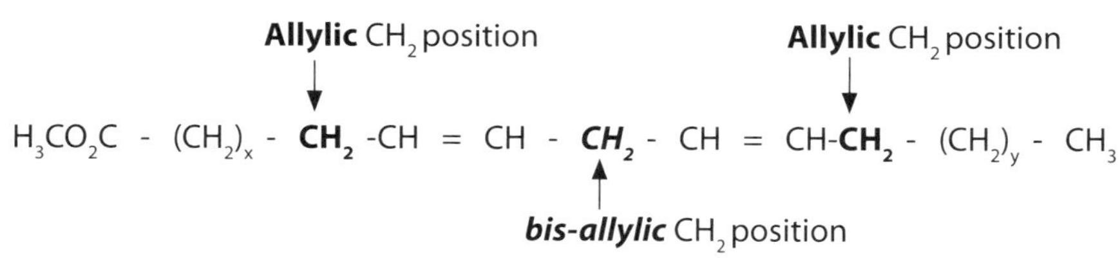

Concentration of the more oxidizable fatty compounds plays a role in a mixture such as biodiesel, although some research shows that small amounts of the more unsaturated fatty compounds in a mixture increase susceptibility to oxidation more than their small amounts would indicate.

Very important factors regarding oxidative stability are external parameters. Presence of light, air (oxygen; also the size of the contact area with air), elevated temperature and extraneous materials (especially metals, which is important for the nature of storage containers for biodiesel; but also peroxides, sensitizers, antioxidants and other contaminants) all are factors catalyzing biodiesel oxidation.

The reactions taking place during oxidation are manifold and complex. Initially, hydroperoxides are formed, which forms the basis of the peroxide value.

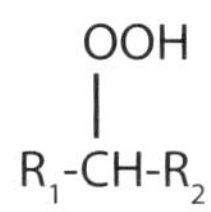

Formally, this is an insertion of molecular oxygen into a C-H bond at a (*bis-*)allylic (the reaction mechanism is actually more complex than just an "insertion"). From that stage on, secondary reactions occur. These secondary oxidation reactions are the formation of free fatty acids, aldehydes and other shorter-chain species as well as double bond isomerization (*cis* to *trans*; rearrangement of double bonds to other positions in the molecule) and polymerization (formation of dimers and other higher molecular weight species). It is beyond the scope of this book and course to discuss the complexities of oxidation in greater detail.

Due to the formation of free fatty acids (both shorter-chain and the same length as well as of higher-molecular weight species), viscosity increases during oxidation. The acid value also increases continuously. The nature of the species formed during the whole oxidation procedure depends on the fatty acid profile and the nature of the oxidation process, which can differ if it is initiated by light or occurs through autoxidation (presence of air). Different extraneous materials can influence the course of oxidation. All these factors result in oxidation experiments often being of limited repeatability and reproducibility, although the mentioned trends and effects are clear.

Lubricity

Lubricity is an important issue for diesel fuels, because in the past diesel engines have relied on the inherent lubricity of diesel fuels to lubricate engine parts, such as fuel pumps and fuel injectors. In ultra-low-sulfur petrodiesel fuels, this lubricating effect has been largely eliminated. Commercial biodiesel, on the other hand, possesses excellent inherent lubricity, a property that is imparted to blends with petrodiesel at around the 2 percent level (B2). Lower blend levels of biodiesel with petrodiesel do not show the same lubricity improvement.

Lubricity is usually determined with an instrument termed the high-frequency reciprocating rig (HFRR) lubricity tester. In this test, a steel ball held down by a 200 g weight rubs against a steel disk immersed in 2 mL of the sample to be tested. The steel ball rubs against the disk for 75 minutes at a frequency of 50 Hz. The most common test temperature is 60°C. The wear scar generated on the disk is used for evaluating the test results. The smaller the wear scar on the ball, the better the lubricity of the test sample. ASTM D6079 describes this test. While not included in biodiesel standards, lubricity is now included in both the American (ASTM D975) and European (EN 590) petrodiesel standards. The specification in ASTM D975 allows for a maximum wear scar of 520 µm, in EN 590 a maximum of 460 µm.

Besides the presence of double bonds somewhat enhancing lubricity, the lubricity of neat fatty compounds depends largely on how oxygen atoms are contained therein. Thus, free fatty acids in the neat form have the best lubricity, with very small to no wear scars. Monoacylglycerols (monoglycerides) also have excellent lubricity in the neat form. The presence of two OH groups plays a significant role in this case. They possess better lubricity in the neat form than methyl esters. Glycerol has excellent lubricity in the neat form but does not act as lubricity-enhancing additive alone in petrodiesel, presumably due to its insolubility in petrodiesel.

Neat fatty esters at additive levels do not enhance lubricity of low-lubricity petrodiesel fuels as much as free fatty acids or monoacylglycerols. However, when adding small amounts of free fatty acids or monoacylglycerols to neat fatty esters and then adding this mixture to low-lubricity petrodiesel fuels, lubricity is enhanced more strongly than by addition of neat esters only. This effect shows that low-level contaminants in biodiesel, such as free fatty acids and monoacylglycerols, which are also limited in standards, are major contributors to biodiesel lubricity. Especially at low blend levels such as B2, these "contaminants" are likely the major lubricity-imparting components of commercial biodiesel.

CHAPTER 25

Process Strategies for Biodiesel Production

In previous discussions we have described the individual processes and equipment that are utilized in biodiesel plants. This chapter deals with the issues associated with integrating the processes into systems that produce biodiesel. It does not advocate for a specific process, although a sample plant is presented. The intent is to provide a means to discuss the variety of approaches that are currently used in biodiesel plants and the trade-offs associated with each. The diagrams that are provided show the major components that are involved but generally do not show valves, sensors or controls. All plants that utilize the concepts discussed here still need to be engineered and equipped with appropriate safety and quality control features.

Figure 32 shows a schematic diagram of a hypothetical multi-feedstock biodiesel plant. The inputs to this plant can be soybeans, soybean oil or yellow grease. Tanks with sufficient capacity are needed to allow deliveries of truck-load or rail car-load quantities of oil and methanol (and possibly sodium methylate). The cost of receiving methanol in barrels or totes is too high for economical biodiesel production.

There should be sufficient tankage to provide at least five days of oil storage to minimize the likelihood of having to shut down the plant due to a shortage of oil. If the plant does not produce its own oil and must bring it in from a distance, then even more on-site storage is recommended. Note that large tanks need to be equipped with spill containment. A SPCC (Spill Prevention, Containment and Control) plan must be in place if the total quantity of oil storage is greater than 1,320 gallons. The truck loading and unloading area should also be equipped with spill containment and control. This can be accomplished with dikes, speed bumps and foam barriers.

Samples of each lot of received material should be collected. When possible, the important specification of each load should be verified before the load is accepted. In any case, samples should be held for 30 days so that if quality problems occur, it can be determined whether the reactants were at fault.

Oil Extraction

Figure 33 shows a process flow diagram for crushing oilseeds such as soybeans or canola. The oilseeds enter the diagram on the right side by passing through an extruder. The extruder compresses the seeds to a high temperature and pressure, which cooks the seed to destroy anti-nutritional elements and breaks up the cell walls to free the oil. High oil content seeds such as canola or sunflower may use a pre-pressing operation before the extruder to remove a portion of the oil.

In some cases, the seeds will be dehulled before entering the extruder. The hulls are primarily fiber, so this increases the protein percentage of the meal. The hulls are also abrasive so removing them extends the life of the extruder. The extruded oilseeds are then fed into the expeller where the oil is pressed out. The meal portion of the crushed seeds is about 82 percent of the oilseed mass for soybeans and about 60 percent for canola. The meal is usually air cooled after pressing and may pass through a grinder to achieve the desired grain size before storage in a bin for transport to its normal use as animal feed.

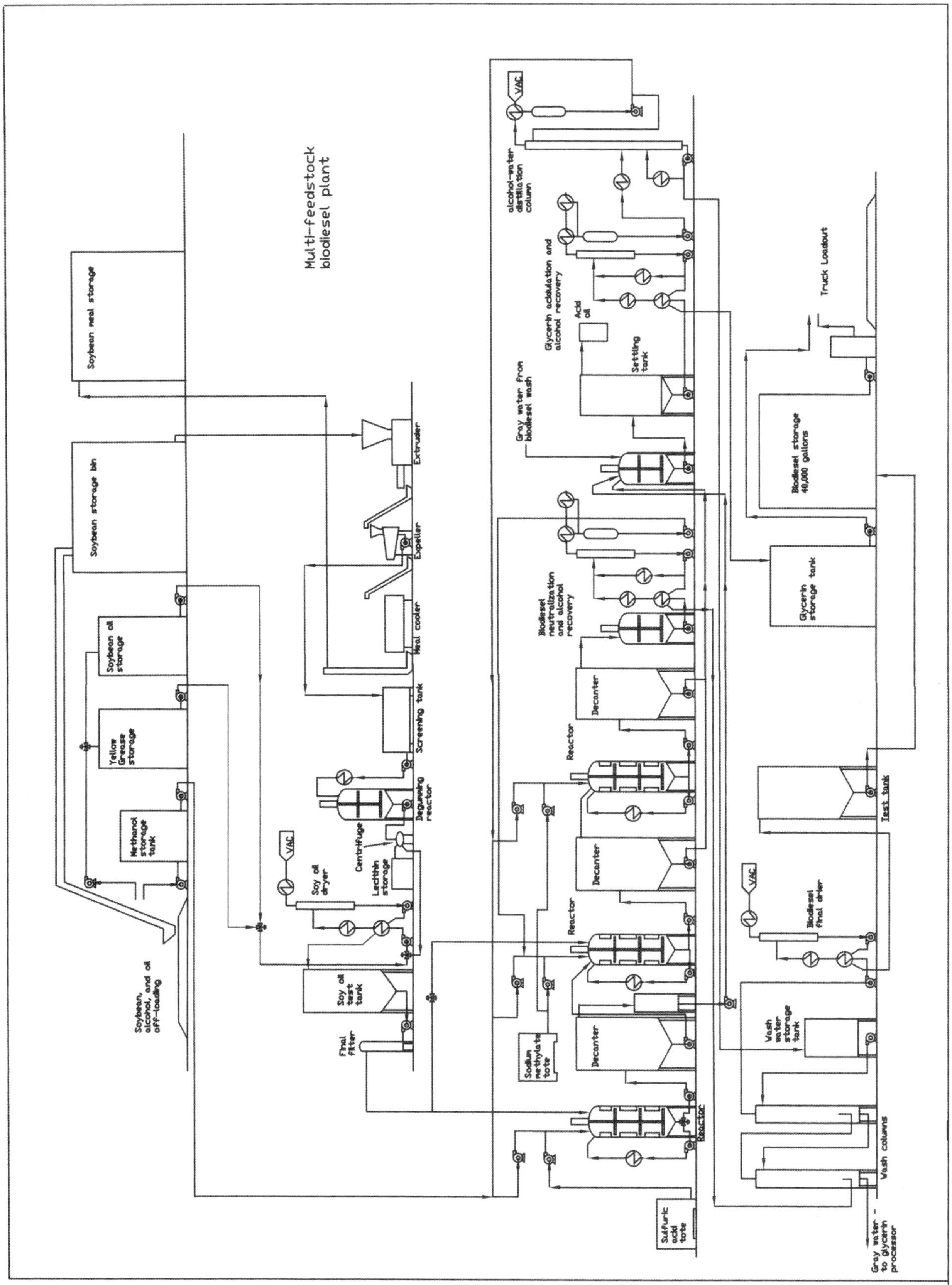

Figure 32. Multi-feedstock Biodiesel Plant.

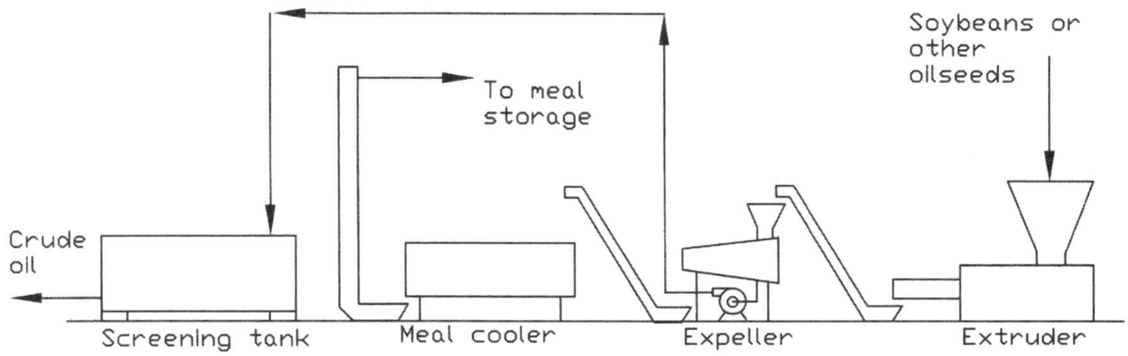

Figure 33. Mechanical Oil Extraction.

Sometimes the oilseeds are directly fed to the expeller, but extrusion helps to increase the fraction of oil that can be extracted from the seeds. Heating the seeds immediately before they enter the extruder or expeller also increases the yield.

Degumming

Figure 34 shows a simple degumming operation for removing the phospholipids, or gums, from crude vegetable oil. Soybean oil and canola oil both have significant amounts of these contaminants, especially with solvent extracted oils. The crude oil is mixed with water at an elevated temperature, which hydrates the phospholipids and produces an insoluble sediment that can be separated with a centrifuge. The phospholipids have value as a source of food-grade lecithin. A portion of the phospholipids need to be treated with dilute phosphoric acid before they can be removed. This will impact the quality of the crude lecithin and may not be necessary depending on the level of these non-hydratable phospholipids. The degummed oil must be dried to remove residual water before it is directed to the biodiesel production process.

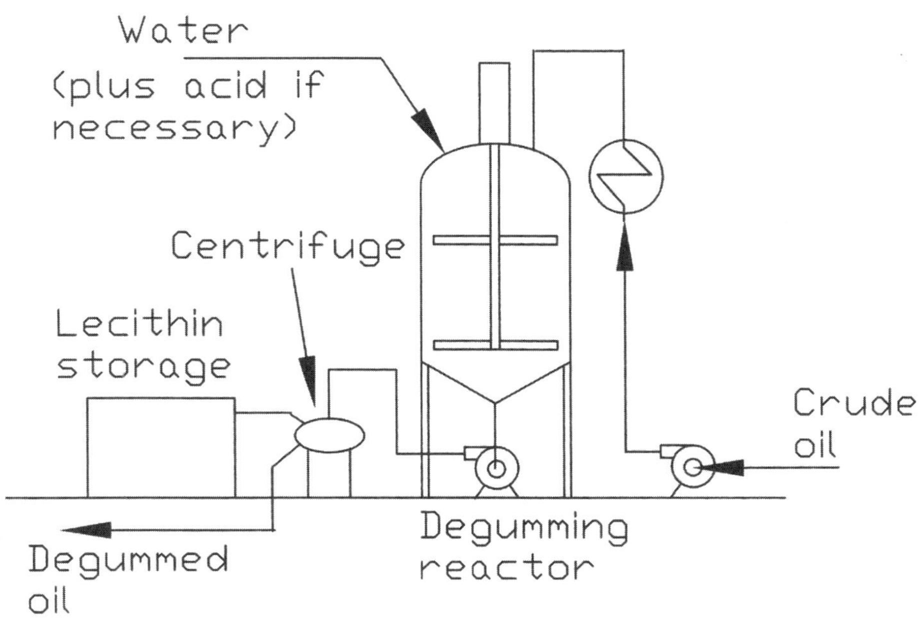

Figure 34. Degumming Crude Vegetable Oil.

Oil Drying

It is a good idea for all incoming oil to be dried. **Figure 35** shows a vacuum drying system for the oil that can be used to remove water from the degummed soybean oil or from yellow grease or purchased soybean oil. It is much better to remove water from feedstocks entering the plant than to have to remove it later or deal with the problems it causes during the process. Oil from rendering plants and crude oil from oilseed crushing typically have 0.5 percent water content. A simple vacuum flash vaporization process with the oil at 240°F and 25 to 28 inches Hg vacuum can reduce the water to less than 0.1 percent, which is a good level to send into either the esterification or transesterification processes.

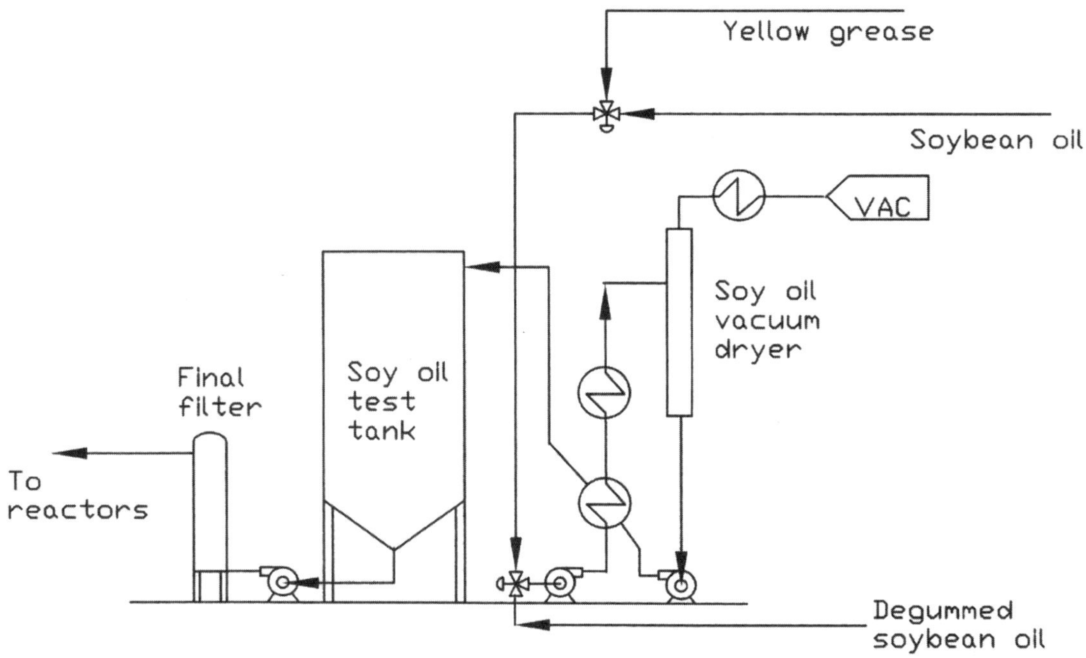

Figure 35. Oil Drying and Filtering.

Oil can often be delivered to the plant in a warm state (110-120°F). If this is available, every attempt should be made to move this oil through the drying system while it is still warm to save on process heat. Similarly, although the diagram shows a test tank, it is desirable to move the dried oil quickly to the next process so that the heat input can be minimized.

It is also recommended that the oil be filtered to make sure that meal particles do not enter the biodiesel processing equipment.

The pump that moves oil from the test tank into the rest of the system is critical because it controls the overall flow rate through the plant. It should be an adjustable metering pump so the plant flow rate can be accurately set and controlled.

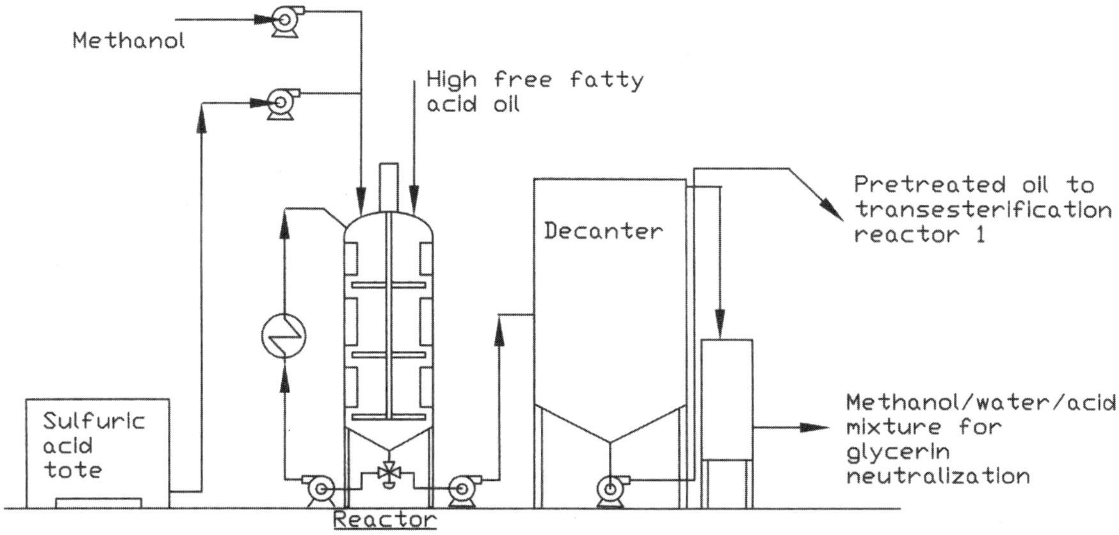

Figure 36. Esterification of Free Fatty Acids.

Esterification

When processing high free fatty acid feedstocks such as yellow grease, it is necessary to pretreat the oil with an acid catalyst to esterify the free fatty acids before putting the oil through the standard transesterification process. The reaction converts the free fatty acids to esters as shown here:

$$CH_3\text{-}(CH_2)_{15}\text{-}COOH + CH_3OH \rightarrow CH_3\text{-}(CH_2)_{15}\text{-}COOCH_3 + H_2O$$

Figure 36 shows the high free fatty acid oil being mixed with methanol and sulfuric acid in an agitated reactor. Since the reaction forms water, the mixture is pumped into a settling tank where a mixture of methanol, water and sulfuric acid rises to the top. This mixture can be removed from the top of the settling tank and used elsewhere in the plant to neutralize and acidulate the glycerin. In some processes, the amount of methanol is less and the methanol, water and sulfuric acid layer forms on the bottom of the oil phase.

The oil phase from the settling tank is now low in free fatty acids and can be pumped to the transesterification reactors. In those cases where the free fatty acids are still too high to progress to transesterification, a second stage of esterification can be used.

Note that this diagram shows the agitation coming from a motor-driven agitator and from a recirculation pump. Both or either method can be used. The external recirculation loop provides an easy way to control temperature by passing the reactants through a heat exchanger.

The esterification process is only needed when processing feedstocks with free fatty acid levels above 5-7 percent. Crude or refined vegetable oils can bypass this step and go directly to the transesterification process.

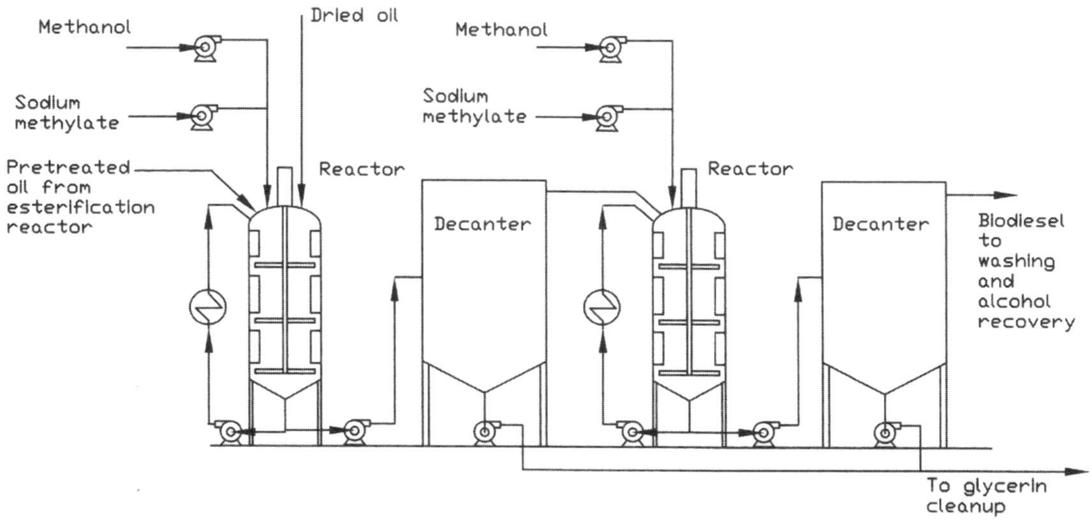

Figure 37. Transesterification Reactors.

Transesterification

The transesterification system shown in **Figure 37** uses two CSTRs. The oil is pumped continuously into the first reactor, along with 80-90% of the total methanol and catalyst. Note that the catalyst is premixed with the alcohol.

The reactants can be pumped out of the first reactor, either by overflowing or to maintain a constant level, and put into the first decanter where the glycerin is separated. The upper phase for the decanter is sent to the second reactor where the other 10-20 percent of the methanol and catalyst are added. A final decanter removes the remaining glycerin.

The incoming oil to this process should be free of water. It should have been dried and, if necessary, pretreated to reduce the level of free fatty acids.

Metering pumps are used to control the relative amounts of methanol and catalyst that enter the reactors.

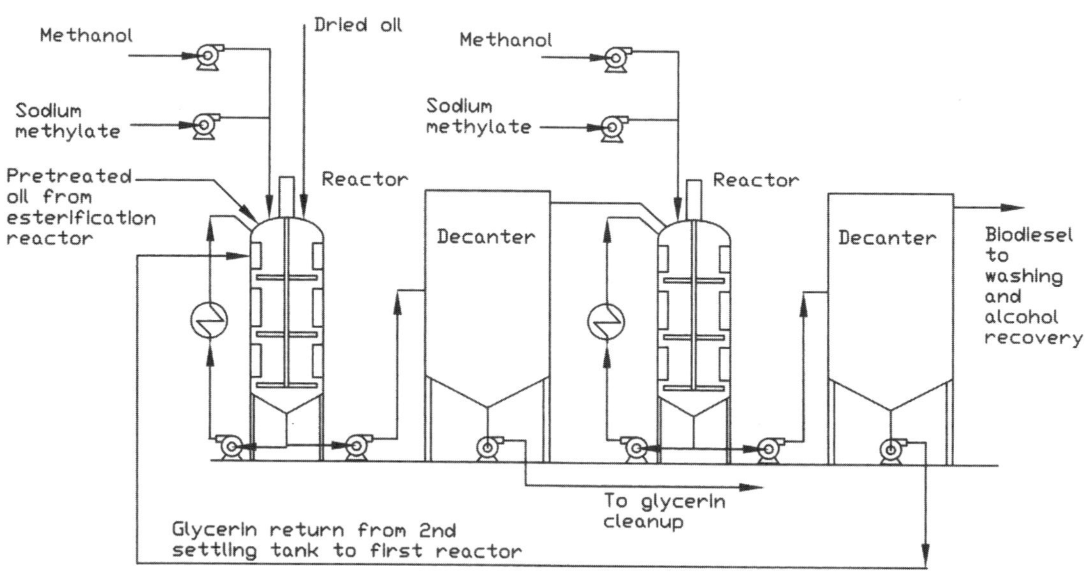

Figure 38. Transesterification with Glycerin Recycle.

Figure 38 shows a variation of the transesterification system with two CSTRs. In this case, the glycerin from the second settling tank is pumped directly back to the first reactor. This glycerin contains methanol and catalyst that can be re-used. The small amount of glycerin that is fed back will raise the glycerin level in the first reactor, which could cause some reaction inhibition due to a high level of the reaction product. This may require the system to be designed with a longer average residence time or a greater excess of methanol. However, advocates of this process contend the savings in catalyst cost and methanol separation makes it cost effective.

The glycerin coming from the second settling tank should be fairly dry. Most of the water that may have entered the reaction with the oil or alcohol or was formed during the reaction will come out with the glycerin from the first settling tank. Because of its water content, this glycerin should not be put back into the reactor, and some outlet needs to be provided for glycerin removal from the system.

Alcohol Recovery from the Biodiesel

Figure 39 shows alcohol removal from the biodiesel. Note that the alcohol recovery process is preceded by an acidulation process. This is intended to split any soap that may be present in the biodiesel into free fatty acids and salt. The free fatty acids will remain with the biodiesel, and the salts will be removed later by water washing. If the level of soap is small, this step may not be necessary; but if soap levels are high (>1000 ppm), then the soap can cause foaming in the flash tank.Removal of the methanol in the flash tank can also cause precipitation of the soap that is held in solution with the biodiesel, and the soap sludge can plug filters and screens.

This diagram shows the recovered alcohol returned directly to the process without drying. Some facilities will pass this alcohol through a distillation column before returning it to the process so water can be removed. If the water in the incoming oil and alcohol is low, then the amount of water in this recovered alcohol should be low. The small amount of water that was present should have been preferentially attracted to the glycerin emerging from the first settling tank and must be dealt with when the alcohol is removed from that stream. If the acidulation reactor is used, then the water added with the acid will be removed with the methanol and needs to be separated by a distillation column.

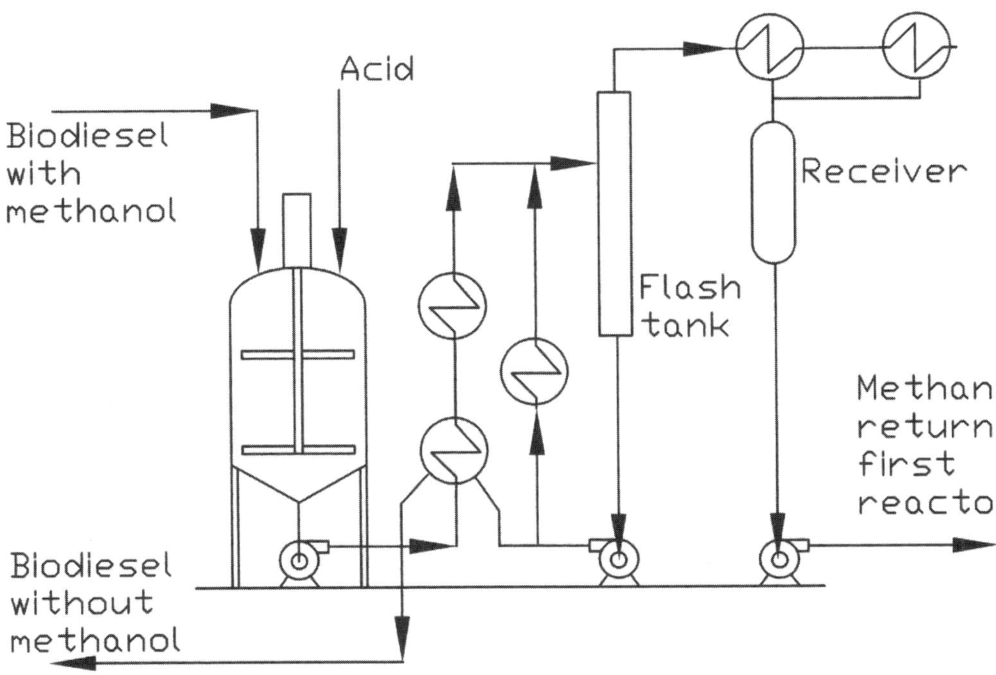

Figure 39. Biodiesel Neutralization and Alcohol Recovery.

Figure 40 shows an alternative transesterification process to the dual CSTR system described earlier. In this case, the reaction occurs in a plug flow reactor that can be visualized as a long pipe with reactants entering at one end and fully reacted product emerging from the other end. The illustration shows only a single reaction coil, but it is easy to see how this might be extended to multiple coils with glycerin removal steps between them. Static in-line mixers can be used to keep the reactants well mixed. A plug flow reactor (PFR) might also be combined with a CSTR with one device performing one stage and the other finishing the process.

An advantage of the PFR is that it is relatively easy to raise the pressure of the reaction. Pipes are low-cost pressure vessels. While the pressure does not directly influence the liquid-phase reaction, it allows operation at higher temperature which provides faster reaction time. Although the higher temperature does allow much faster reaction rates, the pipe lengths needed to get the required reaction times can be quite long, and this can offset some of the advantages of this type of system.

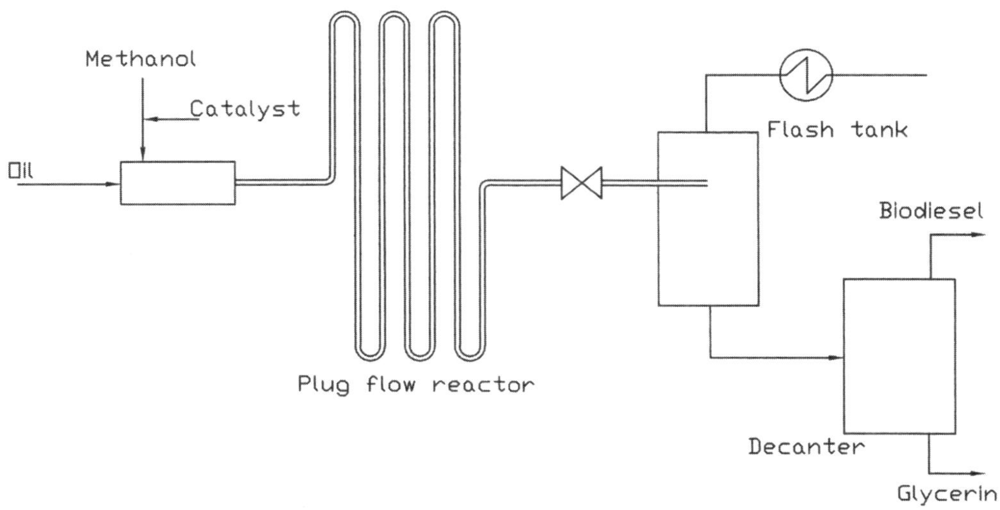

Figure 40. Plug Flow Reactor with Alcohol Recovery.

Since the temperature and pressure in the PFR are already quite high, it is possible to go directly into a flash evaporation process to remove the methanol. The mixture of biodiesel and glycerin emerging from the bottom of the flash tank will separate very quickly, because the methanol is no longer available to act as a phase stabilizer. There is the potential for a reversal of the transesterification reaction, because the methyl esters are in contact with glycerin with a catalyst present. This makes it imperative to separate the biodiesel and glycerin as quickly as possible. There can also be a problem with soap gelling if the temperature of the mixture falls too low. However, if the separated streams are neutralized while the temperature is still high, this should not be a problem.

Glycerin Acidulation and Alcohol Recovery

In **Figure 41**, the glycerin is acidulated to split the soaps and neutralize the catalyst. If available, the stream of methanol/water/acid that came out of the esterification process can be used here. After acidulation, the glycerin goes into a settling tank where the free fatty acid-rich fraction (sometimes called *acid oil* and insoluble in the glycerin) rises to the top. This light fraction can be removed and recycled back to the esterification process. The acid oil should be less than 1 percent of the biodiesel flow, unless the process has excessive soap formation. It is better to remove the methanol after acidulation, because removal of the methanol while the soap is still present creates solid glycerin.

It is common practice to mix other process streams such as the wash water into the glycerin acidulation tank. These other streams also need to be acidulated to split their soaps, and the resulting free fatty acids and salts will be added to those of the glycerin.

Figure 42 shows the alcohol recovery from the glycerin. The first step in the process is an isothermal flash vaporization process which strips the volatile compounds (methanol and water) from the glycerin. This can be accomplished by recirculating the glycerin through the flash tank and then through a heat exchanger to restore the energy loss due to vaporization of the methanol. A jacketed flash tank or bayonet heater in the lower part of the flash tank could also be used to provide this heat input. The alcohol level in the glycerin (30-50 percent) is usually much higher than for the biodiesel (3-5 percent). The glycerin may require multiple passes or internal packing (not shown) to achieve low alcohol levels.

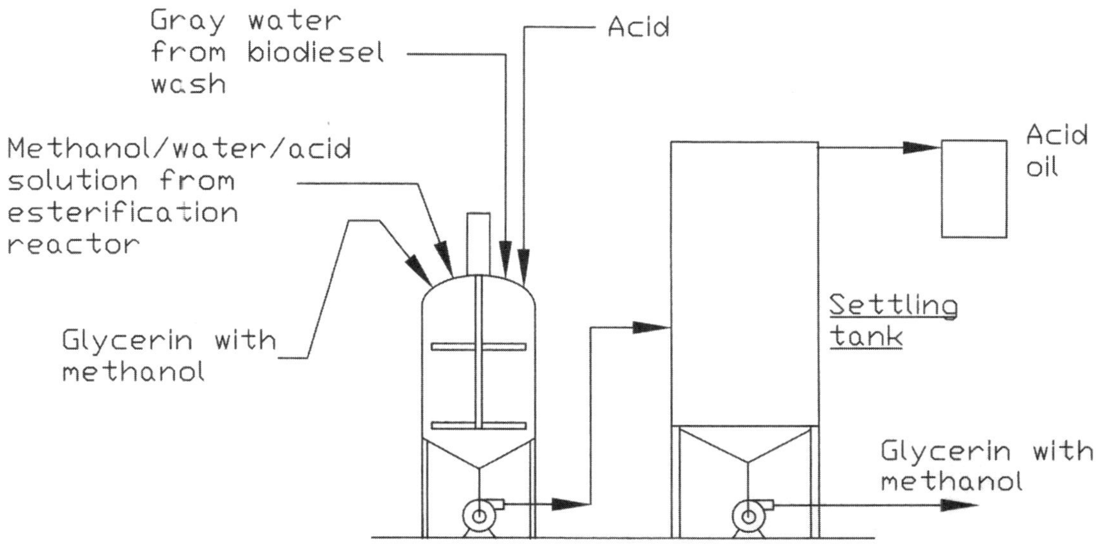

Figure 41. Glycerin Neutralization.

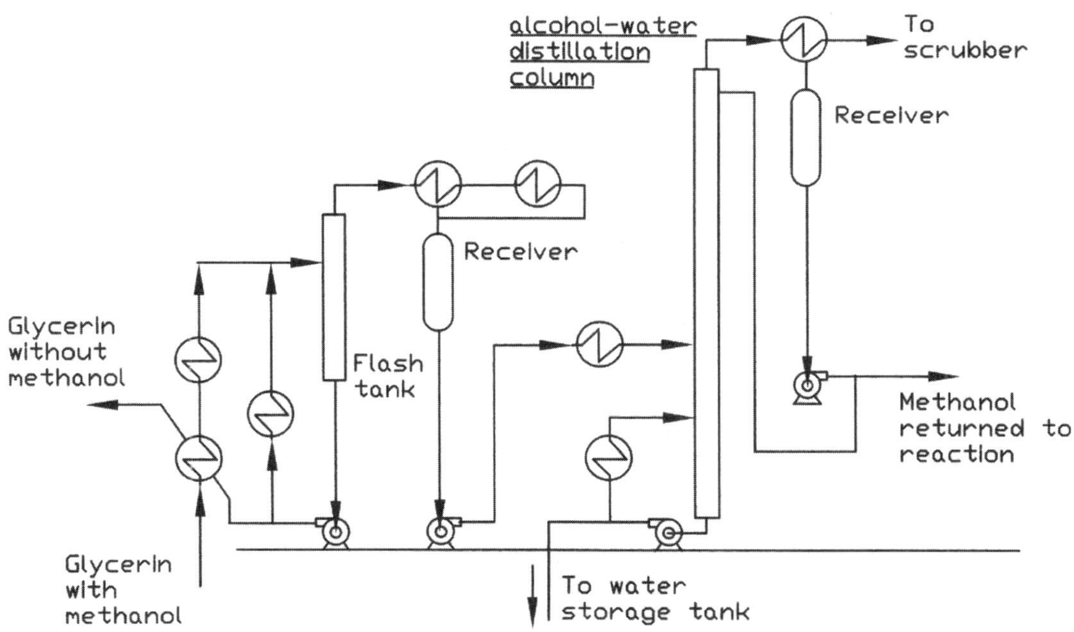

Figure 42. Alcohol Recovery from Glycerin and Alcohol Distillation.

The alcohol recovered from the glycerin will probably be higher in water content than the alcohol recovered from the biodiesel because of the attraction between water and glycerin. A distillation column is shown to separate the alcohol and water. This is one area where there will be significant differences between methanol and ethanol. A simple distillation column is not sufficient to separate ethanol and water, so molecular sieves will need to be used. The water recovered in the distillation can be reused for washing the biodiesel. Some producers have found it to be advantageous to install a grease trap in the water stream emerging from the bottom of the distillation column. This captures the small amount of biodiesel or oil that might come through with the volatiles from the flash vaporization process. This captured material can be returned to the first reactor for further processing.

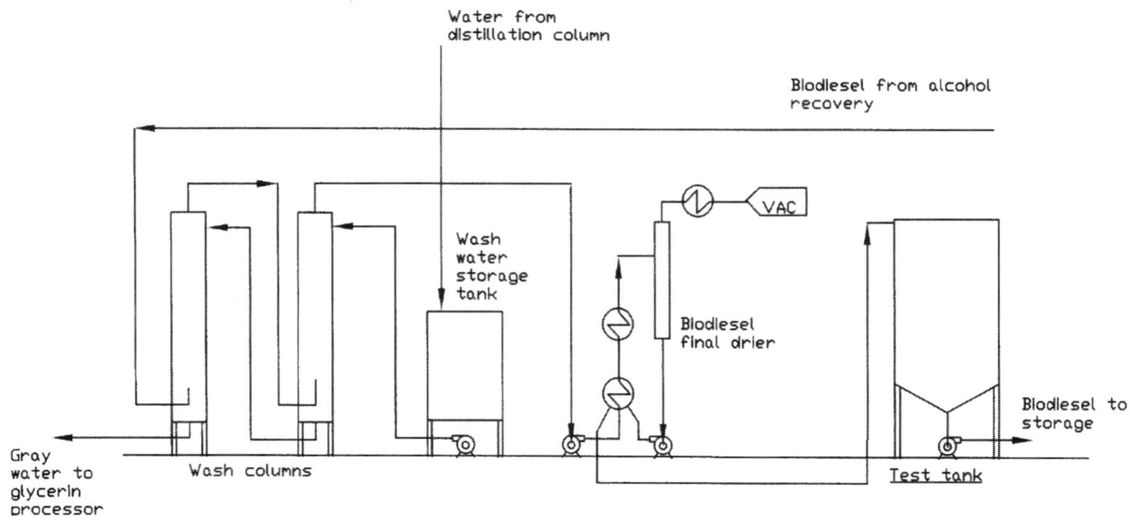

Figure 43. Biodiesel Washing and Final Drying.

Water Washing

Figure 43 shows two stages of water washing of biodiesel with counterflow wash columns. These columns might include baffles or packing to encourage intimate contact between the two fluids. Some agitation might also be used, but emulsion formation is always a concern. Addition of small amounts of hydrochloric or sulfuric acid to the wash water to neutralize residual catalyst and split soaps reduces the likelihood of emulsion formation and minimizes water consumption. The biodiesel leaving the wash columns is dried with a flash evaporator and then sent to a test tank. The fuel can be checked in the test tank before it is sent to the larger storage tank.

Washing might be combined with neutralization as shown in **Figure 44**. In this case, a dilute acid-water solution is mixed with the biodiesel to split some or all of the soaps. Then this water is separated by settling or with a centrifuge. The biodiesel is then contacted with additional clean water that is also separated. The gray water from this second wash stage might be used to dilute the acid for the second stage. The gray water from the first stage might be added to the glycerin acidulation tank so the remaining soaps can be split and the free fatty acids separated and recycled. **Figure 44** shows that the biodiesel and water are mixed in agitated tanks, but in-line static mixers could also be used because the reaction is fast and long residence times are not required.

In both of these systems, the spent wash water is directed to be mixed with the glycerin before acidulation. The acidulation will split the soaps in the wash water, so the fatty acids can be recycled and the salts will stay with the glycerin. The water is removed by the flash vaporization process along with the methanol, so they must be separated with the distillation column. When the water is recovered in this manner, it can be reused for washing, and the plant is considered to have *zero discharge*. Most current large biodiesel plants are designed to be zero discharge plants.

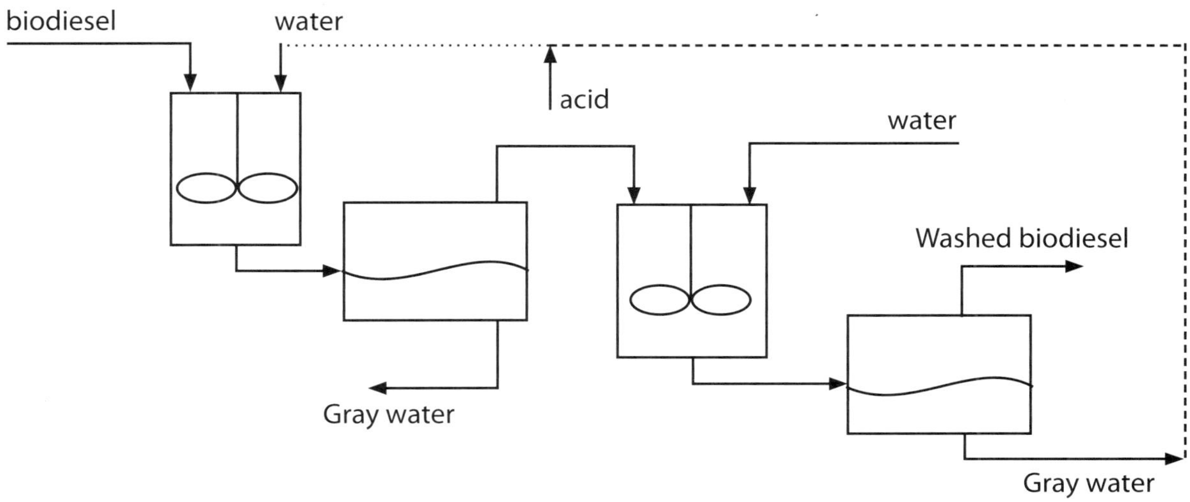

Figure 44. Combined Neutralization

Some biodiesel processors advocate washing the biodiesel before methanol removal instead of removing the methanol first as just described. The two approaches are shown schematically in **Figure 45**. In **Figure 45(a)**, the biodiesel that emerges from the transesterification process after glycerin removal may have water added to it. One variation involves the addition of acid with the water to neutralize the catalyst and split the soaps. Either hydrochloric or sulfuric might be used. Another variation is to add the water before the last stage of glycerin separation to encourage more complete glycerin removal. For all of the cases, a mixture of water and methanol is separated from the biodiesel. Generally, some methanol will remain in the biodiesel. Then the biodiesel is subjected to a flash evaporation stage that removes the remaining methanol and any water that might have carried over from washing.

In contrast, in **Figure 45(b)**, the traditional approach of removing the methanol from the biodiesel before washing is shown. In this case, the methanol that is removed should be quite dry and can probably be reused without the need for distillation. After methanol removal, the biodiesel is washed with water that may contain acid to split the soaps. Then a final drying operation is needed to remove any free or dissolved water.

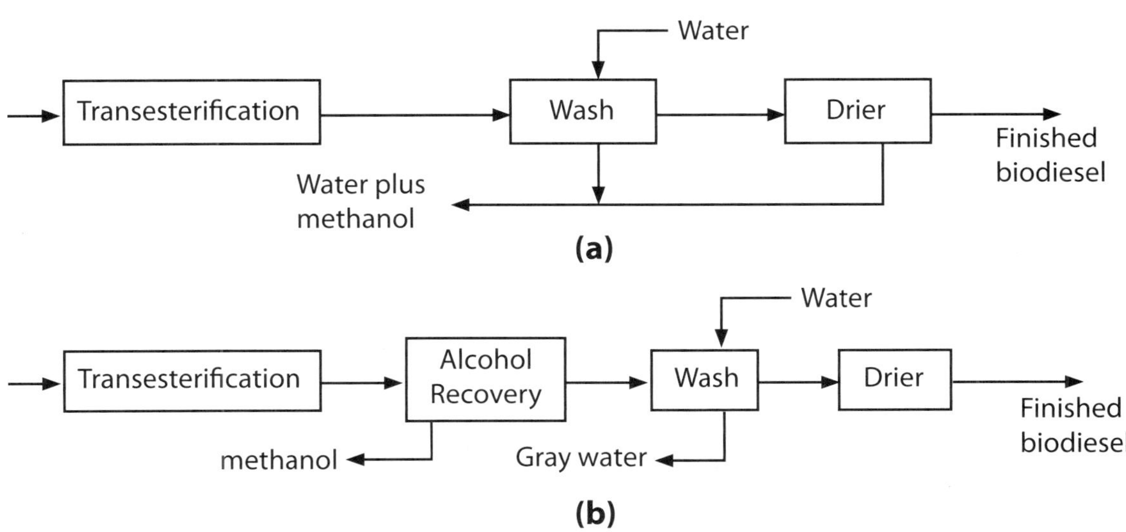

Figure 45. Variations in the Order of Biodiesel Washing and Alcohol Removal.

The first approach has the advantage of eliminating one of the flash drying operations because the residual water and methanol are removed in one process. It also eliminates a major concern that exists when methanol is removed from crude biodiesel. When biodiesel still contains soaps from the transesterification reaction, these soaps are held in solution by the presence of the methanol, which acts as a co-solvent. When the methanol is removed, the soap may precipitate out as a viscous sludge that can plug filters and pumps. When the soaps are split or washed out prior to methanol removal, these solid soap sludges do not have a chance to form.

The second approach has the disadvantage of requiring two flash operations, but the methanol and wash water are kept mostly separate. This minimizes the energy required to separate them providing a considerable energy savings if the recovered methanol does not have to go through a distillation column. However, if the soap levels in the biodiesel are too high, the soap sludges described earlier will be a problem.

One unfortunate consequence of soap splitting in the biodiesel is that the free fatty acids will stay in the biodiesel. While the levels are usually low enough that the fuel can still meet the acid value specification in ASTM D 6751, some producers prefer to keep the acid level as low as possible. If there is concern about these free fatty acids, then the wash system could be expanded into three or more stages, with the acid wash occurring in the middle. The first water wash can remove some of the soap, then the acid wash will split the rest. The final wash will remove the salts, residual acid and methanol, and free glycerin. The risk with this approach is the potential for emulsion formation in the first wash process.

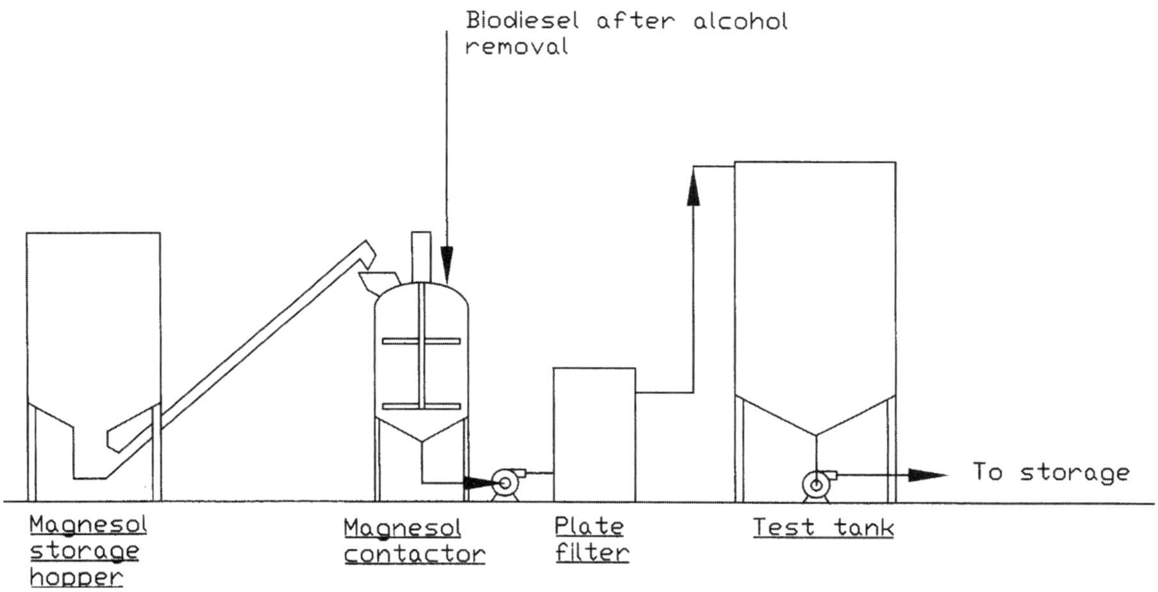

Figure 46. Use of Absorbents as an Alternative to Water Washing.

Alternatives to Water Washing

An alternative to water washing is the use of a solid absorbent to remove the soap. A commercial product called Magnesol (Dallas Group, Jeffersonville, IN) is one example. The Magnesol powder is mixed with the biodiesel after the methanol has been removed. The mixture is agitated at elevated temperature (60°C) for 20 minutes, and then the Magnesol is filtered out. Small amounts of soap can be removed with 0.5 percent Magnesol but higher amounts, as are typical with yellow grease-based biodiesel, may require 2 percent Magnesol. At a 1 percent treatment level, Magnesol (at \$0.75-\$0.82/lb) adds about \$0.05/gallon to the plant operating cost. However, this may be acceptable in areas where water is expensive. The Magnesol is non-toxic and may be disposed of by adding it to animal feed.

Removing the Magnesol from biodiesel requires a pressure plate filter. These units can be expensive, especially if they include automatic filter cleaning.

Note in **Figure 46** that a test tank is shown between the plant and the final product storage. This allows a predetermined amount of the fuel to be tested prior to mixing it with the larger amount of fuel in the main product storage tanks. If the fuel is found to be out-of-spec, it can be reprocessed without contaminating the larger tank. The test tank might hold one day's production, one shift's production or even one hour's production, depending on the frequency of quality testing.

Final Storage and Load-out

Tank trucks typically can haul 6,500-7,200 gallons of biodiesel. This can increase to 10,000 gallons if tow trailers can be used. Rail cars typically hold 21,000 or more gallons. To fill these trucks and rail cars efficiently requires pumps with flow rates of 250-300 gallons per minute. The load-out station should be equipped with a flow measurement device that can accurately measure the biodiesel dispensed. It is also highly recommended that a final filter be installed in the line to ensure the fuel is free of particulates when it leaves the plant.

A truck scale can be used to measure fuel if the buyer and seller agree on the fuel density to be used. The fuel density relationship depends on the fuel temperature. The scale can also be used to measure glycerin, soybean meal or other byproduct sales.

Samples of fuel from each load should be collected and retained for 30 days in case the buyer of the fuel makes a claim against the fuel quality. If this occurs, the fuel sample can be submitted for analysis to provide the seller with the state of the fuel when it left the plant.

SECTION THREE

Workplace Issues

CHAPTER 26

Lab Safety, MSDS and Right to Know

A biodiesel production facility always involves handling hazardous materials such as methanol and strong bases. In most cases, the facility will include a quality control laboratory that will also handle hazardous materials depending on the analytical methods practiced in the laboratory. It is important that anyone considering entering the biodiesel industry be aware of the need for laboratory safety.

This chapter outlines general principles and federal regulations concerning basic laboratory safety, Material Safety Data Sheets (MSDS) and Right to Know requirements.

Most people who work in the laboratory setting find discussions of safety tedious but definitely understand the need for them. All have heard general safety guidelines numerous times. A safe lab environment is best achieved by the application of common sense, common courtesy and keeping things as simple as possible. While some basic lab safety issues will be addressed, it is assumed that anyone working in the lab has been trained in basic safety measures.

Specific federal regulations pertaining to an employer's obligation to laboratory personnel exist under OSHA regulations. These obligations will be outlined.

Laboratories or producers of chemical hazards or wastes also have to follow Community Right-to-Know regulations with regards to public notification of possible environmental concerns.

In this chapter, we will:

1. Review basic laboratory safety practices.
2. Present a synopsis of employer obligations for OSHA compliance and reporting.
3. Describe company requirements under the Community Right-to-Know regulations.

NOTE: Due to the legal and technical nature of these topics, this presentation in no way provides the basis for performing duties as a Facility Safety Officer or Compliance Officer. Specialized training by a certified instructor is required.

OSHA Requirements

Lab safety is covered by the OSHA Occupational Exposure to Hazardous Chemicals in Laboratories Standard (Title 29, Code of Federal Regulations, Section 1910.1450). In the appendix of that section, noted as non-mandatory but widely accepted, reference is made to the National Research Council publication *Prudent Practices for Handling Hazardous Chemicals in Laboratories* concerning chemical hygiene in laboratories. The terms "chemical hygiene" and "lab safety" are essentially interchangeable.

OSHA's Hazard Communication Standard (Title 29, Code of Federal Regulations, Section 1910.1200) requires Material Safety Data Sheets (MSDS) for all chemicals used in an industrial facility. MSDS are provided by manufacturers when they ship chemicals. They contain information that allows users to be aware of the physical properties and possible hazards associated with use of the chemical.

Just as OSHA's Hazard Communication Standard requires employers to keep their employees informed, Right-to-Know requires users and producers of chemicals to keep the public informed of possible hazards. It is required under the Emergency Planning and Community Right-to-Know Act (EPCRA).

OSHA Definition of 'Hazardous Chemical'

The OSHA Occupational Exposure to Hazardous Chemicals Standard applies to all employers engaged in the laboratory use of hazardous chemicals. "Hazardous chemical" means a chemical for which there is statistically significant evidence, based on at least one study conducted in accordance with established scientific principles, that acute or chronic health effects may occur in exposed employees.

OSHA standards as covered in other sections of 29CFR1910 obviously still apply, unless they are superceded by more specific requirements of the same standard. This definition does not apply to uses of hazardous chemicals outside of the laboratory.

Specific requirements of the OSHA Lab Safety Standard include:

- Permissible Exposure Limit - PEL - For laboratory uses of OSHA-regulated substances, the employer shall assure that laboratory employee exposure to such substances does not exceed the permissible exposure limits specified.
- Employee exposure - Employers shall perform initial and periodic monitoring of PELs if there is reason to believe that exposure limits routinely exceed the action level. Employees must be notified of monitoring results.
- Respirators - Where the use of respirators is necessary to maintain exposure below Permissible Exposure Limits, the employer shall provide, at no cost to the employee, the proper respiratory equipment.
- Medical consultation - The employer shall provide all employees who work with hazardous chemicals an opportunity to receive medical attention performed by or under the direct supervision of a licensed physician. This attention shall be provided without cost to the employee, without loss of pay and at a reasonable time and place.
- Records - The employer shall retain a written record of the physician's opinion concerning the exposure.
- Information and training - The employer shall provide employees with information and training to ensure that they are apprised of the hazards of chemicals present in their work area.
- Hazard identification - Hazards shall be labeled, and the MSDS shall be made available to employees.
- Chemical hygiene plan - A chemical hygiene plan (lab safety manual) is required to be developed specifically for each lab. Among its contents should be general safety considerations, standard procedures involving the use of hazards, requirements for adequate ventilation and designation of personnel responsible for the implementation of the plan. Recommendations for preparation of a hygiene plan are given by the NRC.

National Research Council Lab Safety Recommendations

The OSHA Lab Safety Standard lists *Prudent Practices for Handling Hazardous Chemicals in Laboratories* as a non-mandatory recommended reference. It is a publication of the National Research Council. The book is cited because of its wide distribution and acceptance and because of its preparation by members of the laboratory community through the sponsorship of the National Research Council.

However, none of the recommendations given modify any requirements of the Lab Safety Standard. The publication deals with both safety and chemical hazards, while the laboratory standard is concerned primarily with chemical hazards. An outline of its contents is included in the OSHA Lab Safety Standard appendix.

General Principles

- *Exposures* - It is prudent to minimize all chemical exposures. Because few laboratory chemicals are without hazards, general precautions for handling all laboratory chemicals should be adopted, rather than specific guidelines for particular chemicals. As a rule, skin contact with chemicals should be avoided. Methanol, a primary ingredient in biodiesel, can enter the body through the skin and accumulate to toxic levels. Other compounds, such as the strong bases and acids used as catalysts, can cause burns or eye injury if contact occurs.

- *Avoid underestimation of risk* - For work with substances which present special hazards, special precautions should be taken. One should assume that any mixture will be more toxic than its most toxic component and that all substances of unknown toxicity are toxic.
- *Ventilation* - The best way to prevent exposure to airborne substances is to prevent their escape into the working atmosphere by use of hoods and other ventilation devices. Within the production facility, methanol vapors should be captured or contained and vents should be equipped with scrubbers.
- *Chemical hygiene plan* - A mandatory chemical hygiene plan (lab safety manual) designed to minimize exposures is needed. It should be a regular, continuing effort, not merely a standby or short-term activity. Its recommendations should be followed in academic teaching laboratories, as well as by full-time laboratory workers.
- *PEL and TLV* - The Permissible Exposure Limits (PEL) of OSHA and the Threshold Limit Values (TLV) of the American Conference of Governmental Industrial Hygienists should not be exceeded.

Responsibility

The Chief Executive Officer has ultimate responsibility for chemical hygiene within the institution and must, with other administrators, provide continuing support for institutional chemical hygiene. The Chemical Hygiene Officer works with administrators and other employees to develop and implement appropriate chemical hygiene policies and practices. The Laboratory Supervisor has overall responsibility for chemical hygiene in the laboratory. The laboratory worker is responsible for developing good personal chemical hygiene habits.

Laboratory Facility

The laboratory facility should be designed to function as a laboratory, with adequate ventilation, storage space, lab sinks, safety shower and eye wash in addition to appropriate fire safety apparatus. The facility should be kept clean and well maintained. Work conducted and its scale must be appropriate to the physical facilities available.

Chemical Hygiene Plan

The Chemical Hygiene Plan (Lab Safety Manual) should contain procedures covering the following topics:

- *Chemical procurement and storage* - Before a substance is received, information on proper handling, storage and disposal should be known to those who will be involved. Chemical inventories should be separated into compatible groups and appropriate storage provided.
- *Housekeeping, maintenance and inspections* - Formal housekeeping and chemical hygiene inspections should be held semi-annually; informal inspections should be continual. Safety equipment should be inspected regularly (e.g., every 3-6 months).
- *Medical program* - Regular medical surveillance should be established to the extent required by regulations. Personnel trained in first aid should be available during working hours and an emergency room with medical personnel should be nearby.
- *Protective apparel and equipment* - Protective apparel compatible with the required degree of protection for substances being handled, an easily accessible drench-type safety shower, an eyewash fountain, fire extinguisher, respiratory protection, fire alarm and telephone for emergency use should be available nearby.

The Chemical Hygiene Plan (Lab Safety Manual) should contain procedures covering the following topics:

- *Records* - Accident and medical records should be kept. Chemical Hygiene Plan records should document that the facilities and precautions are compatible with current knowledge and regulations. Inventory and usage records for high-risk substances should be kept.
- *Signs and labels* - Emergency telephone numbers, location signs for safety showers, eyewash stations and exits should be posted. Identity labels showing contents should be on all containers (including waste receptacles).
- *Spills and accidents* - A written emergency plan and spill control policy should be established and communicated to all personnel. All accidents or near accidents should be carefully analyzed.
- *Information and training program* - Assure that all individuals at risk are adequately informed about the work in the laboratory, its risks and what to do if an accident occurs.

- *Waste disposal program* - The waste disposal program should specify how waste is to be collected, segregated, stored and transported and include consideration of what materials can be incinerated.

The Chemical Hygiene Plan should also contain the obligatory list of common dos and don'ts:

- Avoid chemical contact with skin and eyes.
- No food or drink in lab.
- Wear appropriate personal protective equipment.
- Use a ventilation hood when necessary.
- Do not work alone.
- Keep the work area clean.
- Don't use damaged glassware.
- Confine long hair and loose clothing.
- Plan work well before beginning.
- Pay attention to what you and others are doing.
- Follow standard procedures carefully when using toxins or allergens.

Material Safety Data Sheets (MSDS)

- Required by OSHA Hazard Communication Standard.
- Provided by chemical supplier.
- Must be made available to employees.
- Must contain certain basic information.

Information Required in MSDS

- *Material manufacturer and identification* - Name and address of manufacturer, name of chemical, Chemical Abstract Service number and date MSDS were prepared.
- *Hazardous ingredients* - Lists the product ingredients which are hazardous.
- *Physical and chemical data* - Boiling point, freezing point, solubility, etc.
- *Fire and explosion data* - Flash point, recommended extinguishing media and special protective equipment needed.
- *Reactivity data* - Chemical compatibility information, product stability over time and light sensitivity.
- *Health hazard* - Threshold Limit Value (TLV) is the recommended maximum exposure concentration. Respiratory or contact hazard information, exposure symptoms, carcinogens, etc.
- *Precautions for safe handling and use* - Disposal information and special procedures for spills or leaks.
- *Control measures* - Personal Protective Equipment (PPE) recommended.

Emergency Planning and Community Right to Know Act (EPCRA)

The Emergency Planning and Community Right to Know Act (1986) regulates property owners and businesses that are involved with hazardous chemicals. There are five basic reporting requirements under EPCRA:

- *Emergency planning* - Required if an Extremely Hazardous Substance (EHS) becomes present at the facility in excess of a Threshold Planning Quantity (TPQ). If TPQ is exceeded, the State Emergency Response Commission (SERC) must be notified and an emergency response coordinator must be appointed at the facility.
- *Emergency release reporting* - If a Reportable Quantity (RQ), as determined by the EPA, of an EHS is released, the operator must report it to the SERC.
- *Submission of MSDS and submission of inventory* - If a Threshold Reporting Quantity (TRQ), as determined by the EPA, becomes present at the facility, the operator must give a copy of the MSDS and chemical inventory forms to the SERC. TRQ of most hazards is 10,000 lbs. TRQ of an EHS is 500 lbs.
- *Submission of toxic chemical release forms* - Operators that manufacture or process 25,000 lbs or otherwise use 10,000 lbs of a toxic chemical in a year are subject to this requirement. They must submit to the EPA an estimate of the quantity of toxic chemicals released into the environment or transported off-site.

CHAPTER 27

Workplace Safety

Safety of your employees is important to the future success of your biodiesel business. In most biodiesel plants there will be more than 10,000 pounds of a hazardous chemical on-site (usually methanol) that brings the plant under federal regulation 29 CFR 1910.119, Process Safety Management of Highly Hazardous Materials. This requires specific programs on:

- Employee participation
- Process safety information
- Process hazardous analysis
- Written operating procedures
- Documented operator training
- Contractor safety training and screening
- Having a mechanical integrity program
- Having a management of change program
- Incident investigation and emergency planning and response

Here, we outline some highlights of addressing workplace safety in your business.

Risk Communication

Risk communication involves the following seven rules:

- *Accept and involve employees* - Communicate any and all possible risks to employees and people involved.
- *Plan carefully and evaluate your efforts* - Always be careful and improve procedures to keep the workplace safe.
- *Listen to concerns* – Encourage employees to voice concerns and make comments.
- *Be honest and open* – This will encourage employees to speak up about concerns.
- *Coordinate and collaborate with other credible sources* – Talk to manufacturers or anyone else involved with the workplace or that can help you improve the workplace.
- *Meet the needs of the media* – If an unfortunate situation should occur, use the media as your resource to prevent and stop possible disasters through communication and education.
- *Speak clearly and with compassion* – When dealing with risk communication, keep in mind the emotions and sensitivity that people may have due to the nature of the topic.

(Source: Seven Cardinal Rules of Risk Communication, U.S. Environmental Protection Agency, Washington, D.C)

Influencing Factors on Risk Perception

When addressing the influencing factors on risk perception from *A Primer on Health Risk Communication Principles and Practices*, U.S. Department of Health and Human Services, the following factors come to mind:

- Risks perceived vs. comparison of risks
- Voluntary vs. imposed
- Under individual's control vs. controlled by others
- Natural vs. manmade
- Statistical vs. catastrophic
- Generated by a trusted source vs. generated from a new source
- Familiar project vs. new project

People are influenced by many different factors. These factors do not necessarily have to be statistical. Just as every individual is different, every person's perception of risk and the magnitude of risk are different. Here are some influencing factors that should be considered when addressing risk communication to effectively explain the risk to your personnel.

When looking at these last three influencing factors, keep in mind you are dealing with perception of the employees and community. If you think the risk is minimal, but your audience believes otherwise, listen to other points of view before you try and impose yours. Then discuss your differences and try to find a middle ground.

(Source: *A Primer on Health Risk Communication Principles and Practices*, U.S. Department of Health and Human Services, Public Health Service, Agency fro Toxic Substances and Disease Registry.)

Questions to Ask

This is a topic we all wrestle with when discussing safety with our employees. So, you may be asking yourself right now, "How do I communicate risks to my personnel?" Here are some suggested questions to ask according to *Communicating with the Public: 10 Questions Environmental Managers Should Ask*, from the Rutgers Center for Environmental Communications:

- Why am I communicating?
- Who am I speaking to?
- What does my audience want to know?
- What do I want to get across?
- How will this be communicated?
- How will I listen?
- How will I respond?
- Who will carry out the plans? When?
- What problems or barriers has our team planned for?
- Have we succeeded?

After you have asked yourself these questions and determined your answers, you are ready to communicate with your personnel. These questions will also help you plan your meeting and how you plan to proceed when faced with some of the audience's questions and reactions to your information.

Workers' Right to Know

Workers have a right to know what they are working with and the possible harm associated with it and that information comes from Material Safety Data Sheets. Originally it was intended for a select group of chemical manufactures, but in 1987, it was extended to all business regardless of size or classification.

Workers' Right to Know is a federal standard and cannot be overturned by state laws and local laws. This has become a more important topic in today's workforce. Workers want to be informed about the risk and safety issues that are present in their workplace. It is easier for them to accept some risk if they know about it when they start the job.

Translation to the Workplace

Workers are protected by Hazardous Communication Standard (HCS). HCS protects any worker exposed to hazardous chemicals and other hazards included in the HCS:

- *Hazards covered* - HCS is not limited to hazardous chemicals and physical hazards (flammability or explosions); HCS also includes health hazards.
- *Right to Know hazards and precautions* - Employees have the right to know of hazards they will be associated with at the workplace and what precautions to use when dealing with hazardous chemicals, physical hazards and health hazards.
- *Information shared* - Sharing information with employees will prevent injury and illnesses due to hazards.
- *Employers' responsibility* - Employers, manufacturers and importers are responsible for informing and training workers about workplace hazards.
- *Employee protection programs* - Once information is shared between employer and employees, development of employee protection programs is highly recommended.

Hazardous Communication Standard (HCS)

OSHA has provided a simple summary of the HCS in a pamphlet entitled *Chemical Hazard Communication*, OSHA Publication Number 3084. It has been helpful to employers. A copy may be obtained from your local OSHA Area Office or by contacting the OSHA Publications Office at (202) 693-1888.

It is recommended that you:

- Obtain a copy of the rule from OSHA.
- Read and understand the requirements. This will entail reading the rule and making sure you understand the requirements. You may have to call OSHA to obtain clarification of any questions you may come up with after reading the rule.
- Assign responsibility for the tasks. This will make someone responsible to see that you meet the HCS mandates. Many times this will be your safety person or manager. That person will have to conduct a hazard assessment, in which, each chemical used in the workplace will have a hazard assessment conducted on it by the employer.
- Provide an inventory of chemicals. A list of chemicals used in the workplace must be made available to employees.
- Provide a system of labeling. Employers must adopt a labeling program for each chemical used in the workplace and make sure it is used and all chemicals are labeled.
- Obtain Material Safety Data Sheets (MSDS) for each chemical. This includes your raw materials, maintenance chemical supplies, office chemicals and any other place chemicals are used or stored. You can obtain these MSDS from the supplier where you bought the chemicals. Also read the MSDS and make sure that you know what safety procedures are needed when handling these chemicals or in case of a spill or contact with an employee.
- Prepare a written program. Employers are required to provide a written program describing the HCS that must be made available to all employees. Make sure to incorporate it into your safety manual.
- Make MSDS readily available to workers. It is required that employers have a MSDS for each and every chemical used in the workplace.
- Conduct training of the workers. Employees are to be trained about MSDS and HCS. Employers are required to demonstrate that employees have been trained.
- Establish procedures to maintain the current program. These procedures should also be in the safety manual and make a person responsible for maintaining the program and all documentation.
- Establish procedures to evaluate effectiveness of the program. This will provide feedback on how effectively your program is being conducted and accepted by the workers.

HCS Hazards

These are some of the common HCS hazards found in the workplace. Evaluate your plant for each of these and look for additional ones that might be specific to your industry:

- Chemical
- Explosion and fire
- Oxygen deficiency
- Noise
- Ionizing radiation
- Safety hazards
- Electrical hazards
- Heat stress
- Cold exposure
- Biological hazards

OSHA's Purpose

This is the mission statement for OSHA: To "assure so far as possible every working man and woman in the Nation [has] safe and healthful working conditions."
(Source: OSHA website @ **www.osha.gov**.)

Lab Safety, MSDS and Right to Know

The following are the regulation authorities that dictate lab safety, MSDS and Right to Know:

- Lab Safety - OSHA regulations and NRC recommendations
- MSDS and OSHA Hazard Communication Standard
- Emergency Planning & Community Right-to-Know Act (EPCRA)

The OSHA Occupational Exposure to Hazardous Chemicals Standard covers lab safety for the Hazardous Chemicals in Laboratories Standard (Title 29, Code of Federal Regulations, Section 1910.1450). In the appendix of that section, *Prudent Practices for Handling Hazardous Chemicals in Laboratories*, by the NRC, is again recommended for reading concerning chemical hygiene in laboratories.

The terms "chemical hygiene" and "lab safety" are essentially interchangeable. The manufacturers provide MSDS when they ship chemicals that contain information that allows users to be aware of the physical properties and possible hazards associated with the use of a chemical. They are required under the OSHA Hazard Communication Standard (Title 29, Code of Federal Regulations, Section 1910.1200).

Just as the OSHA Hazardous Communication Standard requires employers to keep their employees informed, Right to Know requires users and producers of chemicals to keep the public informed of possible hazards. It is required under the Emergency Planning and Community Right to Know Act (EPCRA).

OSHA lab safety requirements and recommendations are given in the previous chapter. They apply and cover workers in lab situations unless they are superceded by more specific requirements of the standard. Please review the list of requirements of the OSHA Lab Safety Standard in the earlier chapter.

General Recommendations of the NRC also are included in the previous chapter and bear reviewing here by readers. Total safety in the work environment, including laboratories, is mandated and is also in the best interests of your company.

The Chemical Hygiene Plan (Lab Safety Manual) should contain procedures covering the following topics: chemical procurement and storage; housekeeping, maintenance and inspections; medical program; protective apparel and equipment; records; signs and labels; spills and accidents; information and training program; and waste disposal program. The previous chapter contains additional information on these, as well as some general good-practices for working with chemicals.

Other Issues

Additional issues will need to be addressed for the safety of all in your workplace. They include:

- Safety and health programs
- Personal protective equipment
- Flammable and combustible materials
- Hand and portable powered tools
- Lock-out/tag-out procedures
- Confined spaces
- Electrical
- Walking-working surfaces
- Hazard communication
- Ergonomics
- Industrial hygiene

Resources

A Primer on Health Risk Communication Principles and Practices. prepared by M. Lum, and T. Tinker, Washington, D.C.: U.S. Department of Health and Human Services, Public Health Service, Agency for Toxic Substances and Disease Registry. 1994, Adapted from Acceptable Risk.

Seven Cardinal Rules of Risk Communication, pamphlet OPA-87-020, drafted by Vincent T. Covello and Frederick H. Allen, U.S. Environmental Protection Agency, Washington, D.C. April 1988.

Communicating with the Public: 10 Questions Environmental Managers Should Ask, by Chess, Hance, and Rutgers Center for Environmental Communications.

CHAPTER 28 Basic Plant Equipment and Operation

In this chapter, we look at the principles behind the primary plant equipment typically used in a biodiesel production facility. Focus will be on reactors, pumps, centrifuges and distillation columns. Although there will be additional equipment in the plant such as settlers, storage tanks, etc., the four classes of equipment included here represent the heart of the process.

Reactors

The reactor is the only place in the process where chemical conversion occurs. Therefore, it dictates what chemical species must be handled downstream of the reactor. Reactors can be placed into two broad categories: batch reactors and continuous reactors.

In the batch reactor, the reactants are charged into the reactor at the determined amount. The reactor is then closed and taken to the desired reaction conditions (temperature, pressure and stir rates). The chemical composition within the reactor changes with time. Once the prescribed reaction time is complete, the chemical contents of the reactor are removed and sent for subsequent processing.

In contrast, continuous reactors have a steady flow of reactants into the reactor and products out of the reactor. Once a continuous flow reactor reaches steady-state operation, the product composition leaving the reactor becomes constant. Two of reactors within the continuous reactor category are continuous stirred tank reactors (CSTRs) and plug flow reactors (PFRs). For CSTRs, the reactants are fed into a well-mixed reactor. The composition of the product stream is identical to the composition within the reactor. Hold-up time in a CSTR is given by a residence time distribution. For PFRs, the reactants are fed into one side of the reactor. The chemical composition changes as the material moves in plug flow through the reactor. The residence time is defined by the length of time required for molecules to travel through the reactor.

The most important considerations within a reactor are the extent of reaction of the reactants, which is known as conversion, and the selectivity of the reaction to the desired products, which is known as the selectivity. Key reactor variables that dictate conversion and selectivity are temperature, pressure, reaction time (residence time) and degree of mixing. In general, increasing the reaction temperature increases the reaction rate and, hence, the conversion for a given reaction time. However, if more than one reaction is taking place, the selectivity to the desired products can be impacted by changing the reaction temperature. In the transesterification reaction, the selectivity of the reaction is not negatively impacted by increasing temperature. Increasing temperature in the transesterification reaction, however, does impact the operating pressure. The reaction is a liquid phase reaction, so the pressure in the reactor must be maintained at a level that keeps the methanol in the liquid phase. Therefore, as the reaction temperature is increased the pressure must also be increased. Another means to increase the conversion in a reactor is to increase the reaction time. For a given temperature, conversion will increase as the reaction time increases. Increasing the reaction time has the negative effect of decreasing the chemical throughput in a given plant or increasing the reactor size for a given chemical throughput. A final very important parameter in the reactor is the degree of mixing or mass transfer. For batch reactors and CSTRs, the degree of mixing is directly related to the amount of energy introduced through the impeller. Increasing the speed of the impeller, which increases the energy input to the reactor, increases the degree of mixing, which improves the performance of the reactor. A threshold exists where additional mixing will not provide any performance enhancement. For PFRs, the degree of mixing is dictated by the design of the reactor and/or catalyst used within the reactor, since no mixing device is present in the reactor that can be independently manipulated.

Batch reactors have several positive features including good mixing characteristics and relative ease of handling homogeneous catalysts as used in the biodiesel transesterification reaction. As mentioned earlier, the reactant concentrations change with time in the reactor as does the product concentration. Therefore, the effect of reaction inhibition by reaction products such as experienced with glycerol can be minimized. However, batch reactors are generally not used in the production of large volume chemicals since it is most efficient to operate the subsequent separation and purification steps in a continuous mode. This batch to continuous mode can be overcome by using multiple reaction vessels sequenced to give effectively continuous operation or by employing a large "surge" vessel after the reactor.

CSTRs, which obviate the primary problem with batch reactors, also provide good mixing characteristics. However, the uniform concentration present in the reactor gives a lower reaction driving force and can also maximize reaction inhibition effects. PFRs or tubular reactors are the most common reactors in the chemical industry. Despite being continuous, the reactant concentration in these reactors changes with distance down the bed to maximize the driving force for the reaction, which is similar to a batch reactor. Achieving good mixing can be more problematic with PFRs, and these type of reactors typically use solid catalysts.

Pumps

While the role of pumps in a chemical plant seems rather mundane, they play a key role in moving chemicals through the manufacturing plant. The most common type of pump in the chemical industry is a centrifugal pump. In many applications, the fluid shear introduced by a centrifugal pump is not a point of concern. However, the shear created by a centrifugal pump can create emulsion problems for the product stream from the biodiesel reactor. The amount of fluid shear imparted by the pump can be significantly reduced by using a positive displacement pump.

There are a number of different types of positive displacement pumps including gear pumps (external and internal) and lobe pumps. External gear pumps generally have two gears with an equal number of teeth located on the outside of the gears, whereas internal gear pumps have one larger gear with internal teeth and a smaller gear with external teeth. The liquid is "pumped" from the action of being carried between the gear teeth followed by displacement as the teeth mesh. In most external gear pumps, one gear drives the other. The internal gear pump has the advantage of fewer moving parts, but it cannot handle liquid streams containing solids or abrasives.

Figure 47. Lobe Pump.

Lobe pumps impart less shear to the fluid being pumped than gear pumps. The lobes in the pump are designed for no contact, so timing gears are required for each lobe. Therefore, the lobe pump has more moving parts than gear pumps. Lobe pumps are best for handling "fragile" and high viscosity fluids at low pressures. As with gear pumps, lobe pumps work by carrying the fluid between lobes followed by displacement as the lobes come together. Since the lobes are designed for no contact, fairly high amounts of slip can occur if the fluid to be pumped has low viscosity.

Centrifugal pumps convert velocity into feet of head where the feet of head is independent of the density of the liquid being pumped. However, conversion of the feet of head to output pressure requires inclusion of the fluid density. For example, 231 feet of water in a column (231 feet of head) exerts a pressure of 100 psig at the base, whereas the same 231 feet of biodiesel with specific gravity of 0.88 would exert a pressure of 88 psig. Therefore, the discharge pressure of the pump when pumping water would be higher than with biodiesel.

The primary components of a centrifugal pump are (1) a shaft, (2) a coupling attaching the shaft to a motor, (3) bearings to support the shaft, (4) a seal around the shaft to prevent leakage, (5) an impeller and (6) a volute, which converts the kinetic energy imparted by the impeller into feet of head. The role of the impeller is to increase the kinetic energy of the liquid. The liquid enters and leaves the impeller at the same pressure. As the liquid flows through the volute, the velocity is decreased with the energy transferred to pressure.

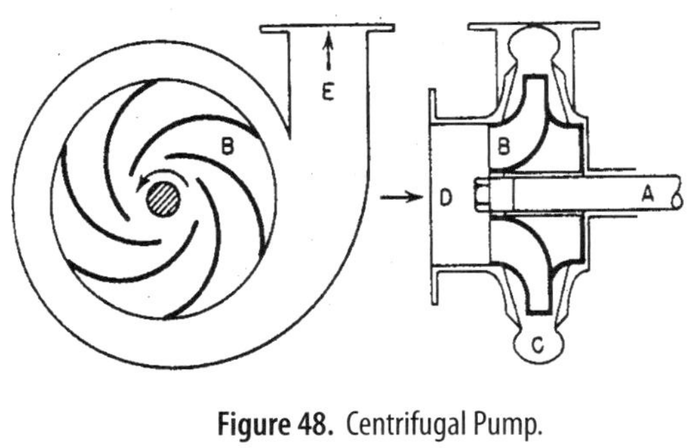

Figure 48. Centrifugal Pump.

While leaking pump seals or impeller wear can be operational issues with pumps, the most common issue with pumps is the loss of suction pressure. Suction pressure is the pressure supplied to the suction side of the pump by the liquid. During loss of suction pressure, which is known as cavitation, the pump discharge pressure and the discharge flow rates become erratic and low. Cavitation is created by vaporization of liquid at the pump suction. This situation can be created be a plug in the suction piping or by starting the pump up too rapidly. In general, positive displacement pumps are more difficult to cavitate than centrifugal pumps.

Centrifuges

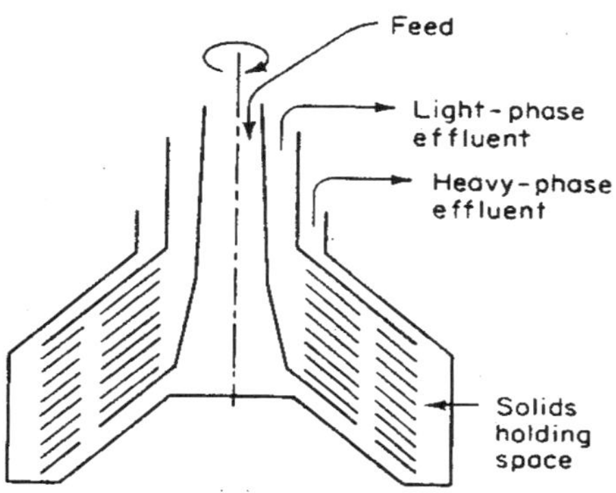

Figure 49. A Centrifuge.

Centrifuges are most typically used to separate solids and liquids, but they can also be used to separate immiscible liquids of different densities. This type of separation can also be achieved using a settling tank. While a settling tank may be cheaper, a centrifuge can be used to increase the rate of separation relative to a settling tank. In a centrifuge, the separation is accomplished by exposing the mixture to a centrifugal force. The more dense phase will be preferentially separated to the outer surface of the centrifuge. A centrifuge generally consists of (1) a bowl containing the mixture, (2) a drive shaft and drive-shaft bearings, (3) a drive mechanism and (4) a casing to segregate the separated products. Centrifuges are very amenable to continuous operation. A range of different centrifuge configurations are available with a common form being disk centrifuges.

The performance of a centrifuge depends on the specifications of the centrifuge as well as the characteristics of the mixture to be separated. While centrifuges are generally rated in terms of mixture-handling capacity, the product quality specifications are the most important consideration. The choice of appropriate centrifuge type and size are predicated on the degree of separation needed in a specific system. The viscosity of the liquid has important implications on the choice of centrifuge size since higher viscosity fluids are more difficult to handle. If a sufficient density difference exists between immiscible fluids, centrifugation is a faster means of separation than settling.

Distillation

Another means of separating chemicals in a fluid mixture is by exploiting the differences in boiling points between the chemicals. If the boiling points are sufficiently different for the chemicals to be separated, such as with water and biodiesel, an evaporator or flash vaporizer can be used for the separation. In the evaporator, the liquid is heated to a temperature in which only the more volatile chemical species will vaporize. As such, the vapor stream leaving the evaporator will be enriched in the more volatile species and the liquid stream from the evaporator will be enriched in the less volatile species. In an evaporator, the separation is accomplished by supplying heat while the mixture is held at a fixed pressure. In contrast, a flash vaporizer first heats the liquid at an elevated pressure. Then, the heated liquid is sent through a flash valve that decreases the pressure. The decrease in pressure causes the more volatile portion of the liquid mixture to vaporize.

An important separation device for miscible fluids with similar boiling points (e.g., methanol and water) is the distillation column. Separation in a distillation column is predicated on the difference in volatilities (boiling points) between chemicals in a liquid mixture. In a distillation column the concentrations of the more volatile species are enriched above the feed point and the less volatile species are enriched below the feed point. The vaporization in the column is driven by heat supplied in the reboiler, which is subsequently removed in the overhead condenser. The temperature in the distillation column is the highest at the bottom and decreases moving up the column. Distillation columns can use either trays or packing.

The degree of separation that can be achieved in a distillation column is dictated by the relative volatilities of the chemicals to be separated, the number of trays or the height of the packing and the reflux ratio. Chemicals with very different boiling points are easier to separate than those with similar boiling points. Increasing the number of trays or the height of packing can increase the amount of separation. Reflux ratio is the amount of condensed overhead vapor that is fed back into the top of the column. Increasing the reflux ratio increases the degree of separation. Increasing the number of trays or height of packing increases the column height, and increasing the reflux ratio increases the column diameter. Either of these increases in column dimensions increases the capital cost of the column.

Trayed distillation columns are the most common. A tray is where contacting occurs between the vapor stream and the liquid stream. The liquid enters the feed tray through the downcomer from the above tray, flows across the tray where is interacts with the vapor, spills over a weir and, finally, flows through the downcomer to the next tray. The vapor flows up through holes in the tray. It is important to have appropriate flow rates of the liquid and vapor to get optimal column operation. If the liquid flow rates become too high, poor contacting occurs with the vapor and extreme flooding of the column can occur. The onset of flooding in a column is easy to observe since the column's pressure drop will increase markedly when flooding begins. If the vapor flow rate is too high, liquid can be entrained and carried over to the above tray leading to poorer separation. In contrast, too low of a vapor rate can lead to weeping of the liquid through the tray, which diminishes the separation. The liquid and vapor flow rates are dictated by the operation of the reboiler and the condenser, which means they are controlled by the energy input and removal.

The materials of construction are an important consideration for the reactor and storage tanks. The demands on the materials used for the storage tanks downstream of the reactor are lower since these tanks contain nearly pH neutral chemicals. The materials of construction for the storage tanks are discussed in the Transportation and Storage chapter. In contrast, the materials of construction required in the reactor must withstand basic conditions for the transesterification reaction or acidic conditions if an esterification approach is used to convert free fatty acids. For the base-catalyzed transesterification reaction, stainless steel is the preferred material for the reactor. However, stainless steel will not be a good choice for use in the acid-catalyzed esterification reaction because stainless steel is subject to attack by acids. Under these conditions, an acid-resistant material such as Hastelloy should be used for the reaction vessel.

CHAPTER 29

Chemical Plant Controls

To produce quality products from a biodiesel production plant, it is necessary to be able to control the production process. While a tremendously important control variable is the properties of the feedstock to the process, this section will only discuss the control of chemical plants within the unit operation of the processes, such as those discussed in the previous section. Two general types of processes can be used in the production of biodiesel: batch or continuous. For batch processes, the process control variables will be temperature, pressure and levels. Temperature and pressure control are critical within the reaction and separation sections of the plants. Level control dictates charging to the reactor, operation of separation devices and product storage. For continuous processes, the process control variables include temperature, pressure, level and flow rate.

It is important to note that process variables, which are monitored and controlled, are not the same as the product specifications. Specifications are properties that are required of the final product such as those given in ASTM D 6751 for biodiesel. Control variables are those process variables that are monitored and adjusted to meet the desired specifications. As such, specifications are generally measured off-line in an analytical laboratory, while the process variables are process conditions that are measured in real time. The process variables are related to the specifications through some type of process model. The process model will typically be provided by the process licensor with subsequent refinement from plant operation experience. Some manufacturing plants will develop their own statistical or empirical models. It should be noted that the process models are typically feedstock dependent.

In discussing process variable measurement instruments, the performance of the instruments can be characterized by their accuracy and/or precision. Accuracy is the ability of an instrument to measure the correct or "true" value of a process variable. In contrast, precision is the ability of an instrument to reproduce a process variable value within a certain interval. As such, it is possible to have either "accurate and imprecise" or "inaccurate and precise" instruments. For day-to-day operation, the ability of a measurement instrument to be precise is generally more important than the need to be accurate. However, when the measurement instrument needs to be replaced, accuracy becomes an important issue.

When considering an instrumental control system for a chemical plant, there are a number of factors that should be considered: cost, precision, reliability and operator interface. It is desirable to have the lowest cost monitoring system that will allow appropriate control of the process. Precision is important to assure that there is reproducibility of measurements. Reliability insures that the system is not prone to failures. Finally, it is desirable to have a monitoring and control system that provides an easy interface with the plant operators.

The most common process variable monitored in a chemical plant is the temperature. The temperature at a specific point is generally measured using thermocouples, which are electrical devices. Thermocouples are made of two different metallic wires. These wires are connected at a junction and an electrical current is established when the junction is heated. The electrical current is proportional to the temperature at the junction. For this reason, the thermocouple must be calibrated to correlate temperature to current. Thermocouples are categorized by the metals used in their junctions, with different metal junction combinations being more appropriate for different temperature ranges. Individual thermocouples are typically not calibrated by the supplier. Instead, manufacturers will typically test a representative portion of the thermocouples they produce. Due to the manufacturing procedure, thermocouples are

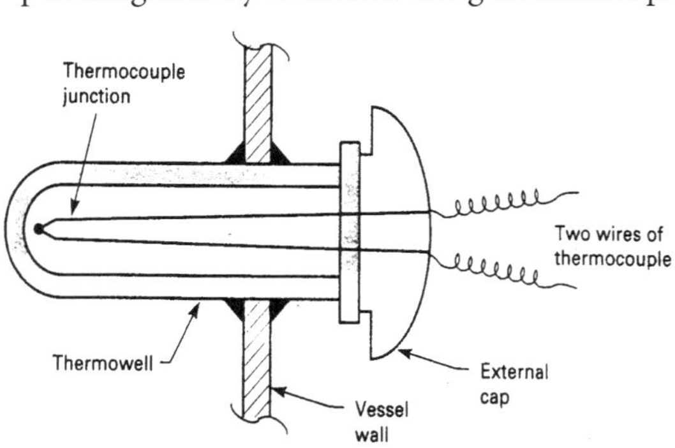

Figure 50. Thermocouple Device.

usually precise (or completely don't work) but can be inaccurate. The thermocouple is typically enclosed within a thermowell. The thermocouple must be fully inserted into the thermowell with the thermowell extended appropriately into the process to obtain a correct reading. If the temperature of a process liquid is being measured it is generally adequate for the thermowell to extend about 2-3 inches into the liquid. Due to inferior heat transfer, the thermowell should extend about 6 inches into a process vapor.

Thermocouples are typically used in processes for both monitoring and controlling the process. When a thermocouple is used as part of a process control loop, the electrical current output from the thermocouple is compared with the current expected for the set point temperature. Deviation of the actual temperature from the set point temperature will cause a change in the output to a heater or cooler that is included in the control loop.

Pressure values can be measured using several different types of devices such as a liquid column, an elastic element and electrical sensing. The most common form of a liquid column device is a manometer in which the liquid height can be observed visually and correlated to the pressure through the density of the liquid. Liquid column devices are rarely used in chemical processes. Elastic element devices measure pressure by determining deformation of an elastic material. This elastic material is usually metallic. One type of elastic element device uses bellows elements. The displacement of the bellows can be correlated to the pressure value.The most common pressure measurement instruments used in chemical processes are electrical sensing devices, which are known as strain gauges. Strain gauges are predicated on the fact that the electrical resistance of conducting solids changes with length and diameter of the solid. The dimensional changes of the solid induced by a pressure causes an increase in the electrical resistance of the conducting solid. An elastic element or electrical sensing device used to measure pressure is called a pressure transducer. The pressure tap for a transducer should be located in the vapor phase of a process. A transducer located in the liquid phase will detect a different pressure at different depths in the liquid.

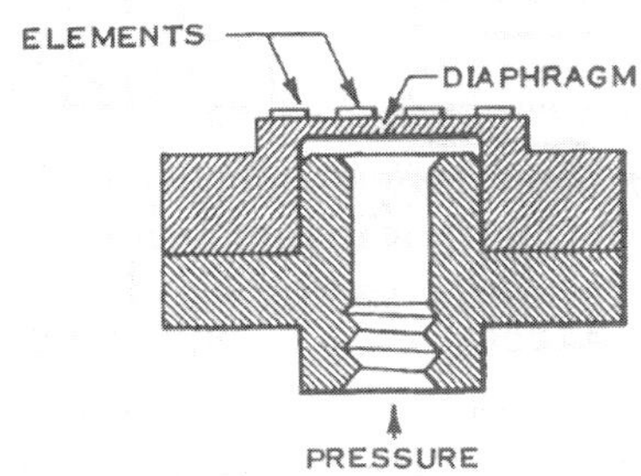

Figure 51. Pressure Transducer.

To use a pressure transducer for process control or in continuous monitoring, the transducer must produce an electrical signal. For the elastic element device, the direct pressure reading due to displacement must somehow be converted to an electrical signal. An advantage of the electrical sensing approach is that these devices directly provide an electrical signal. As with thermocouples, pressure transducers can be used for process control by comparing the measured pressure to a desired set point pressure.

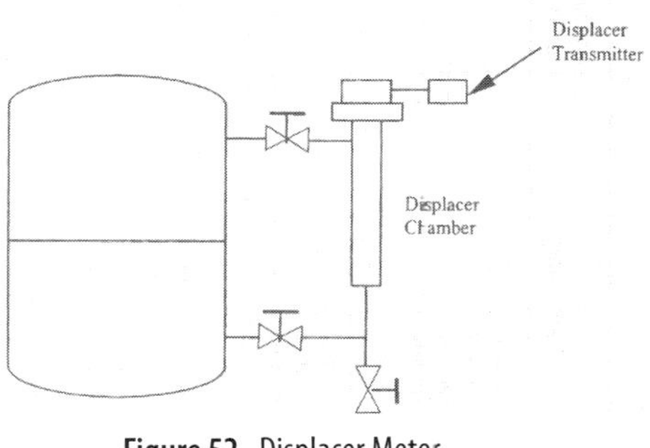

Figure 52. Displacer Meter.

Level measurement can be used to yield mass balance information across the process or loading in a batch reactor. Historically, level measurement devices have typically been displacer-based meters or differential pressure meters. The displacer-based meter is based on buoyancy. The displacer is immersed in a displacer chamber that is located as a side chamber. The displacer is restrained by an elastic element whose motion is proportional to the buoyant force, so the level can be determined by the vertical location of the displacer. Differential pressure meters, which are the most common level indicating devices, measure the difference in pressure between two pressure taps in a vessel. It is important to note that both of these level measurement indicators are liquid density dependent, so that changes in liquid densities can affect their readings. This effect can be particularly important if several liquid phases are present in the

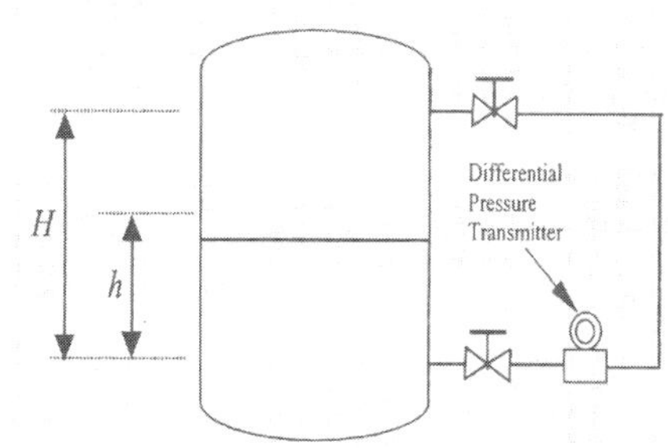

Figure 53. Differential Pressure Meter.

device. For example, the height of liquid in a gauge glass connected to a vessel is not a direct measurement of the liquid level. Recently, level measurement devices have been developed that use signal reflectance from an ultrasonic transducer or radio frequency, which measures the change in impedance between two capacitor electrodes.

In addition to density, there are a number of factors that can influence the reliability of a level measurement. Plugged taps can occur without causing apparent discrepancies in the level measurement device. Excursions to high liquid levels can impact the reliability of the level measurement as can the presence of foam in the vessel.

The measurement and control of level in a vessel generally requires two taps into the vessel. It is important that these taps are properly valved. Since the level indicator can be used to monitor the amount of material in a vessel, it can be used to track chemical inventory within the process as well as to control the charging of chemicals into a batch process. Level measurements can be converted to electrical signals that can be used for control. The most common control loop that includes level measurement is with operation of pumps.

The final process variable that will be discussed is flow rate. Flow rate measurement will generally only be needed if a continuous flow process is used. There are a number of methods for measuring flow rates, but the most commonly used approaches in chemical processes are differential pressure flowmeters and positive displacement flowmeters. Flowmeters can be used in monitoring and control. The most common control loop that includes flow rate is with operation of pumps.

Differential flowmeters measure the difference in pressure between the two sides of a restriction in a confined stream. These flowmeters are based on restrictions imposed by a venturi tube, an orifice plate or a flow nozzle with the orifice plate meter being the most common. The orifice plate hole should be appropriately sized for the flow rate range of interest. If the hole is too large the resulting pressure change and, therefore, accuracy of the flow measurement will be inaccurate. In contrast, if the hole is too small, the flow rate measurement is at the cost of too high of a pressure drop. While orifice plate meters are typically used for flow rate measurement and subsequent control, it should be noted that the meters do have some potential problems. First, this type of flowmeter is highly dependent on the density and viscosity of the liquid. Orifice plates can erode or have blockage, which will lead to inaccurate measurements. The pressure taps can become plugged leading to erroneous flow values.

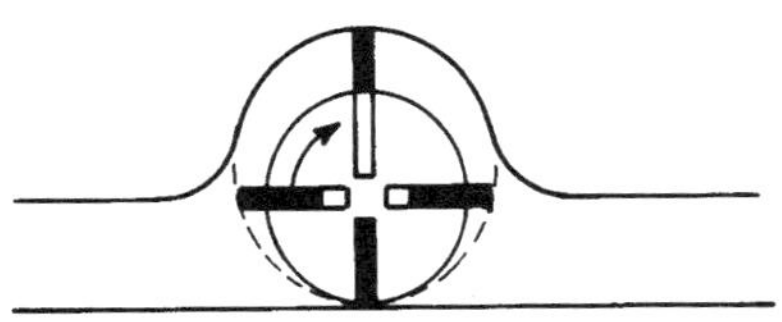

Figure 54. Rotary Flow Meter.

Positive displacement flowmeters measure flow rates using the fluid to displace a measuring device. Most common are rotary displacement flowmeters. In these meters, flow rate is measured by the amount of rotation created by flow of the fluid past a rotary device. An example of a rotary flowmeter is a vane-liquid flowmeter in which a set of vanes mounted on a rotor with opposing pairs rotated in a cylindrical chamber.

The overall monitor and control system for a process can be achieved by individual monitors and controllers located directly at the process. However, it is most common in the chemical processes to have a centralized system for monitoring and controlling. This can be readily accommodated for most processes by a PC-based system. A number of different software options are readily available for these types of systems.

CHAPTER 30

Chemical Plant Instrumentation

A biodiesel manufacturing plant is a chemical plant and as such, will need to have the capability to measure properties of the biodiesel product. Many of the properties of the biodiesel product will be amenable to analysis at the production facility in an associated wet chemistry laboratory. Most chemical manufacturing facilities include a wet chemistry laboratory on the premises. This chapter will discuss the chemical plant instrumentation used to analyze product quality in biodiesel production. The process control instrumentation used in the manufacture of biodiesel will not be discussed here.

The role of the wet chemistry laboratory in a chemical production facility is to monitor product specifications as well as to provide quality control. The equipment required in the wet chemistry laboratory to measure the biodiesel fuel properties is dictated by the specifications delineated in ASTM D6751 and the underlying ASTM standards discussed within it. While the product properties are important, there may also be a need to have capabilities to measure oil feedstock properties if the production philosophy is to accommodate variable feedstocks. For example, if the plant will be run only on soybean oil as the feedstock, the feedstock variability will not be significant. However, if yellow greases such as from animal rendering or spent cooking oil are to be used, then feedstock quality can vary significantly.

While all of the specifications in ASTM D6751 must be met, the specifications can be grouped into several categories. First are processing specifications that include free and total glycerol, kinematic viscosity and flash point. Although these specifications are influenced by the choice of feedstock, they are significantly influenced by the processing of the feedstock, i.e., how the manufacturing plant is operated.

Other quality specifications that can be influenced by processing conditions are water and sediment, copper strip corrosion and acid number. For these specifications, the performance of the manufacturing process and the properties of the feedstocks are both important.

A third specification category can be called property specifications. The specification set includes cloud point, cetane number, distillation temperature and sulfated ash. The intrinsic chemical characteristics of the feedstocks are extremely important for these specifications. For example, higher saturation in the oil feed typically leads to biodiesel fuels with higher cloud points since saturated methyl esters will "freeze" at higher temperatures than unsaturated methyl esters.

The final specification category, which is completely linked to the feedstock properties, is natural contaminants. Included in this category would be the specifications on sulfur and phosphorus.

Of the processing specifications, the most important for monitoring the manufacturing process is the free and total glycerol since this value addresses the extent of the transesterification reaction. While free and total glycerol is a very important product quality test, it is difficult to test for and will likely require the most costly equipment for the wet chemistry laboratory. Nonetheless, it is a test that needs to be available at the manufacturing plant. As discussed previously, there are several possible methods for measuring free and total glycerol. The wet chemistry method is the least costly in terms of equipment since only standard glassware is required. However, this method is both very slow and cumbersome. The gas chromatography method requires a gas chromatograph and is somewhat faster than the wet chemistry method, but in many ways is just as cumbersome. A new method is near infrared (NIR), which requires appropriate IR equipment. This method is quicker and less cumbersome than the other two methods, but the equipment must be appropriately calibrated which requires the necessary expertise.

The additional processing specifications are the kinematic viscosity and flash point. In addition to standard wet chemistry glassware and thermometers, measurement of the kinematic viscosity requires a viscometer. The flash point is measured using a specialized apparatus, so a flash point apparatus must be available to measure this property.

The other quality specification category includes water and sediment, copper strip corrosion and acid number and can reasonably be measured in a standard wet chemistry laboratory. The water and sediment measurement requires standard wet chemistry equipment and a centrifuge. This specification is a product quality specification that relates to both processing and storage conditions, so it must be measured accordingly. The copper strip corrosion can be measured using a test bomb and also requires access to a set of standards. The acid number can be measured using a potentiometric titration cell.

The property specification set, which includes cloud point, cetane number, distillation temperature and sulfated ash, requires specialized equipment that generally does not have to be included in an on-site wet chemistry laboratory. The use of an external laboratory can also be considered for the natural contaminant specifications for sulfur and phosphorus. The values of the property and natural contaminant specifications are strongly related to the characteristics of the feedstock used in the biodiesel facility. If a standard feedstock such as soybean oil is to be used, then these two specification sets can readily be handled by external testing. However, if variable feedstocks are to be used it might require some additional test facilities for some of the specification measurements to be available at the plant or to have availability to the tests at an external laboratory.

There are a number of external laboratories that will measure the necessary properties to determine whether a biodiesel sample meets specifications. While not an exhaustive list and not meant as an endorsement of any specific laboratories, three laboratories that offer the full range of diesel fuel measurements are:

Phoenix Chemical Laboratory, Inc.
3953 Shakespeare Ave.
Chicago, IL 60647-3497
(773) 772-3577

Southwest Research Institute
Petroleum Products Research Department
6220 Culebra Road
San Antonio, TX 78228-0510

Williams Laboratory Services
1090-A Sunshine Road
Kansas City, KS 66115
(913) 621-3603

These laboratories can measure all the values required in biodiesel, as specified by ASTM D6751 except for the free and total glycerol. Only Williams Laboratory Services performs this test.

The fatty acid profiles of oils and biodiesel as well as the free fatty acid content of an oil can be measured by the following laboratories:

Williams Laboratory Services
1090-A Sunshine Road
Kansas City, KS 66115
(913) 621-3603

Woodson-Tenent Laboratories
3507 Delaware Ave.
Des Moines, IA 50313
(515) 265-1461

Representative costs for the various tests are in the following table. These values represent the minimum values given as of December, 2002. The actual prices are negotiable based on the number of samples and the number of tests to be run.

ASTM No.	Test	Cost per sample
D93	Flashpoint	$25
D130	Copper corrosion	$15
D445	Kinematic viscosity	$20
D613	Cetane number	$65
D664	Acid value	$43
D874	Ash, sulfated	$83
D1160	Distillation (vacuum)	$145
D2709	Water & sediment	$19
D4751	Phosphorus	$60
D5453	Sulfur	$30
D5773	Cloud point	$35
D6584	Free & total glycerol	$89

Table 32. Representative Costs of Standard Tests.

From a quality control perspective, it is important to note that all of the specifications listed in ASTM D6751 are off-line measurements. In fact, essentially all of the specifications in ASTM D6751 are too slow for on-line measurement. Therefore, a key question becomes how is product quality monitored during production when the quality measures are all "after the fact" measurements.

The discrepancy between specifications and real time measurements to determine product quality is a common problem in the chemical/fuel industry. Typical process variables that can be measured on-line include temperature, pressure, fluid level, residence time, and flow rates. These on-line measurements are critical for tracking the progress of the manufacturing process, but they must be correlated to the final specifications, which are the ultimate determination of product quality. Process models are used to provide the correlation between on-line measurements and final product quality. Process models can exist in a number of different forms including statistical models and empirical models. Statistical models are developed by measuring product quality values relative to a range of process variable combinations. The model results from a statistical correlation between product quality values and process variables. Empirical models correlate the process variables to product quality values using chemical models.

Effective quality control monitoring combines the on-line monitoring of key process control variables and off-line monitoring of product quality. As discussed above, process models are used to track the on-line variables relative to product quality. Process models will typically be supplied by the process licensor although the models will commonly be refined by the operations at an individual plant. Since the process models will almost always be empirical, the models will likely be feedstock dependent. These models will be used as part of the process control strategy for the manufacturing process. Process control will be discussed in more detail in a later chapter.

In summary, a biodiesel manufacturing facility will need to have a wet chemistry laboratory associated with it. Testing equipment will be needed for some of the product specifications while other product property values can be measured using external laboratories. A potential set of equipment to consider for the wet chemistry laboratory includesstandard glassware, viscometer, flash point apparatus, centrifuge, copper strip corrosion test bomb, potentiometric titration cell and equipment for free and total glycerol testing (gas chromatograph or wet chemistry method).

CHAPTER 31 Biodiesel Transportation and Storage

There are several key properties that need to be considered in the transportation and storage of biodiesel. These properties include exposure temperature, oxidative stability, fuel solvency and material compatibility. Exposure temperature refers to the temperature at which the biodiesel is being stored or transported. Oxidative stability concerns the ability of the biodiesel to be stored for extended periods without degradation. Fuel solvency refers to the solvent properties of the biodiesel. The interaction of the biodiesel with materials of construction for storage tanks, seals and gaskets is the material compatibility. Each of these properties will be discussed in more detail and within the context of storage.

All types of diesel fuels can gel at low temperatures. The temperature required to induce gelling (freezing) in biodiesel is generally higher than that of conventional diesel fuel and depends on the composition of the biodiesel. In particular, saturated methyl esters will have higher freezing points than unsaturated methyl esters. The composition of the biodiesel is, of course, dictated by the oil feedstock used for its production.

As with gelling, all diesel fuels have the potential for oxidative stability issues. Oxidative stability concerns the degradation of diesel fuel due to reaction of the fuel with oxygen and catalysts. As such, the reactivity of the fuel is important. This reactivity can be related to the presence of C=C (olefinic) bonds in the fuel with increased content of the C=C bonds correlated to decreased oxidative stability of the diesel fuel. The increase in instability of a given diesel fuel molecule is generally directly proportional to the increase in the number of C=C bonds in the molecule (i.e., a molecule containing two C=C bonds has half the stability of a molecule containing one C=C bond). The oxidative stability of a diesel fuel can be estimated using the iodine number (ASTM D 1510). The longer term stability of a diesel fuel can be evaluated using an accelerated stability test (ASTM D 2274). Oxidative degradation of diesel fuel is manifest by higher acid numbers, increased viscosity and the formation of gums and sediments.

While biodiesel is only a mild solvent, it does have higher solvency properties than diesel fuel. Due to this property, residual sediments in diesel storage tanks or vehicle fuel tanks can be solvated by biodiesel. The solvent properties of biodiesel are mitigated by the use of biodiesel in blends with diesel fuel. Typically, 20 percent or less blends of biodiesel in diesel will almost completely dilute the solvency effect. However, this solvency effect should be considered if pure biodiesel is stored in a tank that was previously used for standard diesel.

The material compatibility is also an important property to consider with biodiesel. It is best to use stainless steel or aluminum equipment in the processing of biodiesel. Oxidation and sediment production in either biodiesel or diesel can be initiated by contact with brass, bronze, copper, lead, tin and zinc. Biodiesel can have compatibility issues with some polymers. Shown in **Table 33** is compatibility of biodiesel with a number of polymers relative to the polymer's compatibility with standard diesel. Biodiesel is compatible with the seals, gaskets and adhesives commonly used today. However, there may be issues with pre-1993 seals, gaskets and adhesives.

Polymer	Relative to standard diesel
Fluorosilicon	Hardness little change, swell +7%
Nitrile	Hardness −20%, swell +18%
Nylon 6/6	Little Change
Polypropylene	Hardness −10%, swell +8-15%
Polyurethane	Hardness little change, swell +6%
Polyvinyl	Much worse
Teflon	Little change
Tygon	Worse
Viton A401-C	Little change
Viton GFLT	Little change

Table 33: Biodiesel Material Compatibility.

When storing biodiesel, the temperature that is experienced by the biodiesel must be closely monitored so as to avoid the formation of crystals. These crystals are formed by the solidification (freezing) of the biodiesel. Crystal formation is undesirable since the crystals can plug fuel lines and fuel filters. To avoid the formation of crystals, biodiesel should be stored at a temperature at least 15°F higher than the pour point of the fuel. For most pure biodiesel storage temperatures of 45-50°F are generally adequate. Even in most cold climates, underground storage of pure biodiesel commonly provides the necessary storage temperature to prevent crystal formation. If pure biodiesel is stored above ground in cold climates, the storage system needs to include additional features such as insulation, agitation capability and/or heating capability. Blends of biodiesel with standard diesel can frequently be stored at lower temperatures. These blends still need to be stored at least 15°F above pour point temperature for the blend. The formation of crystals is reversible as heating of the biodiesel can be used to melt the crystals. Alternatively, the crystals can be filtered from the biodiesel.

Stability of the biodiesel is an important attribute if the biodiesel is to be stored for a prolonged period. Poor stability can lead to increasing acid numbers, increasing fuel viscosity and the formation of gums and sediments. Information about the stability of the stored biodiesel can be achieved by monitoring the acid number and viscosity. Storage stability of biodiesel has not been extensively examined relative to its composition. Therefore, the current best practice involves not storing biodiesel or biodiesel blends for more than six months. The stability of stored diesel fuels can be enhanced by using antioxidants. Examples of acceptable antioxidants include t-butyl hydroquinone (TBHQ), Tenox 21 and tocopherol. Water contamination must be minimized in the stored fuel as it can lead to biological growth in the fuel. If water contamination occurs, biological growth can be mitigated by using biocides.

When selecting tanks for the storage of biodiesel several factors should be considered. Acceptable materials of construction for biodiesel storage tanks include aluminum, steel, Teflon and fluorinated polyethylene or polypropylene. If the tanks to be used have been previously used for the storage of standard diesel fuels, the tanks should be cleaned prior to use for biodiesel storage. Tanks should be selected to minimize the possibility of water contamination.

An appropriate choice of storage tanks must also be considered for feedstock storage tanks. As with biodiesel storage tanks, the acceptable materials of construction for the feedstock storage tanks are aluminum, steel, Teflon and fluorinated polyethylene or polypropylene. Biodiesel plants based on a feedstock of fats must provide heated storage tanks to keep the feedstock liquid. For vegetable oil feedstocks, the need for heated storage tanks is dictated by the geographical location of the biodiesel plant.

Several factors should be considered when handling biodiesel. While biodiesel contains no hazardous materials, it should still be stored and used in well-ventilated areas. Storage of biodiesel should not be near heat, spark or flames sources. If stored in drums, proper drum handling procedures should be followed, which includes no drum puncturing, dragging or sliding. It is recommended that PVC-coated gloves as well as safety glasses or goggles should be used when handling biodiesel.

Typical forms of transportation vessels for biodiesel are rail cars, tank trucks and drums. The choice of vessel is dictated by the amount of biodiesel being transported. When transporting biodiesel, the transportation vessels should be clean and dry. The preferred transportation vessels would be ones dedicated to biodiesel. While dedicated biodiesel vessels may be possible with drums, it is less likely with tank trucks or rail cars. When using tank truck or rail cars, it is important to specify that the vessel should be steam cleaned and *blown dry* prior to use if the vessel is not routinely used for biodiesel transport. In addition, the vessels must conform to the same materials of construction used in storage. Seals, gaskets and adhesives present in the transfer systems should comply with the compatible materials discussed above. Due to its pour point, pure biodiesel is difficult to ship in cold weather. Methods to transport biodiesel in cold weather include hot biodiesel in tank cars that are rapidly delivered, solidified biodiesel in tank cars that are equipped with steam coils, 20 percent biodiesel blends with winterized diesel and 50 percent biodiesel blends with kerosene (diesel No. 1) which has a pour point of 0°F.

Biodiesel is commonly used in blends with standard diesel fuel. As discussed above, use of biodiesel blends can improve the thermal tolerance of the fuel by decreasing the pour point. Therefore, blending may be considered at the biodiesel manufacturing site prior to storage. Several blending techniques can be employed. Biodiesel has a higher specific gravity than diesel fuel (0.88 versus 0.85), so blending can be achieved by splashing the biodiesel on the top of diesel, this technique is known as splash blending. Alternatively, agitated blending can be used. Agitated blending uses direct mechanical agitation or blending induced by pump the biodiesel/diesel fuel mixture back and forth into a tank truck. While water contamination is undesirable due to the potential to promote biological growth in stored blends, biodiesel/diesel fuel blends will not separate upon exposure to water.

CHAPTER 32

Equipment and Process Planning as Related to Feedstocks and Processes

About half of the biodiesel plants in the United States and Europe have been designed to use clean, high-quality seed oil and methanol. The remainder use recycled restaurant oils and some animal fats. The range of acceptable free fatty acids ranges from less than 4 percent to as high as 50 percent. Most technology providers offer technologies that can handle up to 20 percent free fatty acids. Here, we will consider some of the issues involved in using a variety of feedstocks, including different oils and fats, different alcohols and different catalysts. We also will look at some process variations, particularly the issue of batch versus continuous flow.

Of the facilities that use vegetable oils, most use refined oils. Some facilities could use unrefined oils depending on the technology. Alternative feedstocks require changes to the production process. The important parameter to be considered is the free fatty acid level. Crude soybean oil will usually have a free fatty acid level less than 1.0 percent. If the oil is refined, the free fatty acid level will approach zero. However, recycled restaurant waste oils will have a free fatty acid level of 1 percent to as high as 20 percent, with the most frequent range being 2-10 percent. The low end of this range corresponds to a recently changed oil and the high end is an oil that may have been stored for a considerable time before it is processed. Rendered animal fats will have free fatty acids between 5 percent and 30 percent, mostly depending on the time of year. In the winter, when animal carcasses cool quickly with little decomposition, the free fatty acid will be low while in hot summers, the free fatty acid can be quite high. Finally, greases from restaurant grease traps and float grease from sewage or wastewater treatment plants can have free fatty acid levels ranging from 40-100 percent.

High free fatty acid feedstocks cannot be processed to biodiesel using the conventional alkaline-catalyzed transesterification. As shown in the reaction below, the alkaline catalyst, sodium hydroxide in this case, reacts with the free fatty acids to form soaps.

$$R-\overset{\overset{\Large O}{||}}{C}-OH + NaOH \longrightarrow Na^{+}\ -\ ^{-}O-\overset{\overset{\Large O}{||}}{C}-R + HOH$$

These soaps can cause two separate problems. First, the soaps act as an emulsifier to prevent the glycerol from separating from the biodiesel. Second, if the glycerol does separate, then the addition of water during the washing process can also create an emulsion that is hard to break. In either case, the plant may end up with a batch of emulsified product that will require acid treatments to break down before the batch can be reprocessed.

There are a variety of ways the high free fatty acid material can be treated. One technique that is familiar to many in the oil seed processing industry is free fatty acid removal using caustic stripping. This involves adding a sodium hydroxide-water solution of the oil and deliberately making soap. Then, the water-soap solution is separated by centrifugation. This is frequently used during soybean oil refining to remove the free fatty acids and a product called *soapstock* is produced. Soapstock is generally considered to be a low-value product, but some research has shown that it is possible to convert it to biodiesel. In any case, caustic stripping usually involves removing the free fatty acids and directing them to a separate product stream. An alternative process is to use an acid, such as sulfuric acid, as a catalyst for transesterification but the acid-catalyzed reaction of triglycerides and methanol is very slow, typically taking two to three days to reach completion. However, acids are very effective catalysts for the conversion of free fatty acids to methyl esters. This process takes only about an hour at 60°C. This gives rise to a two-step approach to processing high free fatty acid feedstocks. First, use an acid catalyst to convert the free fatty acids to esters, and then use an alkaline catalyst to convert the remaining triglycerides to methyl esters. Most plants that process high free fatty acid feeds use some variation of this concept.

When a two-step process is used, water management becomes the technological challenge. Acid esterification of fatty acids into biodiesel creates a byproduct of water in the feedstock mix that creates two problems: (1) it reduces the yield of fatty acids to biodiesel and leaves too many fatty acids behind in the feedstock and (2) the water will also reduce yields in the transesterification step. The processor needs to design the process so that water of reaction is removed during the acid esterification step and that the material that proceeds into the transesterification step has a low enough water content.

Ethanol

Ethanol is the most common alternative alcohol for producing biodiesel. Other alcohols such as isopropyl and butyl have also been used but to a much smaller extent. The University of Idaho has done most of their process development with ethanol and probably has the largest experience base with this alcohol. Ethanol has the advantage that, like soybean oil, it is renewable because it is easily made from corn. This allows the argument to be made that the biodiesel is made entirely from renewable feedstocks. Unfortunately, ethanol is more expensive than methanol and about 44 percent more ethanol is required for the reaction. However, the process yield is greater because the extra weight of ethanol is converted to biodiesel. The excess ethanol from the reaction is also hard to recover because water and ethanol form an azeotrope so it is hard to return 100 percent ethanol to the process. Additional processing with molecular sieves or cyclohexane distillation is required at higher expense. Finally, transesterification with ethanol is more prone to soap formation and thus requires tighter process controls than with methanol.

Other Catalysts

Besides sodium hydroxide, there are several different catalyst options. Sulfuric acid, mentioned above, and other strong acids can be used for transesterification but are very slow and thus not commonly used except for pretreatment of free fatty acids. The acids do have the advantage that they do not make soap with free fatty acids.

Sodium hydroxide for transesterification has the strong advantage that it is inexpensive, about \$0.40-\$0.42/lb. Most recipes for biodiesel production specify from 0.5 to 1.0 lb of sodium hydroxide to make 100 lbs of biodiesel. If free fatty acids are present, it will be necessary to add additional sodium hydroxide to compensate for the loss of catalyst to soap.

Potassium hydroxide is also commonly used in commercial biodiesel production. It is generally observed to be somewhat more effective than sodium hydroxide, it dissolves more easily in the alcohol, and it has the added advantage that when the catalyst is removed from the glycerol at the end of the process, it yields a fertilizer (potash).

Sodium hydroxide is mixed with methanol before the alcohol and catalyst is combined with the feedstock. When these two chemicals are mixed, they form a product called sodium methoxide. Water is also formed during the reaction, which tends to inhibit the reaction. Using pure sodium methoxide produced with a water-free process can provide a more effective catalyst although at a higher cost. Experience at Iowa State University shows that sodium methoxide is at least five times more active than sodium hydroxide. An additional advantage of sodium methoxide is that it can be purchased premixed with methanol as a 25 or 30 percent concentrated solution. These liquid concentrates are more convenient to dispense.

Process Variations – Batch vs. Continuous Flow

Processing biodiesel in batches tends to be favored by small plants. This approach is more flexible as it allows the process parameters to be adjusted for each batch so it is relatively easy to compensate for differences in feedstock characteristics. The equipment needed for batch processing tends to be less expensive since all of the operations can be performed at atmospheric pressure in tanks. The disadvantage of batch processing is that the physical size of the plant tends to scale directly with the capacity of the plant. To double the capacity of the plant requires tanks that are twice as large.

Continuous flow processing is favored by larger plants. It uses utilities and other resources in a continuous manner at a lower peak rate, which usually is less expensive. Continuous flow plants usually operate at high temperatures (at least 65°C) to shorten the processing time. They also tend to avoid gravity separation processes which usually involve long times and large tanks. Centrifugal separators are often used to do separations very quickly while taking very little plant space. However, the cost of the separators is very high. The equipment for continuous flow processing does not usually require a lot of space and can be scaled up to provide increased capacity without taking up a corresponding increase in space.

Both batch and continuous flow processing can provide a high quality product and successful plants have been developed using both approaches.

SECTION FOUR | Developing a Business

CHAPTER 33 Identification and Development of Markets

Marketing opportunities for biodiesel are segmented along at least five different lines. The segments are:

1. Application – light duty vehicles, heavy-duty vehicles, off-road vehicles, heating oil, back-up power generation, marine, fixed operations and mining.
2. Blend level – B2 to B5 (lubricity), B20 (EPAct and possible state mandates) and B100.
3. Environmental performance – emissions reduction, biodegradability and non-toxic.
4. Statutory niches – fleet vehicles, military, over-road vehicles, protected land areas and waterways.
5. Aesthetic – acceptable odor, renewable and biodegradable.

The factors effecting biodiesel's entry into all of these segments include the fuel price and availability, legislative mandates and incentives, environmental regulations and consumer awareness. The development of a marketing strategy for an individual producer requires an assessment of the local demands for diesel fuel, choice of what segment(s) to address and development of strategic partnerships to produce, transport and sell the product.

The instructional goals of this chapter are to:

1. Introduce the different ways to segment the markets for biodiesel fuels.
2. Review the size of the total diesel fuel market and the most important factors that impact each application segment.
3. Suggest some strategies for selecting market segments and marketing strategies appropriate to different locations and circumstances.

Examining the B100 Markets

The key characteristics that promote the B100 market include the facts that it is 1) non-toxic, 2) biodegradable, 3) renewable, 4) has substantially reduced emissions, compared to petrodiesel, 5) has a non-objectionable odor, and 6) can be used and stored in equipment designed for use with biodiesel with little modification in warmer climates. Each of these attributes can be an advantage in different applications. In addition, B100 has been recognized as an "alternative fuel," offering benefits for its use as a 20 percent blend in EPAct fleet applications.

The major application segments for B100 include national parks, marine, underground mining and off-road vehicles.

Biodiesel is being used in more than 20 national parks across the country. Each park has a slightly different focus (B100, B20, transportation, power, marine), but the sheer versatility of the fuel has been noted. As a result, state park and recreation managers are becoming more interested in biodiesel, particularly in areas frequented by the public.

Biodiesel has had success in the pleasure boat market because of its more favorable odor, both fuel and exhaust, compared with petrodiesel. This has been particularly true in the sailboat market, where fuel cost is a relatively minor consideration. The biodegradability and the demonstrated lower toxicity of biodiesel in spill events is both an aesthetic and an economic advantage in marine applications. The fuel biodegrades rapidly in the aquatic environment where fines for spills can be as high as $10,000 per day. Numerous tour boats in Florida, Hawaii and California now use biodiesel to improve the enjoyment of their clientele.

The marine market for diesel fuel is broken into segments as shown in **Table 34**. There has been a high acceptance rate for B100 in the recreational sector, particularly for sailboats. A 5 to 10 percent penetration in this area would give a 7-15 million gallon-per-year market. Public transit ferries are another potential market but will likely require environmentally based mandates. The commercial marine markets are unlikely to be receptive to B100 for cost reasons.

Application	Diesel market
Sailboats	4 million gpy
Other recreational boats	132 million gpy
Public transit ferries	22 million
Commercial marine	2400 million gpy

Table 34. Marine Market Size.

The reduction in particulate emissions and potentially lower health risks from biodiesel are important factors in penetrating the underground mining market. The dominant factor in the market, however, is the fuel cost. The engines are very large and run continuously. High efficiency particulate traps have kept petrodiesel competitive in meeting emission standards. The current consensus is that B100 is probably too costly to play a significant role in underground mining applications, but testing with B20 and other blends is on-going.

There are several off-road vehicle segments: airports, farming, logging, industrial and construction. Off-road vehicles and stationary engines are a significant diesel market totaling about 2.1 billion gallons per year. Biodiesel is finding acceptance in areas such as parks, resorts and forests because of biodegradability, lower emissions and odor properties. Renewability is an advantage, particularly in areas emphasizing eco-tourism.

The off-road market most likely to see a significant penetration of B100 is in airports in air quality non-attainment areas. A 10 percent penetration in this market represents about 1.8 million gallons of B100 annually. By contrast, a 1 percent penetration into the construction market represents a 0.5 million gallon per year market. Farming is a market segment with high interest and motivation, but not for B100. There is a consistent interest for B2 to B10 blends in the soybean growing regions of the U.S.

Examining B20 Markets

The B20 blend was originally chosen as an optimum between reductions in exhaust emissions and fuel cost. B20 provides about a 14 percent decrease in PM 10 (10 micron particulate matter) emissions, a 9 percent decrease in CO and a 7 percent decrease in hydrocarbons, compared with petrodiesel. B20 also causes about a 2 percent increase in NOx.

Biodiesel is listed as an alternative fuel under the terms of the Energy Policy Act (EPAct) of 1992, as amended by the Energy Conservation Reauthorization Act of 1998 (ECRA). Fleet users have the option to purchase and use 450 gallons of biodiesel in vehicles in excess of 8,500 GVW in lieu of purchasing an alternative fueled vehicle.

The primary application segments for B20 are:

- Qualifying fleet vehicles under EPAct.
- Non-tactical military vehicles and GSA vehicles for fuel purchased under CID A-A-59693A.
- Municipal transit fleets and school buses (although not included under EPAct).

There is extensive documentation of B20 performance in light duty and heavy duty trucks and in buses. This documentation includes engine performance and fuel economy, emissions testing and documentation of Tier 1 and Tier 2 health effects.

A key factor in favor of selection of B20 as an alternative fuel is that there is no investment required for new infrastructure to switch to biodiesel. There have been a number of successful trials of B20 by the military and municipal transit systems. The public and market awareness is increasing.

There may be an opportunity for developing a B20 automobile market, similar to that in Europe and in the voluntary E10 market in the U.S., but the number of diesel powered automobiles in the U.S. is declining. Heavy trucks that fall within the EPAct requirements are candidates for B20 use. Buses, in theory, should be a substantial market. However, in the urban bus fleet there is substantial competition with CNG powered buses, in spite of the documented lower life-cycle costs for B20 fueled buses. Also, B20 is likely too expensive for the long distance bus lines. **Table 35** shows the overall sizes for these potential B20 markets.

Application	Diesel use, GPY
Automobiles	900 million
Light trucks	1,600 million
Heavy trucks	25,900 million
Buses	1,300 million
Fleet vehicles	220 million

Table 35. Potential Market Size for B20.

Examining B2 to B5 Markets

The low blend option, in the range of B2 to B5, is essentially the use of biodiesel as a diesel fuel additive to enhance lubricity. Low-level blends will also reduce emissions although the reductions will be proportional to the blend level, so the reductions may be small. The mandated transition to low sulfur diesel fuels has opened the door to biodiesel's largest potential market, the over-the-road trucking industry. The B2 blend combines improved lubricity performance with a minimal increase in fuel price.

The low blend approach appears to offer the greatest opportunity for significant entry into the major use markets for diesel fuel. The heavy truck market is about 26 billion gallons per year. Railroad engines use 3.4 billion gallons of diesel per year and off-road vehicles account for an additional 2.1 billion gallons. Even a 10 percent penetration of biodiesel into these markets at the B2 level will require 62 million gallons per year of B100 fuel. The recent mandate for the use of B2 in all heavy trucks in Minnesota has drawn attention from a number of other states.

Summary

Biodiesel markets are segmented by application and by blend level. The driving forces behind the expansion of biodiesel markets are environmental mandates and superior performance. The cost of the fuel continues to be a negative factor opposing more rapid acceptance of biodiesel. The most likely market, however, is the low (B2 to B5) blends that carry minimal cost impact, but significant performance enhancement in a huge market.

CHAPTER 34

Business Plan Development

Many individuals considering entering the biodiesel field are looking to start new businesses. A necessary first step for all new businesses, especially if external funding is sought, is the development of a business plan. Individuals already engaged in successful businesses will be familiar with the material in this chapter. Here it lends special attention to those aspects that relate to biodiesel companies.

It is extremely important to the success of your operation that you, the owner/CEO, write the business plan. Often, the process of writing a business plan is more beneficial than the plan itself. It is also a living document that changes as the company evolves.

Components of a Business Plan

The business plan should consist of the following 11 sections, each title beginning on a new page:

1. Request Page
2. Table of Contents
3. Executive Summary
4. Business Description
5. Management
6. Market Analysis
7. Marketing Plan
8. Product or Services
9. Manufacturing Plan
10. Financial Data
11. Supporting Documents

A business plan may be a document written to persuade a lender or lenders to provide capital for your venture. As such, it is an essential management tool for your business. The plan may serve as the implementation plan for a strategic plan. The business plan outline in this handout applies to an entrepreneur or a businessperson seeking money for a new business start-up or a business expansion.

10 Key Points in Writing a Business Plan

1. Be honest. Do not be overly optimistic or try to hide limitations or weaknesses.
2. Write in easy to understand terms. Avoid jargon and terms that are unfamiliar to people outside of your industry.
3. Describe your company's image. You need to convince the reader you understand all aspects of the business.
4. Provide the reader with an understanding of your business and how you will use the loan.
5. Evaluate the company's management team. This is a major focus of the plan. Point out the strengths and weakness and how you are going to address these weaknesses.
6. Answer these three strategic planning questions: Where are we now? Where do we want to be? How do we get there?
7. Quantify your market, sales, production and cost data. Do not generalize. Be specific. Use data to help tell the story.
8. Begin each major section on a new page with the appropriate title, for example, <u>Marketing Plan</u>.
9. The actual content of the business plan will vary depending on the nature and complexity of the business, the stage of development and the type of financing needed.
10. The business plan may be used as a sales document. The content and quality of the plan should be representative of your company.

Business Request Page

When writing the Business Request Page you should include your business description with general information such as your company name, address and contact information. If you are requesting financing, include the requested dollar amount.

It is necessary to include the terms and predicted timeline required. Be specific about the purpose of the loan. Also include how money will be spent, the type of collateral, dollar amount and type of equity you are proposing. As owner, how much cash will you invest, or what type of assets will you contribute? Finally, list the contact person at the firm who is responsible for the proposed business plan.

Executive Summary

The Executive Summary should contain a brief synopsis of all business plan components. It gives the readers an overall summation of the company and highlights the main points of the plan.

Included will be a business description containing the following information:

- Name
- Starting date
- Location
- Biodiesel plant description (type of process, feedstock, glycerol processing, such as 80 percent raw, 95 percent technical grade, or 99.7 percent kosher grade)
- Significant company history
- Business goals
- Type of company (corporation, LLC, etc.)

The main points to make include the following. They will be explained in more detail later.

- Management expertise: Discuss the key persons involved in the business with summaries of their relevant expertise or past businesses.
- Market analysis and strategy: Who are the customers? Where are they located? What market niche will you serve? Who is your competition? What is the market (sales) potential? How will you sell your products?
- Targeted market and demand: Discuss the main markets you are targeting and the demand in those markets.
- Product/services: Give a description of the product or service. What differentiates your product from existing products? What features of your product will give you a competitive edge in the marketplace? What is the product's current state of development: do you need further R&D; do you have blueprints but no prototypes; is a prototype built and ready for production? What types of protection do you need: patents, trademarks or copyrights?
- Description and stage of development: What is your current stage of development? Describe each stage and projected timeframe to achieve each one.
- Manufacturing plan: What are the materials, supplies and equipment you will need to manufacture your product? How are you going to accomplish this, and what steps must you follow? How does this relate to the biodiesel process you are going to use and its related feedstock?
- Financial information: Financial information should be included if the business plan is for the purpose of borrowing money. Describe the sources of funds and a breakdown of the amount each source is supplying. Give an explanation of why the money is needed and how it will be used. Finally, show your proposed repayment plan for these funds along with a breakeven summary.

If it is a turnaround situation for the company, describe the steps you will be taking to accomplish this task. It is important that goals are set for evaluating the success in making the transition and bringing the company financials into the positive.

Business Description

The business description illustrates the current status of the business and its future direction by providing information on the demographics of the company. The business description should include basics such as name and type of business (business organization, i.e. partnership, etc). It should go on to provide details that identify your unique competitive advantage. Include the following information:

Company history: Describe the company's development and/or information about how your idea developed.

About the industry: Discuss the biodiesel industry and provide any information that may be relevant to your business.

Legal structure: Describe the organization of the business, accounting for the legal and tax ramifications.
Employment: Determine the number of employees needed and consider employees' age distribution. Briefly discuss employees' qualifications to do the work based on training, education and/or experience.

Mission statement: A business mission statement should include the company's philosophy and values on serving its customers for current and future products.

Business goals: Discuss the current status of the company and where it wants to be (company goals) in the future. State goals quantitatively and analyze your company in terms of strengths, weaknesses, opportunities and threats (SWOT).

Specific line of products/services: When describing the specific line of products/services, it is important to keep in mind who your customers are and how the products/services being offered affect their needs and wants. Channels of distribution, seasons of operation and even hours of operation will impact your customers.

Management Skills

These six skills or traits should be focused on in selecting management personnel:

1. Ability to identify and develop business strategies
2. Ability to organize and maximize efficiency
3. Able to coordinate all activities
4. Ability to understand and adjust business plan
5. Ability to delegate
6. Ability to control and supervise the business

The most critical part of a business plan is the management section. The management of a business faces many diverse circumstances and challenges on a daily basis. The management team must be able to stay on top of the changing markets and adapt to them while running the business in an efficient manner. It is important that management personnel work as a team to delegate tasks, remain in control and plan for the future of the business while meeting the demands of today.

Hence, it is the management team that provides the ability to take the business plan from paper to reality. When writing this section, be honest, but neither modest or boastful.

The first five areas to address in the management section are:

- Key personnel
- Management team
- Reporting relationships
- Directors/advisors
- Staffing plan

Key personnel: List the key management personnel and their duties and responsibilities. Resumes should be included in the supporting documents to show management has the experience and skills needed to manage the company. Emphasize past successes and the current role in the business

Management team: Explain how the diversities and similarities (in education, training, experience, etc.) among key personnel will create a management team that will lead the business to success.

Reporting relationships: Define responsibilities of the officers. Also define the reporting relationships. Organizational charts give a clear picture of the company's management structure and who is responsible for which division or tasks. Include salary structure and ownership share information.

Directors/advisors: Determine the form of the business and define who makes up the board of directors and outside advisory services.

Staffing plan: Discuss management needed in the organization, how you will fill key slots, hiring plans and the date positions will be filled. Also consider the ability to house more employees for potential growth.

The next five areas to address in the management section are:

- Business organization
- Ownership
- Management duties
- Investment
- Competitive advantage

Business organization: Define your business organization.

Ownership: State the names of stockholders and shareholders.

Management duties: Discuss the managers or private firms who will handle management duties. Include internal control systems for accounting, inventory and management information reporting systems.

Investment: Discuss the amount of money invested by the owners.

Competitive advantage: A successful business will tailor how products and services are offered to meet the demands of customers, giving the business a competitive edge by using the management team's experience and/or skills.

Market Analysis

Market analysis is needed for the products that will be produced by your company. While biodiesel is probably the main product (it is used as an example in the following discussion), your company may also be selling glycerol and other byproducts. When you consider market analysis you should ask yourself the following questions:

- Who are the biodiesel customers? Determine and understand the preliminary biodiesel market area and identify potential target markets.
- What biodiesel products will they buy? Gather information regarding the biodiesel market area and target markets.
- What quantity of product is needed at this time? Analyze information to project sales.
- What is the best way to get the product/service to the customers? Study data available from the industry. Pay particular attention to transportation. Plant location relative to the feedstock and the major markets will be an important factor in the selling price.
- What are the future prospects for the business? Set a business course.

The market section will answer questions and considerations like these by discussing a marketing mix: product/service, distribution channels, price and promotion. This section looks at the elements in the marketing mix in comparison to the industry and, more specifically, against your competition.

The market analysis will help you answer these questions by leading you to:

1. Focus on target markets.
2. Discuss the demographics of potential customers and why they want your product.
3. Look at the willingness of people to buy your product.
4. Show that your business will be able to survive and grow.

Customers

Identify customers and potential customers. Discuss demographic information about customers such as age, sex, income, type of work and where they are located. Determine customers by targeting the market you are seeking. This will help focus your sales effort.

Target Markets

According to Iowa Small Business Development Centers, a target market is the prime user of current products and/or service that are similar to what you will bring to the marketplace. In this case, the target market is the user of diesel fuel, further broken down into different categories of use.

Identify who the customers are now and potential customers, also referred to as targeted markets.

There are three components of a target market analysis:

- Use preliminary market data to identify the potential target market. Preliminary market information can come from a variety of sources; gather as much as you can.
- Identify where the target market is located using geographic boundaries and/or characteristics.
- Identify who the target market is by demographics and socioeconomic characteristics.

After determining your target markets, gather as much information about your markets and the customers as you can. Don't be afraid to invest time to track down the information. It will help you decide how you are going to capture the markets.

An important concern for the biodiesel industry is whether the primary market is individual consumers or organizations. This will be affected by the nature of any governmental subsidies and mandates that may be in place. If the market has a major consumer component, there are additional items to consider. Look for variables in your potential market. Those include age, occupation, education, family cycle (intergenerational use) and whether commercial, noncommercial or agriculture users.

Such variables may offer some insight about who makes up the target market. Family life cycle, for example: If the wife is a customer, will her husband be a customer, too? Or if one generation is a customer, will future generations (or other members) of that family be loyal customers as well?

Other variables may include, but are not limited to: modes of transportation used by the potential market, competitors and types of business customers (i.e. commercial or noncommercial).

If the market is composed mainly of organizations, the analysis should focus on the potential from the following groups:

- Federal government fleets that must meet EPAct requirements.
- States that are pushing mandates for biodiesel, school boards that are concerned about student exposure to diesel fumes as the students wait to either board buses or unload from buses.
- Cities with pollution concerns (biodiesel would have to have a proven NOx reducer additive).
- Fleet operations that want a green fuel.
- Underground mines that need to reduce particulate emissions.

Market Area

When we talk about market area analysis, we are talking about:

- Gathering information regarding the market area and target markets.
- Obtaining population figures for present and future.
- Checking the census bureau for information on age, income, sex, occupation and education.
- Identifying the transient populations (if applicable).

For population figures for the present and future you may want to check past population numbers. Look for increases and decreases to indicate booms or busts within the specific area. Use census information and websites to find this information. Also, economic development offices/agencies in your area or the area where your market is located can be great resources.

In your analysis, check for future community developments that could affect major changes in sales. Identify and understand competitors (biodiesel fuel and other alternative fuel or diesel fuel companies). Determine how much is spent on product/service (diesel fuel) in the area through city retail locations or commercial distribution terminals. Determine cost, selling price and benefits of competitive products.

This will help you identify what your customers have been buying or are going to buy, and from whom they are going to buy.

Market Size and Trends

Discuss market in current unit and dollar size. Include future industry growth and trends. Support with documentation.

In the area of market trends, relate:

- Economic trends that affect the business. Information on economic trends will reflect where the industry is now in relation to age, direction and business cycle.
- Positive trends versus negative trends.
- Information on regulatory trends.
- Trends with competitive fuels or fuel technologies.
- Feedstock trends.
- Technology trends.
- Government price supports.

Information on economic trends will reflect where the industry is now in relation to age, direction and business cycle.

Examples of trends include size, ages, areas and increases and decreases.

Explore whether customers buy larger quantities or smaller. Is the target market younger than five years ago? Is a particular age group's spending down from a year ago? Is there more or less spending in a certain locations? Are rural agencies having an increase in their contract clientele?

Positive economic trends need to be considered in relation to negative trends. Example: Store A is located in a down-town area and sells a variety of clothing to teen-agers and young adults. An economic trend that downtown spending is falling would be a negative trend for this store. However, if teen-age spending is up, this positive trend may outweigh the effects from the negative trend of a decrease in downtown spending.

Economic trend information can be found in a variety of places; you can find general trends for the country to industry specific trends. Sources include the U.S. Department of Commerce, Census Bureau Data Annual Survey and current industrial reports and trade journals.

Regulatory trends can have a profound impact on the biodiesel industry. They affect how fast the industry can grow and the climate in legislative bodies of the government toward biodiesel and renewable fuels in general.

Trends with other competitive fuels or fuel technologies will help you understand how biodiesel fits in the market-place with respect to them. These trends can give you an insight to the financial viability of biodiesel production in your area.

Feedstock trends will affect the cost structure of your business plan. Historical data, geographical influences and supply/sales trends can help identify the best feedstock to use in your specific case.

Technology trends need to be considered to determine future needs in the area of biodiesel production and use. These trends will help identify new markets as technology is developed for equipment and uses of biodiesel.

Competition

Identify competitors and where they are located. Discuss competitors' annual sales volume, market share, strengths and weaknesses. Discuss key differences of your company and product compared to the competition and your product price compared to competitors' prices.

Estimated Market Share and Sales

Estimate market share and sales by including projected unit and dollar sales and the anticipated area of sales.

You can determine projected sales by any of the following methods:

- Calculate the average number of people per business in your area.
- For more information, you may want to establish a test market.
- Survey your potential market.
- Research trade and government publications on market size and distribution.

Surveys give direct information from the potential target market, which can be helpful in your analysis. Test markets and surveys will provide information on the current market position and on how your product compares to that of the competition.

Product Distribution and Sales

Discuss how this type of product is sold in the marketplace. Discuss distribution and your plans compared to customary practices.

Analyze this information. To gain the necessary information on product distribution and sales:

- Talk with similar businesses in other (but similar) communities about the market area.
- Ask manufactures, suppliers and distributors for information about a possible market in your area of interest.

You will develop a better understanding of the demand for your product/services. This also will help you to determine the quantity of product(s) needed at this point in time.

Competitive Advantage

Discuss the competitive advantages of your company and product compared to the overall industry and competition.

Ask yourself these three questions to help you determine your competitive advantage:

- What makes your company unique?
- What is it that draws customers to you instead of the competition?
- How do you plan to keep your competitive edge so customers keep buying from your business?

These can be the hardest and most important questions to answer. They are the questions the banker or lending institution will be asking. Your findings and answers will address why you think the customers will purchase your product over the competition and ultimately lead you to establishing projected sales.

Strengths/Weaknesses/Opportunities/Threats

Analyze strengths and weaknesses of your product line and company versus competitors' product line and company.

Be honest and try to be as objective as possible when comparing your company to that of the competitor. Any weakness you point out should be addressed in your business plan. How you are going to improve them?

Marketing Plan

The Small Business Development Center (SBDC) of Iowa suggests that you ask yourself the following questions when writing your marketing plan:

- Is there a real customer need?
- Can you get a price that gives you good margins?
- Is your company a credible source for the product?
- Does the biodiesel product or service produce a clear benefit significantly better than the competition?
- Is there a cost-effective way to get the message and product to the customer?
- How and where are you going to sell your biodiesel product?

The marketing plan section of your business plan demonstrates you know how and where to sell your product. For biodiesel, you will want to cover three main areas in a marketing plan:

- Marketing goals and objectives – What you plan to do with the marketing effort within certain time frames? What are sales forecasts?
- Overall marketing strategy – What market niches will you serve? Discuss how customers will be identified and sold to; pricing strategy and policy versus competitors' policies; service and warranty policies; how you will capture customers from competitors; and credit terms. Separate information by each goal and give realistic timetables for each.
- Sales methods – Discuss advertising and promotion policy; sales force management; sales staffing (manufacturing representatives or company sales force); sales area; distribution and sales methods (factory direct, dealers or wholesalers); and how the sales methods will ensure projected sales levels.

Three additional areas to discuss in your written plan are:

- Test markets – Determine acceptance/demand for product. Offer for a limited time.
- Marketing budget – Show the budget with dollars allocated for advertising, travel, sales bonuses and commissions, promotional materials, trade shows and samples.
- Key assumptions – How will you keep up with what anticipated changes?

Provide results of completed test marketing or outline the plan to conduct test marketing. Test markets may be used to determine acceptance/demand by producing a limited amount of product and/or offering services for a limited time. This will help you determine the interest in your product and/or service and help you identify future customers. A recent example: Hershey's Chocolate decided to conduct a market analysis on a possible new product. It offered two different types of Hershey's Kisses – Dark Kisses and Creamy Kisses – for an eight-week time period across the United States. Hershey's is now using the information gathered to determine if there is a market for these new products.

Future Business Prospects

Once you address the current plans or state of your business markets, you should address the future prospects by asking: Is your target market the same tomorrow as yesterday, or even today?

Once you have stated an answer, move on to explore these options:

- Expanding the target market.
- Possible contracts.

Contracts lock in a select market for a certain time frame. Will your market renew those contracts?

Continuing to talk about the future business prospects, you can identify market share and its changes by looking at the following:

- People in the area.
- Dollars per person spent in the area.
- Number of related businesses in the area.
- Square footage of related local businesses.
- Locations of businesses.

These are things to keep in mind through changes over time. You want your business to maintain a competitive edge, so keep in mind what difference a change in customers brings and what is changing with your competitors.

Will it be helpful to include a market survey in your marketing plan? If so, what type of survey will you conduct? Briefly state the rationale for the survey, such as a secondary way to get information out about your company, determining needs in the area, price likely to be paid, etc.

Surveys can help you ascertain what is changing and what is happening with both your customers and competitors. You will need to select the type of survey to be run (telephone, mail and/or personal). Provide people you are surveying with information about your project, the reason for the survey, the amount of time the survey will take and the value of the survey to you. Be brief and specific.

Products and Services

There are six areas to address under the section of the business plan designated for products and services. The first three are:

- Description of complete product line or services offered. What is the quality of goods and services?
- Legal protection. Do you have the necessary patent, copyright, legal and technical considerations covered?
- Comparison to competitors' products or services.

Here you will provide a clear picture of your product or service. Whenever possible, use photographs, sketches, digital images or any other visual aids to present your product/service in the clearest manner. Include marketing materials used to promote the product. State the purpose of your product or service.

Show an understanding – as well as actual documentation if you have it – of the necessary patents, copyrights and/or trademarks that are in place, needed or applied for.

What are the similarities between your and your competitors' operations and products. Be clear.

The remaining three points to include in the products and services section of your plan are:

- Regulatory agency requirements. Demonstrate you know these and explain how you plan to comply.
- Competitive advantage – Determine your unique competitive advantage versus the competition. What makes your product/service stand out?
- Customer benefits – Discuss the benefits the customer will receive from your product and/or service. Also include benefits customers receive by a relationship with you. Show how your staff members' knowledge, experience or attitudes, for example, are a benefit to the customer.

Manufacturing Plan

The objective of the Manufacturing Plan section is to demonstrate that you know how to produce the product, and you can produce a quality product in quantities in a sufficient amount of time to meet customers' demand. By covering the following points you will help to convince the reader of your business development plan that you can do this.

- Facility
- Location – site selection
- Production

Facility

Include dimensions of the facility you are building along with a plant layout sketch. It may be a good idea to plan for capital improvements. Minimum economic size for a plant will depend on feedstock type and the extent of processing to be conducted on-site. The minimum economic size for a glycerol refining plant may be much larger than what is needed to process the glycerol produced by a minimum size biodiesel plant.

Location

Include the city, county, state and address where the plant will be located. Plant location is important to reduce transportation costs. If using soybean oil, for example, the plant should be located near a crushing or extraction facility. If using yellow grease, the plant location should be in an area with sufficient population to generate the feedstock for the plant.

Production

Discuss how this plan for a product will transform into a reality. Discuss your production methods and the state of art of your production process. Explain what equipment is needed. Discuss how the manufacturing operation will help make the business a success. Give your timeline to make product. Detail your production capacity versus sales needs. Include product literature on key pieces of equipment you will purchase and why you are buying that piece. Discuss how you will produce or provide the service or product. This will be a lengthy section, displaying your knowledge base. Include specific information.

Other topics to address in your manufacturing plan are:

- Staffing
- Inventory
- Quality
- Raw material
- Environmental issues

Staffing: Describe the size of workforce you will need. Discuss the skills needed in the ideal worker and possible training programs to build upon skills and talents of workers.

Inventory: Outline your inventory policy for raw materials and finished goods. Discuss how you plan to keep track of materials to avoid over-ordering or experiencing a lack of essential materials during crucial production timeframes. You need to include plans for providing the working capital to maintain the finished goods inventory.

Quality: Detail quality control, quality assurance plans and plans to implement ISO 9000 or BQ 9000.

Raw material: Explain what raw materials are needed and sources of supply. List key suppliers.

Environmental issues: Discuss environmental issues and any anticipated environmental factors. Discuss your environmental compliance program in detail.

The final four areas to address in your manufacturing plan are:

- Condition of production assets
- Manufacturing process advantages
- Government requirements
- Key assumptions

Condition of production assets: What is the condition of your production facility and equipment? Are you using state of the art equipment?

Manufacturing process advantages: State any unique competitive advantage you have in the manufacturing process arena. How are you doing it differently from your competitor, and why are they not following in your footsteps?

Government requirements: Discuss how you intend to comply with governmental agencies' requirements, such as OSHA for safety and air quality permits for the state and federal government.

Key assumptions: What are the key assumptions you used in developing your manufacturing plan?

Financial Data

Here, you want to illustrate how you can make a profit, have positive cash flow and be able to repay any debt. The following materials and points, which you will provide in your written plan, will aid you in completing this section.

- Sources and applications of funding
- Equipment list
- Pro forma balance sheet
- Breakeven analysis
- Income projections

Sources and applications of funding: Explain where the funding is coming from and what it is being used for.

Equipment list: List all equipment and, if possible, include the model and serial numbers.

Pro forma balance sheet: This is self-explanatory. Be sure to include notes and explanations.

Breakeven analysis: Use standard formulas to demonstrate and discuss at what point you will break even and what variables bring you to that point.

Income projections: These sheets should include a three-year projection with notes of explanation. The first year should be done on monthly projections, and the remaining years should be done on quarterly or annual basis.

The next areas to consider are:

- 12-month cash flow
- Cost of goods sold
- Audited financial statements
- Collateral
- Aging accounts payable and accounts receivable (for an existing business)
- Personal financial statements

12-month cash flow: Along with the spreadsheets, include notes of explanation.

Cost of goods sold: Make a detailed schedule for the cost of goods sold.

Audited financial statements: Include audited financial statements for the past three to five years.

Collateral: Make a list of collateral offered to secure a loan(s).

Aging accounts payable and accounts receivable: For existing businesses, financial data can be obtained from accounts payable and accounts receivable information already on file.

Personal financial statements: Owners with 20 percent or more ownership should keep a personal financial statement.

Personal financial statements should be done for each owner.

The following pages show examples of a Balance Sheet and an Income Projection Sheet.

Balance Sheet

CURRENT ASSETS	
Cash	$________
Accounts Receivable	$________
Inventory	$________
TOTAL CURRENT ASSETS	$________
PROPERTY & EQUIPMENT	
Land & Buildings	$________
Fixtures & Equipment	$________
Vehicles	$________
TOTAL	$________
Less Accumulated Depreciation	$________
NET PROPERTY & EQUIPMENT	$________
OTHER ASSETS	
Licenses	$________
Goodwill	$________
TOTAL ASSETS	$________
CURRENT LIABILITIES	
Notes Payable	$________
Accounts Payable	$________
Accrued Expenses	$________
Taxes Owed	$________
Current Portion, Long-term Debt	$________
TOTAL CURRENT LIABILITIES	$________
LONG-TERM DEBT	
Loans	$________
TOTAL LIABILITIES	$________
STOCKHOLDER EQUITY	
Capital Stock	$________
Paid-In Capital	$________
Retained Earnings	$________
TOTAL NET WORTH	$________
TOTAL LIABILITIES & NET WORTH	$________

Income Projections

	Qtr 1	Qtr 2	Qtr 3	Qtr 4	Yr 20xx	Yr 20xx
SALES						
Cost of Sales						
Material						
Labor						
Supplies						
Gross Profit						

Operating Expenses

Officers Salaries						
Office/Mgr's Salaries						
Payroll Expenses						
Supplies (office)						
Advertising						
Travel						
Depreciation						
Freight/Postage						
Insurance						
Utilities						
Accounting/Legal						
Real Estate & Pers. Taxes						
Telephone						
Repairs & Maintenance						
Misc. Expenses						
Auto/Truck Expenses						
Rent or Lease Expenses						
Product Testing						
Total Expenses						
Operating Profit						
Interest Expense						
Other Income						
Other Expense						
Net Profit (Before Income Taxes)						
Estimated Income Taxes						
NET PROFIT						

General Supporting Documents

Some general supporting documents to put into your plan include:

- Letters from future customers – These also can be contracts or letters of intent to do business with you.
- Plant layout – A general plant layout should be included. This will show that you have done enough engineering or development work to know that the building size for your business is adequate.
- Product brochures, promotional materials – These materials should have been developed and used to describe your product to the potential customer market. They don't have to be fancy at first, but need to contain information about your business and products.
- Resumes – Include resumes of all upper management personnel.
- Market research data – Include documented evidence of market growth, trends and statistics.
- Patent or trademark information – Include all legal documentation.
- Letters of financial commitment – Documents should show financial commitments from primary investors for initial capital or working capital, etc.

Critical Risks and Problems Supporting Documents

In the Supporting Documents section, include critical risks and problems that you think may occur. Identify and discuss major risks and problems that you think you may encounter. Feedstock price swings and pricing differential between soy oil and yellow grease are potential risk factors. Glycerol pricing and the potential for a glut in the glycerol market are also risks.

Discuss risks including, but not limited to the company, industry, personnel, market, product and financing. Indicate which assumptions or potential problems are most critical and how your plans reduce the impact of unfavorable developments. Address any weaknesses of your company that were mentioned in previous sections of your business plan.

Additional documents that might fit in these areas are:

- Price cutting by competitors (including loss of mandates, loss of or changes in subsidies or changing competitive advantage of other producers as a result of state subsidies, etc.).
- Any potentially unfavorable industry-wide trends.
- Design or manufacturing costs in excess of estimates.
- Sales projections not achieved.
- Product development schedule not met.

You also may want to include information relating to:

- Difficulties or long lead times encountered in the procurement of parts or raw materials.
- Difficulties encountered in obtaining bank credit lines because of tight money (or because product market is considered risky).
- Larger than expected innovation and development costs to stay competitive.
- Availability of trained labor.

You may encounter difficulty in obtaining an uncommon raw material or have longer delivery times if it is transported from a far distance. Describe how you will deal with this situation or a similar one and still meet customer needs.

In times of a growing economy, money can be in tight supply. This might mean higher than normal interest rates or difficulty in finding banks that are willing to work with new start-up companies. Biodiesel markets may be considered risky because they may depend on mandates or subsidies. Typically, debt financing is difficult to obtain and equity financing makes up a larger share of financial commitments.

Some processes or industries require state-of-the-art technology to stay competitive. The computer industry, for example, is one that has a high cost for research and development.

The labor market might be tight, and it will be hard to find good workers. This can affect the ability to hire the workforce that you need to efficiently work at your plant.

Resources for Information

Some examples of useful resources for additional information on writing and developing a business plan are given here. Comparable organizations should be available in most states.

- Iowa State University Extension
- Center for Industrial Research and Service (CIRAS)
- Manufacturing Extension Partnership (MEP)
- John Pappajohn Entrepreneurial Centers
- Small Business Development Centers (SBDC)
- U.S. Small Business Administration (SBA)

Additional resources that often may be accessed over the Internet include:

- State Department of Economic Development
- State Workforce Development
- Secretary of State
- State Department of Revenue and Finance
- Local agencies in your area (Economic Development, City Council, etc.)
- USDA and Value Added Agriculture Extension
- National biodiesel organizations

Sources Used in this Chapter

- Small Business Development Center
- Iowa Manufacturing Extension Partnership (IMEP)
- Center for Industrial Research and Service (CIRAS)
- Iowa State University Extension
- National Renewable Energy Laboratories
- *How to Write a Business Plan*, by Verl Anders, CIRAS

CHAPTER 35

Plant and Operational Costs

In starting a business and writing a business plan, one always has to understand the cost of the project, the cost of operating the plant and how you will be able to make a profit. It can be helpful to have "ballpark" or "rule of thumb" numbers to use for the initial stages of project development as long as you know they are rough estimates. They can help you understand the scope of the project and start to put the task of financing and fundraising in perspective.

The biggest problem with estimates is that they are just that – estimates – and not firm numbers. Many assumptions go into producing those numbers. In many cases, they are average numbers with a conservative slant. Once they guide you to refining your business plan and plant size, you should seek out an engineering company to develop accurate numbers that you can then use for obtaining financing.

The engineering company can customize the cost of building a biodiesel plant for your project, taking into account your particular situation and resources. It can evaluate any savings from existing buildings and infrastructure that your selected site can provide. By designing your plant specifically for your location and size, the firm can utilize any synergies that may exist between the technology and available resources.

Figure 55 is a graph of the estimated costs of a biodiesel plant installed for various production capacities. This graph was composed with information gathered from various plant installations and calculations. The budgetary installed costs include pretreatment of feedstock where the free fatty acid (FFA) is above 2.0 percent. The modular costs are for a prefab plant built as component modular with modules assembled at the plant site. The cost per gallon for the smaller modular plants is almost equal to the cost per gallon of the larger plants, therefore minimizing or eliminating the economies of scale based on equipment. The cost per gallon for a 30 mm gpy 10 percent FFA plant is around \$0.73 per gallon based on installed equipment, compared to that of a 5 mm gpy modular 10 percent FFA plant at \$0.87 and a 15 mm gpy modular 10 percent FFA plant at \$0.74. This shows that the costs per gallon of the larger plants are not as different as those of the smaller plants when based on equipment cost.

The actual cost of the major pieces of equipment or what is referred to as "basic equipment" is estimated at 20-25 percent of the installed cost. If you refer to the cost of all equipment, which includes valves, fitting, piping, instruments, gages, utility generation equipment and other such items, then the cost is 33-40 percent of the installed cost.

One way to reduce the cost of equipment is to purchase it on the used equipment market, if you can find the pieces of equipment needed in the condition, material of construction and capacity that you need. This becomes more difficult in designs for larger biodiesel plants. If you purchase used equipment, you should add extra money in the maintenance budget to cover repairs of the used equipment.

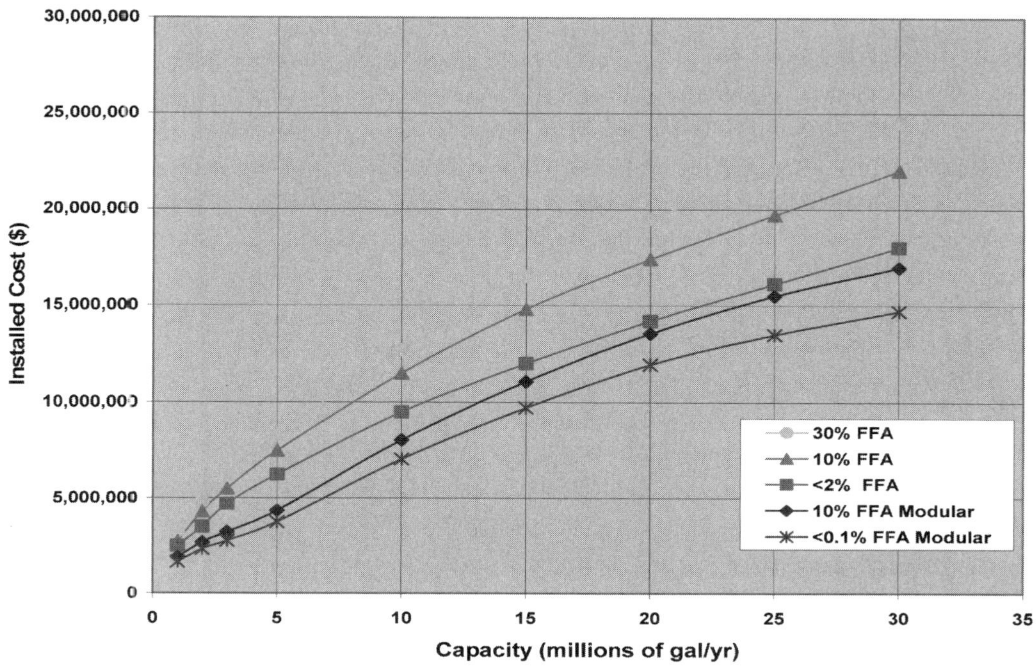

Figure 55. Biodiesel Plant Budget Installed Cost.

After obtaining the equipment list from the engineering company, you can multiply the equipment number by an installation factor of one and one-half to two times the cost of the equipment to cover the installation of those pieces of equipment. It is best to use a total list of the equipment needed for a biodiesel plant as opposed to just one of the basic pieces of equipment. If the cost of all the equipment is around \$5 million, then the installation of that equipment would run an additional \$7.5-\$10 million for a total cost of equipment and installation of around \$12.5-\$15 million. This total figure also will include all buildings, scales and other infrastructure.

The cost can vary if you contract out some of the construction and/or take on the construction management position internally. The plant site also has an effect on the cost, based on necessary earth moving, required regulations and existing infrastructure.

One of the biggest mistakes that groups can make is to look at the equipment or installed cost and then decide to increase the capacity of the plant to achieve perceived economies of scale. They might look at the installed cost and determine that for less than double the original cost of the plant, they can double the original planned size of the biodiesel plant. For example, a 5 million gallon per year site erected biodiesel plant is estimated at \$7 million. A 10 million gallon per year plant is estimated at \$9 million. It might seem to make sense to build the larger plant and take advantage of economies of scale.

The key questions you have to ask are, "Can we sell those extra 5 million gallons per year of product?" or "Can we obtain a continuous supply of feedstock with low enough transportation cost to keep the plant running 24 hours a day, seven days a week?" Once you build a plant over 5 million gallons, it will usually be a continuous plant, although we are seeing continuous plants as small as 3 mm gpy. If a plant is built to run continuously, profits will be lost every time you shut down. Also, continuous plants are designed to run at capacity, and to run it at much lower rates can cause problems and a loss in efficiency. Therefore, the money you might have saved by using the economies of scale and building a much larger plant can be lost if the markets cannot support the product being produced or the feedstock can not be obtained consistently and at a low transportation cost. For example, running a 30 mm gpy plant at less than capacity can be more expensive to operate than a smaller plant running at capacity, and therefore, produce biodiesel at a cost above that of a smaller plant and erase any economies of scale.

A third question to ask before being lured into building a larger plant is, "Can we raise the additional capital needed to build the larger plant?" Often, the last million dollars needed to complete a project are the most difficult dollars to find. In an industry like biodiesel that is just starting to grow and is not well established, the risk perceived by investors and banks is higher than with a well-established industry. Just as it was hard to raise money for the ethanol plant projects a few years ago, the biodiesel industry is still new enough that building a plant to produce more capacity than you have known markets for is considered very high risk, and it becomes difficult to obtain financing.

The best size plant to build should be dictated by the current and proposed market size, finances and a host of other factors. It should not be based on the tempting thought of gaining additional capacity with a proportionally lower incremental cost of equipment. With the emergence of modular plants, it is possible to build a smaller plant and as the market grows, add an additional modular to increase the plant production capacity.

If expansion in a few years is being considered, then the plant can be designed in certain areas that make sense to build extra capacity into the design at a minimal cost. This will take a good engineering company to understand where larger capacity equipment can be installed and operated at lower flow rates, and not affect the quality of product or efficiencies in operation. These areas are limited. It may be better to just add additional equipment when the market demand requires an expansion.

One final point to consider on the size of equipment and potential of building too large a plant is that rather than increasing size, it may be advantageous to build a second plant at a different location. This might enable you to take advantage of transportation and energy issues. Usually, the larger the plant, the farther you have to transport the feedstock and products. It might make financial and economic sense to build a second plant near another source of feedstock and new market customers.

The cost of operation to produce biodiesel is composed of feed, chemical, energy, labor, depreciation, maintenance, loan interest, testing, marketing and other overhead costs. Each category of cost is influenced by the location of the plant, contracts for products, supplies, transportation, chemicals and labor. Each of these costs can be managed and/or negotiated, but in the end, if you don't make a profit, you will not be in business long.

If you look at the cost of production of biodiesel, you will see that the cost per gallon of biodiesel produced is most impacted by the feedstock cost. As you can see in pie charts included in the Biodiesel Production Cost Summary for a 5 mm gpy soybean oil plant in **Figure 56** and for a 30 mm gpy soybean oil plant in **Figure 57**, the cost is around 77.2 percent and 80.6 percent, respectively of the cost to produce biodiesel. This is why it is important to obtain the best feedstock at the lowest cost to minimize the cost of production.

Soybean Oil 5 mm gpy Graphics

Biodiesel Production Cost Summary

Customer: CIRAS Iowa State University
Case: Product cost with $0.22/lb Soybean Oil feed (5 mm gpy plant size)

Cost Data Category		Inputs
Plant Capacity (gal/year)		5,000,000
Feedstock Type		Soybean Oil
Feedstock Cost ($/lb)	$	0.220
Feedstock FFA Content (%)		0.5%
Methanol Cost ($/gal)	$	1.150
25% Sodium Methoxide Cost ($/lb)	$	0.706
Hydrochloric Acid Cost ($/lb)	$	0.131
50% Sodium Hydroxide Cost ($/lb)	$	0.088
Natural Gas Cost ($/1,000,000 btu)	$	10.00
Electricity Cost ($/KWh)	$	0.050
Transportation Cost on Feedstock ($/gal)	$	0.076
Average Labor Cost ($/hr)	$	25.00
Days of Operation		330
Labor Required per Shift		2.5
Hours of Operation per Day		24
Complete Plant Cost Installed ($)	$	3,745,000
Capital Invested by Customer ($)	$	1,872,500
Yearly Maintenance Cost		3.88%
Administration	$	125,000
Estimated Biodiesel Marketing Fee ($/gal)	$	0.03

Biodiesel Production Cost		Cost/gal ($)	Cost/liter ($)	% of Total
Cost of Feedstock		1.672	0.442	73.9%
Cost of Transportation of Feedstock		0.076	0.020	3.4%
Cost of Acid		0.011	0.003	0.5%
Cost of Base Catalyst		0.103	0.027	4.6%
Cost of Sodium Hydroxide		0.001	0.000	0.1%
Cost of Methanol		0.122	0.032	5.4%
Cost of Heat Energy		0.022	0.006	1.0%
Cost of Electricity		0.004	0.001	0.2%
Cost of Labor		0.099	0.026	4.4%
Depreciation		0.069	0.018	3.0%
Cost of Maintenance		0.029	0.008	1.3%
Cost of Administration and Overhead		0.025	0.007	1.1%
Cost of Marketing		0.030	0.008	1.3%
	Total	**2.263**	**0.598**	**100.0%**

Cost of Goods Sold		**2.111**		

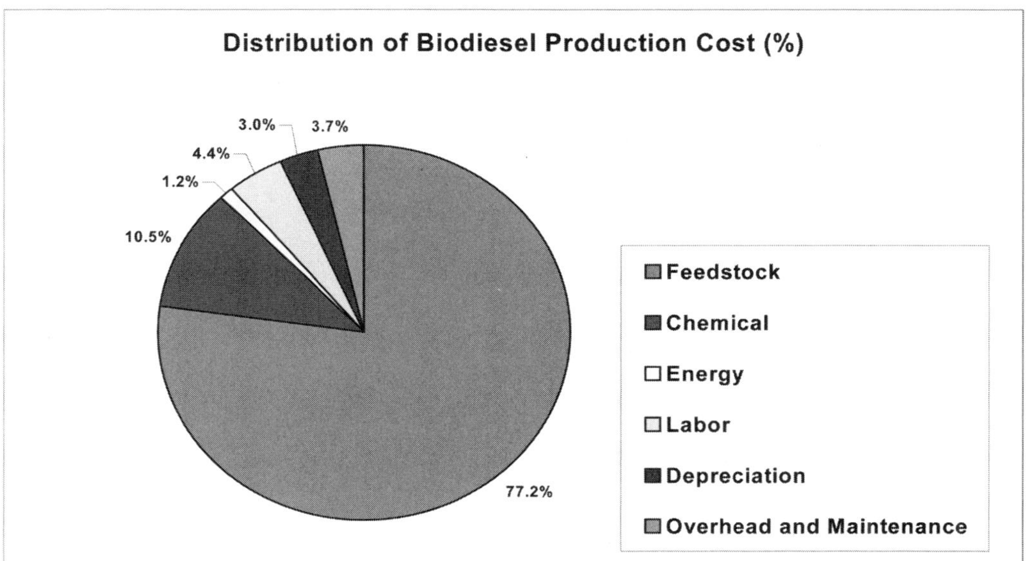

Figure 56. Biodiesel Production Cost for 5mm gpy Soybean Oil Plant.

Soybean Oil 30 mm gpy Graphics

Biodiesel Production Cost Summary

Customer: CIRAS Iowa State University
Case: Product cost with $0.22/lb Soybean Oil feed (30 mm gpy plant size)

Cost Data Category		Inputs
Plant Capacity (gal/year)		30,000,000
Feedstock Type		Soybean Oil
Feedstock Cost ($/lb)	$	0.220
Feedstock FFA Content (%)		0.5%
Methanol Cost ($/gal)	$	1.150
25% Sodium Methoxide Cost ($/lb)	$	0.706
Hydrochloric Acid Cost ($/lb)	$	0.131
50% Sodium Hydroxide Cost ($/lb)	$	0.088
Natural Gas Cost ($/1,000,000 btu)	$	10.00
Electricity Cost ($/KWh)	$	0.050
Transportation Cost on Feedstock ($/gal)	$	0.076
Average Labor Cost ($/hr)	$	25.00
Days of Operation		330
Labor Required per Shift		4.0
Hours of Operation per Day		24
Complete Plant Cost Installed ($)	$	21,600,000
Capital Invested by Customer ($)	$	10,800,000
Yearly Maintenance Cost		3.88%
Administration	$	175,000
Estimated Biodiesel Marketing Fee ($/gal)	$	0.03

Biodiesel Production Cost		Cost/gal ($)	Cost/liter ($)	% of Total
Cost of Feedstock		1.672	0.442	77.1%
Cost of Transportation of Feedstock		0.076	0.020	3.5%
Cost of Acid		0.011	0.003	0.5%
Cost of Base Catalyst		0.103	0.027	4.8%
Cost of Sodium Hydroxide		0.001	0.000	0.1%
Cost of Methanol		0.122	0.032	5.6%
Cost of Heat Energy		0.022	0.006	1.0%
Cost of Electricity		0.004	0.001	0.2%
Cost of Labor		0.026	0.007	1.2%
Depreciation		0.066	0.017	3.0%
Cost of Maintenance		0.028	0.007	1.3%
Cost of Administration and Overhead		0.006	0.002	0.3%
Cost of Marketing		0.030	0.008	1.4%
	Total	**2.168**	**0.573**	**100.0%**
Cost of Goods Sold		**2.038**		

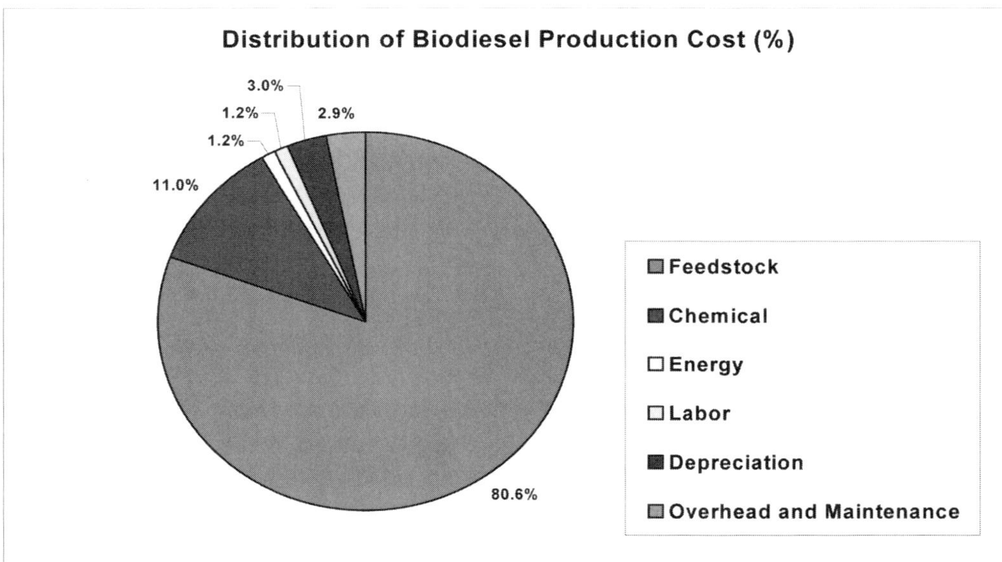

Figure 57. Biodiesel Production Cost for 30 mm gpy Soybean Oil Plant.

In these samples of biodiesel production cost summaries, soybean oil is used as a feedstock because of its availability in the United States. In the examples, a 10-year average of soy oil cost per pound of $0.22 is used for the cost of the feedstock. Other countries use the most plentiful feedstock, which in Canada and Europe is rapeseed and in Malaysia is palm oil. Sometimes the oil from oil seed crops has to have some pretreatment to remove waxes or reduce the FFA.

An alternative feedstock is animal oil or fats, which require additional equipment cost to process the high FFA content of that feedstock. For example, yellow grease can have a FFA of 15 percent, where white grease (animal oil) could be around 1-5 percent FFA at a higher cost penalty of 3-4 cents for that lower FFA. Following are sample biodiesel production cost summaries for 5 mm gpy and 30 mm gpy plants built to use 70 percent soybean oil and 30 percent animal feedstock of around 5 percent FFA content and an average cost of $0.16 per pound (see **Figures 58 and 59**).

Another area to consider is whether the use of oil or grease feedstock qualifies for the government incentives and at what incentive rate. A distinction might be made in legislative circles on the definition of "recycled." It may affect how incentives are applied and for what amount.

Soybean Oil/ White Grease 5 mm gpy Graphics

Biodiesel Production Cost Summary

Customer: Monona County Biodiesel
Case: Product cost with $0.22/lb Soybean Oil feed & $0.16 White Grease (5 mm gpy plant size)

Cost Data Category			Inputs
Plant Capacity (gal/year)			5,000,000
Feedstock Type & Percent	Soybean Oil Crude and Degummed		70%
Feedstock Cost ($/lb)		$	0.220
Feedstock FFA Content (%)			0.5%
Feedstock Type & Percent	White Grease		30%
Feedstock Cost ($/lb)		$	0.160
Feedstock FFA Content (%)			2.0%
Methanol Cost ($/gal)		$	1.15
25% Sodium Methoxide Cost ($/lb)		$	0.706
Hydrochloric Acid Cost ($/lb)		$	0.131
50% Sodium Hydroxide Cost ($/lb)		$	0.088
Natural Gas Cost ($/1,000,000 btu)		$	10.00
Electricity Cost ($/KWh)		$	0.050
Transportation Cost on Feedstock ($/gal)		$	0.08
Average Labor Cost ($/hr)		$	25.00
Days of Operation			330
Labor Required per Shift			2.5
Hours of Operation per Day			24
Complete Plant Cost Installed ($)		$	4,345,000
Capital Invested by Customer ($)		$	2,172,500
Yearly Maintenance Cost			3.88%
Administration		$	125,000
Estimated Biodiesel Marketing Fee ($/gal)		$	0.03

Biodiesel Production Cost	Cost/gal ($)	Cost/liter ($)	% of Total
Cost of Feedstock	1.530	0.404	70.9%
Cost of Transportation of Feedstock	0.076	0.020	3.5%
Cost of Acid	0.015	0.004	0.7%
Cost of Base Catalyst	0.109	0.029	5.1%
Cost of Sodium Hydroxide	0.002	0.001	0.1%
Cost of Methanol	0.122	0.032	5.7%
Cost of Heat Energy	0.032	0.008	1.5%
Cost of Electricity	0.004	0.001	0.2%
Cost of Labor	0.099	0.026	4.6%
Depreciation	0.080	0.021	3.7%
Cost of Maintenance	0.034	0.009	1.6%
Cost of Administration and Overhead	0.025	0.007	1.2%
Cost of Marketing	0.030	0.008	1.4%
Total	**2.158**	**0.570**	**100.0%**
Cost of Goods Sold	**1.990**		

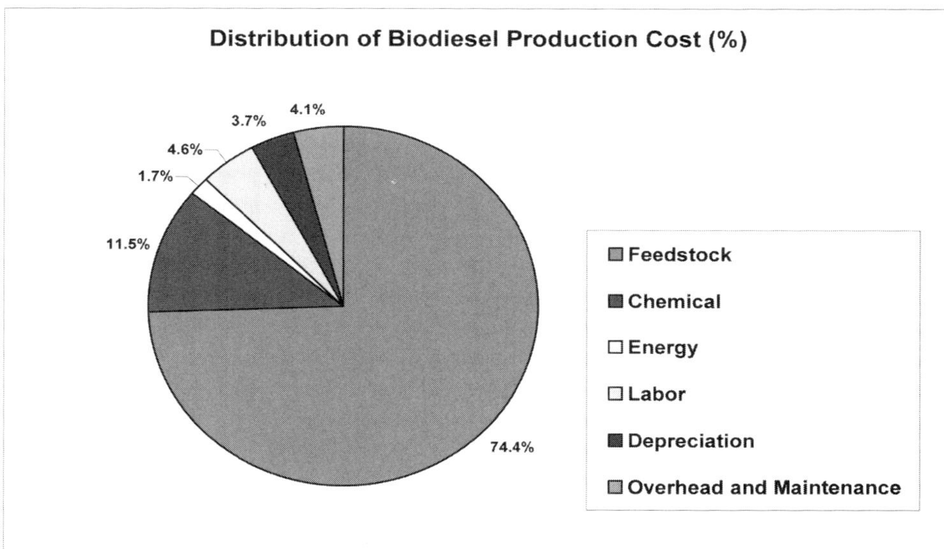

Figure 58. Biodiesel Production Cost Summary for 5 million gpy Soy Oil and Animal Fat Biodiesel Plant.

Soybean Oil/ White Grease 30 mm gpy Graphics
Biodiesel Production Cost Summary

Customer: Monona County Biodiesel
Case: Product cost with \$0.22/lb Soybean Oil feed & \$0.16 White Grease (30 mm gpy plant size)

Cost Data Category			Inputs
Plant Capacity (gal/year)			30,000,000
Feedstock Type & Percent	Soybean Oil Crude and Degummed		70%
Feedstock Cost (\$/lb)		\$	0.220
Feedstock FFA Content (%)			0.5%
Feedstock Type & Percent	White Grease		30%
Feedstock Cost (\$/lb)		\$	0.160
Feedstock FFA Content (%)			2.0%
Methanol Cost (\$/gal)		\$	1.15
25% Sodium Methoxide Cost (\$/lb)		\$	0.706
Hydrochloric Acid Cost (\$/lb)		\$	0.131
50% Sodium Hydroxide Cost (\$/lb)		\$	0.088
Natural Gas Cost (\$/1,000,000 btu)		\$	10.00
Electricity Cost (\$/KWh)		\$	0.050
Transportation Cost on Feedstock (\$/gal)		\$	0.08
Average Labor Cost (\$/hr)		\$	25.00
Days of Operation			330
Labor Required per Shift			4.0
Hours of Operation per Day			24
Complete Plant Cost Installed (\$)		\$	25,600,000
Capital Invested by Customer (\$)		\$	12,800,000
Yearly Maintenance Cost			3.88%
Administration		\$	175,000
Estimated Biodiesel Marketing Fee (\$/gal)		\$	0.03

Biodiesel Production Cost	Cost/gal (\$)	Cost/liter (\$)	% of Total
Cost of Feedstock	1.530	0.404	74.1%
Cost of Transportation of Feedstock	0.076	0.020	3.7%
Cost of Acid	0.015	0.004	0.7%
Cost of Base Catalyst	0.109	0.029	5.3%
Cost of Sodium Hydroxide	0.002	0.001	0.1%
Cost of Methanol	0.122	0.032	5.9%
Cost of Heat Energy	0.032	0.008	1.5%
Cost of Electricity	0.004	0.001	0.2%
Cost of Labor	0.026	0.007	1.3%
Depreciation	0.078	0.021	3.8%
Cost of Maintenance	0.033	0.009	1.6%
Cost of Administration and Overhead	0.006	0.002	0.3%
Cost of Marketing	0.030	0.008	1.5%
Total	**2.064**	**0.545**	**100.0%**
Cost of Goods Sold	**1.917**		

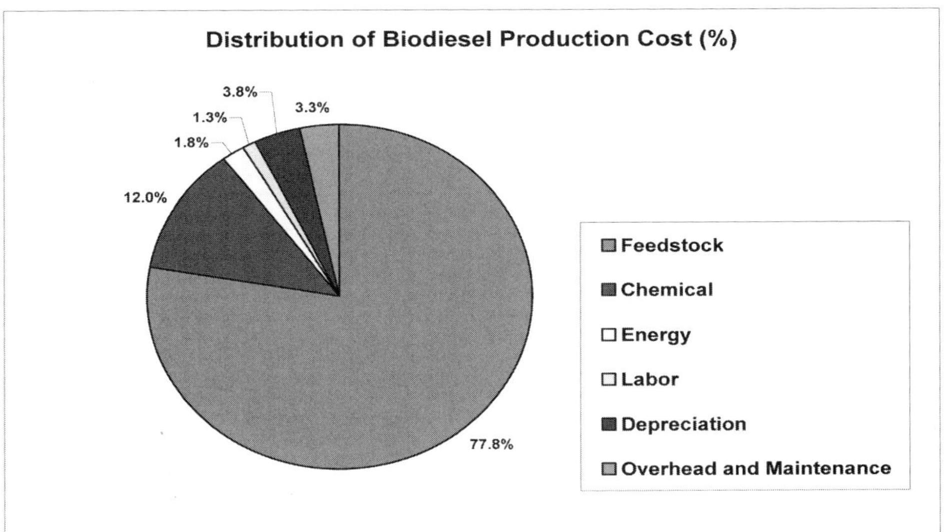

Figure 59. Biodiesel Production Cost Summary for 30 million gpy Soy Oil and Animal Fat Biodiesel Plant.

These two cost summaries give you an idea of what the cost of producing biodiesel may be for a certain plant design and cost of energy, chemicals and feedstock for both a 5 and 30 mm gpy plant size. Many variables affect the cost of production; therefore, these numbers should be used only as a guide, and actual numbers should be calculated with your specific set of circumstances and feedstock.

Feedstock is the largest component of the cost of producing biodiesel. The next largest cost of production is that of all the chemicals used to produce biodiesel. This is closely followed by the cost of depreciation for a large size plant and labor for a small size plant. Many times producers forget about this cost of depreciation, not realizing they will have to replace the equipment someday. The cost of energy is a small percentage of overall costs. With the cost of labor being such a small percentage of the overall cost per gallon produced, it does not make sense to try and operate a plant with less than two persons per shift.

Many of the costs are controlled by the chemical reaction and do not change with the size of the plant on a per-gallon basis. A 5 million gpy plant has the same cost of chemicals and energy per gallon of biodiesel produced as a 30 million gpy plant. Even the cost of feedstock per gallon of biodiesel produced is the same for both plants.

The energy costs are a factor of the plant design and are usually calculated by the engineering company or technology provider. The electrical usage can be determined by adding all the motors and any electric heaters together to obtain kilowatts per hour of operation. Then by finding out the cost of electricity in dollars per kilowatt hour ($/KWh) for your area, you can calculate the electricity used per gallon of biodiesel produced for a certain technology or plant design and size. A similar calculation can be made for the heat needed, usually using natural gas as the fuel for the boiler.

Where the cost will vary with the size of the plant is the depreciation, labor and overhead. The depreciation is the same percentage of the equipment cost for any size plant. But since the equipment cost for a larger volume plant is less when compared to the volume of biodiesel produced as a result of the economies of scale, the per-gallon allocated amount is less for a larger volume plant than with a smaller one when the plant is run at capacity.

Labor is like depreciation in that it increases very little with a large increase in plant size. Two persons per shift can run a large continuous reactor just as easily as they can run a small one. With very large plants, dedicated persons for such tasks as loading and unloading might be necessary and therefore add to the labor cost.

The overhead is an arbitrary number and does not expand in the same proportion as capacity. This number should be broken down into more detail and can change very fast with such functions as marketing and administrative tasks.

In the earlier examples, the operational cost for a biodiesel plant on a per-gallon basis for the same size plant that uses all soybean oil when compared to plant using 70 percent soybean oil with 30 percent animal oil or fat is $0.10 higher per gallon. This also holds true for both a small and large plant size because the feedstock consisting of animal oil or fats is usually less expensive than oil seed feedstock.

Offsetting the lower feedstock cost in an animal oil plant are the higher chemical, depreciation, energy and maintenance costs. Additional equipment to handle the higher free fatty acid results in higher depreciation and maintenance costs, and they are calculated as a percentage of the equipment cost.

CHAPTER 36

Sensitivity Analysis

Sensitivity analysis is a tool used to view historical price data and trends for key inputs and outputs from the production process. In the case of biodiesel, these sensitivity analyses will usually cover the feedstock (usually soybean oil in the United States), natural gas, methanol and biodiesel. The futures market and past closing prices on the open market were used to compile these analyses.

The feedstock is the major cost for producing biodiesel, and therefore will have the most detail in the sensitivity analysis and appropriate charts. Keep in mind that these sensitivity analysis charts should only be used as a guide and can in no way guarantee or predict the future pricing of the represented inputs and outputs from the plant.

As part of the analysis, the cost of production and financial information including interest on borrowed money, depreciation, etc., are used to produce the sensitivity analysis tables. The cost of production is used to generate the net profit or loss associated with a particular feedstock price and selling price of biodiesel. To determine the operational cost for a plant in the size range of 5 million gallons per year, information from a number of engineering and design firms was collected and compiled. The basis for these cost estimates are listed in **Figure 60**.

The sensitivity analysis table addresses the sensitivity effect of the soybean oil price on the profit stated as Earnings Before Income Tax, and Amortization (EBITA) (see **Table 36**). By stating the profit as EBITA the variations of income tax and amortization are removed from the analysis, making it easier to show the effect of soybean oil price and biodiesel selling price on a neutral basis.

Since biodiesel is not traded on the commodity market, there are no standard means to track the average historical biodiesel selling price prior to 2003. To arrive at a biodiesel selling price prior to 2003, one possible method to use is to assume that biodiesel would sell for the historical price of No. 2 diesel fuel plus the equivalent of 1 cent per percent of biodiesel that would be blended with the petroleum diesel. This means that 100 percent biodiesel would run $1.00 higher (100 percent times one cent per percent or $1.00) than the historical selling price of No. 2 diesel fuel on the commodity market.

The 1 cent per percent higher price for each percent of biodiesel in the blend is a current method of estimating the selling price for a blend biodiesel with diesel fuel on the open market. This means that a B2 blend of biodiesel and diesel fuel (a blend consisting of 2 percent biodiesel and 98 percent diesel fuel) would sell for the normal selling price of No. 2 diesel fuel plus $0.02. This is a good estimate of the selling price of biodiesel without using any government incentives, like the blender's credit or producer credit. It also presents a conservative view of the possible selling price of biodiesel blends.

The down side of using the 1 cent per percent higher price for each percent of biodiesel in the blend for the selling price is that the markets that would accept the use of a higher price diesel fuel blends are limited. To appeal to more markets, especially the over-the-road trucking industry, the biodiesel/petroleum blend fuel will have to be priced at or below petroleum diesel fuel. This was one of the reasons for the blender's credit, to eliminate the difference between biodiesel petroleum blends and straight petroleum diesel fuel. But since the producer does not get this credit, it was not considered in these calculations or the sensitivity analysis.

Biodiesel Production Cost Summary

Customer: CIRAS Iowa State University
Case: Product cost with $0.22/lb Soybean Oil feed (5 mm gpy plant size)

Cost Data Category	Inputs
Plant Capacity (gal/year)	5,000,000
Feedstock Type	Soybean Oil
Feedstock Cost ($/lb)	0.220
Feedstock FFA Content (%)	0.5
Methanol Cost ($/gal)	1.15
25% Sodium Methoxide Cost ($/lb)	0.706
Hydrochloric Acid Cost ($/lb)	0.131
50% Sodium Hydroxide Cost ($/lb)	0.088
Natural Gas Cost ($/1,000,000 btu)	14.000
Electricity Cost ($/KWh)	0.080
Transportation Cost on Feedstock ($/gal)	0.04
Average Labor Cost ($/hr)	25.00
Days of Operation	350
Labor Required per Shift	2.0
Hours of Operation per Day	24
Complete Plant Cost Installed ($)	7,000,000
Capital Invested by Customer ($)	7,000,000
Yearly Maintenance Cost	2.88
Administration	75,000
Estimated Biodiesel Marketing Fee ($/gal)	0.03

Biodiesel Production Cost	Cost/gal ($)	Cost/liter ($)	% of Total
Cost of Feedstock	1.672	0.442	73.0%
Cost of Transportation of Feedstock	0.040	0.011	1.7%
Cost of Acid	0.011	0.003	0.5%
Cost of Base Catalyst	0.108	0.029	4.7%
Cost of Sodium Hydroxide	0.001	0.000	0.1%
Cost of Methanol	0.122	0.032	5.3%
Cost of Heat Energy	0.031	0.008	1.3%
Cost of Electricity	0.007	0.002	0.3%
Cost of Labor	0.084	0.022	3.7%
Depreciation	0.128	0.034	5.6%
Cost of Maintenance	0.040	0.011	1.8%
Cost of Administration and Overhead	0.015	0.004	0.7%
Cost of Marketing	0.030	0.008	1.3%
Total	**2.290**	**0.605**	**100.0%**

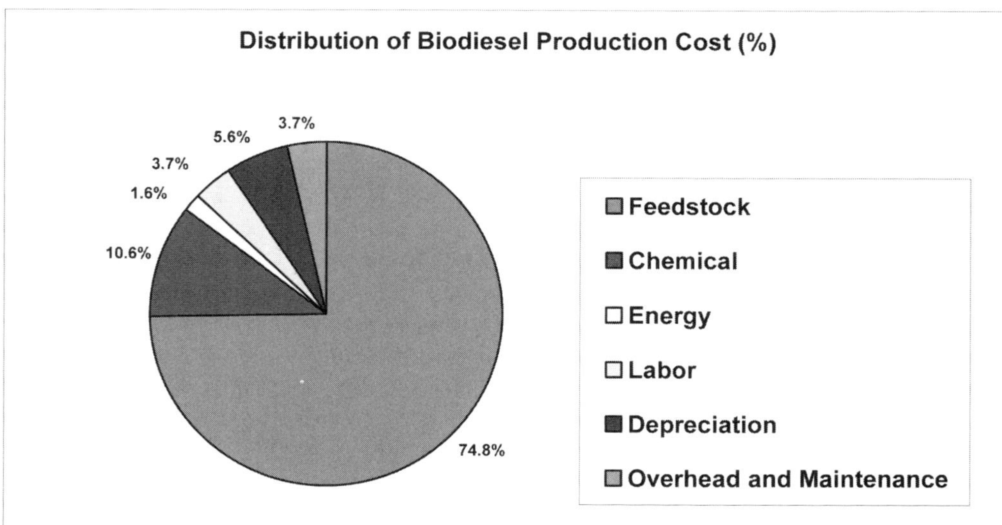

Figure 60. Biodiesel Production Cost Summary.

Biodiesel Produced from Soybean Oil EBITA Sensitivity Analysis with No Producers Credit

B100 Biodiesel Price *** $/gal

Soybean Oil Price $/lb	$1.50	$1.60	$1.70	$1.80	$1.90	$2.00	$2.10	$2.20	$2.30	$2.40	$2.50	$2.60	$2.70	$2.80
$0.12	($220,396)	$279,604	$779,604	$1,279,604	$1,779,604	$2,279,604	$2,779,604	$3,279,604	$3,779,604	$4,279,604	$4,779,604	$5,279,604	$5,779,604	$6,279,604
$0.14	($984,196)	($484,196)	$15,804	$515,804 2001*	$1,015,804 2000*	$1,515,804	$2,015,804	$2,515,804	$3,015,804	$3,515,804	$4,015,804	$4,515,804	$5,015,804	$5,515,804
$0.16	($1,747,996)	($1,247,996) 1999*	($747,996)	($247,996)	$252,004	$752,004	$1,252,004	$1,752,004	$2,252,004	$2,752,004	$3,252,004	$3,752,004	$4,252,004	$4,752,004
$0.18	($2,511,796)	($2,011,796)	($1,511,796) 2002*	($1,011,796)	($511,796)	($11,796)	$488,204	$988,204	$1,488,204	$1,988,204	$2,488,204	$2,988,204	$3,488,204	$3,988,204
$0.20	($3,275,596)	($2,775,596)	($2,275,596)	($1,775,596)	($1,275,596)	($775,596)	($275,596)	$224,404	$724,404	$1,224,404	$1,724,404	$2,224,404	$2,724,404	$3,224,404
$0.22	($4,039,396)	($3,539,396) 1997*	($3,039,396)	($2,539,396)	($2,039,396) 2003*	($1,539,396)	($1,039,396)	($539,396)	($39,396)	$460,604	$960,604	$1,460,604	$1,960,604	$2,460,604
$0.24	($4,803,196) 1998*	($4,303,196)	($3,803,196) 1996*	($3,303,196)	($2,803,196)	($2,303,196)	($1,803,196)	($1,303,196)	($803,196)	($303,196)	$196,804	$696,804	$1,196,804	$1,696,804
$0.26	($5,566,996)	($5,066,996) 1995*	($4,566,996)	($4,066,996)	($3,566,996)	($3,066,996)	($2,566,996)	($2,066,996)	($1,566,996)	($1,066,996)	($566,996)	($66,996)	$433,004	$933,004 2005*
$0.28	($6,330,796)	($5,830,796)	($5,330,796)	($4,830,796)	($4,330,796)	($3,830,796)	($3,330,796)	($2,830,796) 2004*	($2,330,796)	($1,830,796)	($1,330,796)	($830,796)	($330,796)	$169,204
$0.30	($7,094,596)	($6,594,596)	($6,094,596)	($5,594,596)	($5,094,596)	($4,594,596)	($4,094,596)	($3,594,596)	($3,094,596)	($2,594,596)	($2,094,596)	($1,594,596)	($1,094,596)	($594,596)
$0.32	($7,858,396)	($7,358,396)	($6,858,396)	($6,358,396)	($5,858,396)	($5,358,396)	($4,858,396)	($4,358,396)	($3,858,396)	($3,358,396)	($2,858,396)	($2,358,396)	($1,858,396)	($1,358,396)
$0.34	($8,622,196)	($8,122,196)	($7,622,196)	($7,122,196)	($6,622,196)	($6,122,196)	($5,622,196)	($5,122,196)	($4,622,196)	($4,122,196)	($3,622,196)	($3,122,196)	($2,622,196)	($2,122,196)

*Annual average of Soybean Oil Price and $1.00 over annual average Diesel No. 2 price for each year.

	1995	1996	1997	1998	1999	2000	2001	2002	2003	2004	2005
Soybean Oil	$0.268	$0.238	$0.233	$0.258	$0.176	$0.150	$0.145	$0.183	$0.236	$0.286	0.2751**
#2 Diesel	$0.546	$0.667	$0.616	$0.454	$0.552	$0.904	$0.791	$0.735	$0.891	$1.191	$1.759
Biodiesel***	$1.55	$1.67	$1.62	$1.45	$1.55	$1.90	$1.79	$1.74	$1.89	$2.19	$2.76

** Based on only 4 months

*** Biodiesel price is based on No. 2 Diesel annual average price plus $1.00, which is equivalent to 1 cent per percent biodiesel content

Table 36. Biodiesel Produced from Soybean Oil EBITA Sensitivity Analysis.

Since 100 percent biodiesel is pure biodiesel with no diesel fuel in it, the price of the biodiesel is assumed to be the equivalent of a dollar more than the selling price of No. 2 diesel fuel. Since the biodiesel improves the diesel fuel properties, especially in low sulfur diesel fuel, there is normally not any reason to blend the biodiesel with No. 1 diesel fuel, which is a higher priced petroleum diesel fuel. The major reason to use a biodiesel blend with No. 1 diesel fuel as opposed to that of No. 2 diesel fuel is for lower gelling temperatures in the winter when consumers usually switch to No. 1 diesel fuel anyway.

The shaded area in **Table 36** highlights the area where profit is negative based on a given soybean oil purchase price and biodiesel selling price, based on EBITA. Therefore, one would like to operate the plant in the unshaded area to meet financial obligations based on EBITA. This of course does not take into account any other factors that might make the operation of the plant more profitable, such as government incentives.

The years listed in **Table 36** indicate the average annual soybean oil price coupled with the average annual biodiesel price based on No. 2 diesel fuel historical numbers for the indicated year. None of these numbers are corrected for the present value of money nor are any external forces or influences that may have had an effect on these numbers known.

The calculated historical price for biodiesel used in **Table 36** was based on the diesel fuel prices shown in **Table 37**, which is comprised of the average annual price for No. 2 diesel fuel over the past 11 years. The No. 2 diesel fuel in the category of Sales for Resale was used in the calculations for the biodiesel price. The Sales for Resale category was used since it represents sales of refined petroleum products to purchasers who are other-than-ultimate consumers – wholesale sales. Explanations of all the categories are shown in **Table 42** at the end of this chapter. The other categories shown in **Table 37** can be used to explore sensitivity of niche markets for diesel fuel No. 2.

The historical data for soybean oil prices is shown for the period of 1991 through the first four months of 2005 in **Table 38**. The average monthly price is shown for each month for each of the years, and the average for that year is listed in the last column. Only the past 11 years of annual averages were used to conduct the sensitivity analysis in **Table 36**. This information was obtained from commodities market records and the USDA.

The monthly average price for soybean oil ranges from a low of $0.1238 per pound or 12.38 cents per pound to a high of $0.3466 per pound or 34.66 cents per pound. This means that the daily averages or spot prices could vary even more. The overall average for the 11-year period is $0.2196 per pound or 21.96 cents per pound. This number can be used in the financial calculation for projected profits based on historical data. The overall trend in soybean oil prices shows that it follows a cycle that seems to span a 10-year time period reaching a high every 10 years. Keep in mind that this is a look at historical pricing of soybean oil and may not represent the future if conditions change in the soybean supply market.

The historical No. 2 diesel fuel prices can be further broken down into monthly averages. The monthly averages in the category for Sales for Resale Type No. 2 Diesel Fuel Average Monthly Prices are shown in **Table 39**. These monthly prices can be used in conjunction with the 11-year historical numbers for the average monthly price of soybean oil to generate a frequency analysis.

No. 2 Distillate Average Prices by Sales Type

(Cents per Gallon Excluding Taxes)

Product/ Sales Type	1995	1996	1997	1998	1999	2000	2001	2002	2003	2004	2005
No. 2 Distillate											
Sales to End Users, Average	67.3	79.3	75.3	59.9	67.8	104.4	94.8	87.4	105.8	133.9	188.8
Residential	86.7	98.9	98.4	85.2	87.6	131.1	125	112.9	135.5	154.8	204.8
Commercial/Institutional	59.3	71.4	67.4	52.5	60.5	96.2	86.1	79.2	97	126.3	179.4
Industrial	64.4	77.1	71.2	55.6	63.9	99.1	90.7	82.6	100.3	131	180.8
Through Retail Outlets	67	78.8	74.5	59.3	68.5	103.6	94.3	86.2	104.4	134.8	193.7
Other End Users	67.3	79.3	75.3	60	68.7	105	96	87.8	102.1	133.3	193.2
Sales for Resale	53.6	66	61.1	45	53.8	90.1	78.5	72.8	89.1	117.8	173.2
No. 2 Diesel Fuel											
Sales to End Users, Average	63.6	75.7	71.4	56.2	65.4	100.6	91.2	83.7	100.8	131.6	188.1
Commercial/Institutional	58.8	71.1	66.6	51.6	60.5	95.8	85.7	78.9	96.1	126.8	180.1
Industrial	64.9	77.5	71.5	56.2	64.5	99	90.7	82.6	100.4	131.9	186
Through Retail Outlets	67	78.8	74.5	59.3	68.5	103.6	94.3	86.2	104.4	134.8	193.7
Other End Users	67.7	79.6	75.7	60.4	69.4	105.2	96	87.9	102.2	133.7	193.6
Sales for Resale	54.6	66.7	61.6	45.4	55.2	90.4	79.1	73.5	89.1	119.1	175.9
No. 2 Diesel Fuel, Low-Sulfur											
Sales to End Users, Average	64.5	76.7	71.9	56.5	66.3	101.4	91.7	84.1	101.4	132.5	189.9
Commercial/Institutional	59.8	72.3	67	51.5	61.7	96.8	86.8	80.1	97.2	128.4	182.2
Industrial	66.3	79.1	72.2	55.7	64.3	97.1	89.1	81.9	99.7	131.8	188.6
Through Retail Outlets	67	78.8	74.5	59.3	68.5	103.6	94.3	86.2	104.4	134.8	193.7
Other End Users	68.4	80.6	76	60.1	70.4	106.2	94.9	85.8	100.9	132.9	192.9
Sales for Resale	55.1	67.3	61.9	45.7	55.7	90.9	79.4	73.8	89.5	119.7	176.8
No. 2 Diesel Fuel, High-Sulfur											
Sales to End Users, Average	61.4	73.2	69.8	55.5	62	98.1	89.2	82.2	98.6	128.1	181.5
Commercial/Institutional	56.8	68.7	65.9	51.8	58	93.8	83.7	76.7	94.1	123.7	176
Industrial	63.6	75.9	70.8	56.7	64.8	101.7	93.3	83.9	101.8	132.2	182.1
Other End Users	66.7	78.4	75.3	60.8	67.6	103.9	98	91.3	104.4	135.4	194.8
Sales for Resale	52.4	63.9	60.2	43.7	51.9	87.5	77.1	71.2	87	114.6	170

Source: Energy Information Administration and the US Department of Energy

Source: Energy Information Administration, Forms EIA-782A, "Refiners'/Gas Plant Operators' Monthly Petroleum Product Sales Report" and EIA-782B, "Resellers'/Retailers' Monthly Petroleum Product Sales Report"

Table 37. No. 2 Distillate Average Prices by Sales Type

U.S. Soybean Oil: Prices

Soybean Oil, Crude, Decatur: Average Cents Per Pound
1991 - 2004, By Month

Year	Jan.	Feb.	Mar.	April	May	June	July	Aug.	Sept.	Oct.	Nov.	Dec.	Average
1991	21.50	21.60	22.20	21.50	20.20	19.60	19.00	20.20	20.40	19.50	18.70	18.90	20.28
1992	18.70	18.80	19.70	19.00	20.10	20.70	18.80	17.80	18.20	18.30	20.10	20.50	19.23
1993	21.20	21.40	21.00	21.20	21.10	21.30	24.10	23.40	20.90	22.98	25.37	28.09	22.67
1994	29.91	28.84	29.03	27.94	29.10	27.60	24.50	24.50	26.11	27.06	29.84	30.61	27.92
1995	29.04	28.15	28.30	26.30	26.00	26.78	27.60	26.56	26.26	26.56	25.41	24.76	26.81
1996	23.52	23.49	23.60	25.70	26.50	24.95	24.10	23.99	23.92	21.95	21.81	21.60	23.76
1997	22.45	22.41	23.29	23.17	23.68	22.97	21.89	22.06	22.88	24.31	25.73	25.08	23.33
1998	25.10	26.51	27.09	28.10	28.30	25.80	24.90	24.00	25.10	25.20	25.20	24.00	25.78
1999	22.90	20.00	19.50	18.80	17.90	16.50	15.29	16.50	16.79	16.08	15.63	15.30	17.60
2000	15.63	15.09	16.21	17.52	16.75	15.65	14.70	14.34	14.24	13.50	13.37	13.12	15.01
2001	12.53	12.38	13.90	13.53	13.53	14.21	16.49	17.08	15.46	14.38	15.23	15.10	14.49
2002	14.82	14.15	14.75	15.31	15.98	17.69	19.12	20.61	20.32	20.75	23.00	22.60	18.26
2003	21.50	21.20	21.56	22.40	23.17	22.90	21.80	20.40	23.20	27.40	27.76	29.54	23.57
2004	30.34	33.05	34.66	34.19	32.68	30.07	28.05	25.98	25.87	23.23	22.95	21.79	28.57
2005†	20.46	20.70	23.60	23.09	23.38								27.81

† Preliminary. Source: USDA, ERS Oil Crops Outlook, June 2005.

Table38. U.S. Soybean Oil Prices.

Sales for Resale Type No. 2 Diesel Fuel Average Monthly Prices

(Cents per Gallon Excluding Taxes)

Product/ Sales Type	1995	1996	1997	1998	1999	2000	2001	2002	2003	2004	2005
No. 2 Diesel Fuel											
Sales for Resale											
January	51.1	57.0	70.7	50.9	37.4	78.5	91.2	55.2	90.5	97.2	132.2
February	51.5	58.8	68.5	49.3	36.2	86.2	86.5	57.3	108.7	97.9	140.4
March	52.0	63.0	63.4	46.9	44.1	85.6	78.8	67.5	103.4	101.8	160.0
April	55.4	70.7	62.8	49.3	49.4	80.4	83.3	71.5	87.2	108.4	164.8
May	56.4	67.5	61.3	48.3	48.4	82.2	90.3	71.3	80.3	112.9	153.5
June	53.3	59.9	57.4	44.6	50.9	83.0	85.8	69.0	81.8	107.9	168.3
July	52.2	60.8	56.6	43.5	56.9	84.2	76.3	71.8	84.6	116.0	172.3
August	54.9	65.6	59.6	42.2	62.0	92.5	81.5	76.4	89.6	125.0	191.0
September	56.3	72.3	58.6	46.2	65.3	105.3	84.7	84.2	81.6	133.8	214.5
October	55.3	75.9	62.6	46.2	65.3	104.5	72.0	86.3	87.8	153.7	234.4
November	57.0	74.1	62.5	42.4	70.0	103.7	62.4	79.5	87.4	143.3	184.5
December	58.2	71.8	56.4	36.7	70.9	94.6	55.5	83.1	90.1	128.5	177.1
Annual Average	54.6	66.7	61.6	45.4	55.2	90.4	79.1	73.5	89.1	119.1	175.9

Source: Energy Information Administration and the US Department of Energy

Source: Energy Information Administration, Forms EIA-782A, "Refiners'/Gas Plant Operators' Monthly Petroleum Product Sales Report" and EIA-782B, "Resellers'/Retailers' Monthly Petroleum Product Sales Report"

Table 39. Sales for Resale Type No. 2 Diesel Fuel Average Monthly Prices.

In analyzing the monthly historical No. 2 diesel fuel prices representing the sales to wholesale customers, one can see the minimum average monthly price over the 11-year period was $0.3620 per gallon or 36.20 cents per gallon in 1999. The maximum average monthly price was $2.3440 per gallon or 234.40 cents per gallon in 2005. The trend over the 11-year history is that petroleum prices are increasing.

By combining the soybean oil average monthly price with the average monthly biodiesel price calculated from the historical diesel fuel prices, the frequency of each possible combination of soybean oil price with biodiesel price was created as shown in **Table 40**. It shows the percentage of that frequency and can give the reader a sense of how often these various combinations occur. Additional columns had to be added because the monthly No. 2 diesel price averages reached new lows and highs with respect to the annual averages.

The shaded area in the table indicates at what combination of soybean oil price and biodiesel selling price the operation of a 5 million gallon per year plant shows a loss. The frequency of that occurring based on historical prices is also seen in **Table 40**. By adding all the percentages together, the historical data indicated that a 5 million gallon per year biodiesel production plant would operate with a profit in 21 percent of the time over the past 11 years. This does not take into account use of government incentives. Taking into account the producer credit would increase the total profitability frequency to 29 percent. Taking into account other government incentives would increase the frequency of making a profit even higher. Keep in mind that this is only true if the same external influences that impact the various pricing scenarios occur in the future as did in the past.

Although this sensitivity analysis was conducted on a 5 million gallon per year plant, the sensitivity analysis is the same for the feedstock and biodiesel selling price for larger biodiesel plants. The profitability range is, however, different for each plant size because the operational costs are different. For example, the frequency of the various combinations of soybean oil expense and biodiesel selling price is the same for any size plant, but the profitability range is different as shown in the **Table 41**, which represents the frequency percentages and profitability range for a 30 million gallon per year biodiesel plant. As you can see, the total of all percentages in the profitable range is 33 percent for this size plant. Of course this only true if the plant is operating at capacity.

The remaining sensitivity charts show methanol and natural gas based on the average annual price for the period of January 2000 to April 2004 (see **Figures 61 and 62**). These sensitivity analyses represent a small part of the overall cost of making biodiesel. Therefore, they do not represent a high risk due to price fluctuations in the markets. The historical costs for these items are also highly variable as indicated by the graphs, but the effect on profits is minimal when compared to the feedstock.

The sensitivity analysis of other purchased items like methanol and natural gas can be used to determine the cycles the pricing might experience. Also, this information on price sensitivity can be used to determine if long-term contracting would be beneficial. In the case of natural gas, the low cost curve indicates the long-term contracts, and in a volatile market like natural gas supply, it can be very cost effective to obtain contracts for supply.

Biodiesel and Soybean Oil Historical Price Frequency Percent*

B100 Biodiesel Price *** $/gal

Soybean Oil Price $/lb	$1.40	$1.50	$1.60	$1.70	$1.80	$1.90	$2.00	$2.10	$2.20	$2.30	$2.40	$2.50	$2.60	$2.70	$2.80	$2.90	$3.00	$3.10	$3.20	$3.30
$0.12	0%	0%	0%	0%	0%	2%	0%	0%	0%	0%	0%	0%	0%	0%	0%	0%	0%	0%	0%	0%
$0.14	0%	0%	2%	2%	2%	3%	2%	1%	0%	0%	0%	0%	0%	0%	0%	0%	0%	0%	0%	0%
$0.16	0%	1%	3%	5%	4%	2%	0%	0%	0%	0%	0%	0%	0%	0%	0%	0%	0%	0%	0%	0%
$0.18	0%	2%	0%	1%	2%	0%	0%	0%	0%	0%	0%	0%	0%	0%	0%	0%	0%	0%	0%	0%
$0.20	2%	0%	0%	1%	2%	2%	0%	0%	0%	1%	1%	0%	0%	0%	0%	0%	0%	0%	0%	0%
$0.22	1%	0%	3%	3%	3%	2%	1%	1%	0%	1%	1%	0%	0%	0%	0%	0%	0%	0%	0%	0%
$0.24	2%	0%	8%	2%	2%	0%	0%	0%	0%	0%	0%	2%	2%	0%	0%	0%	0%	0%	0%	0%
$0.26	2%	5%	5%	2%	0%	0%	0%	0%	0%	2%	0%	0%	0%	0%	0%	0%	0%	0%	0%	0%
$0.28	0%	5%	0%	0%	0%	2%	0%	0%	1%	0%	0%	0%	0%	2%	2%	1%	0%	1%	0%	1%
$0.30	0%	1%	0%	0%	0%	1%	1%	1%	0%	0%	0%	0%	0%	0%	0%	0%	0%	0%	0%	0%
$0.32	0%	0%	0%	0%	0%	0%	0%	1%	0%	0%	0%	0%	0%	0%	0%	0%	0%	0%	0%	0%
$0.34	0%	0%	0%	0%	0%	0%	2%	1%	0%	0%	0%	0%	0%	0%	0%	0%	0%	0%	0%	0%

Percent of the time the combination of soybean oil expense and biodiesel selling price produced a profit based on monthly averages was 21%

* Shaded area indicates a 5 mm gpy biodiesel plant operating at a loss with not including any government incentives
** Based on only 4 months
*** Biodiesel price is based on No. 2 Diesel annual average price plus $1.00, which is equivalent to 1 cent per percent biodiesel content

Table 40. Biodiesel and Soybean Oil Historical Price Frequency as a Percent for 5 mm gpy Plant.

Biodiesel and Soybean Oil Historical Price Frequency Percent*

B100 Biodiesel Price *** $/gal

Soybean Oil Price $/lb	$1.40	$1.50	$1.60	$1.70	$1.80	$1.90	$2.00	$2.10	$2.20	$2.30	$2.40	$2.50	$2.60	$2.70	$2.80	$2.90	$3.00	$3.10	$3.20	$3.30
$0.12	0%	0%	0%	0%	0%	2%	0%	0%	0%	0%	0%	0%	0%	0%	0%	0%	0%	0%	0%	0%
$0.14	0%	0%	2%	2%	2%	3%	2%	1%	0%	0%	0%	0%	0%	0%	0%	0%	0%	0%	0%	0%
$0.16	0%	1%	3%	5%	4%	2%	0%	0%	0%	0%	0%	0%	0%	0%	0%	0%	0%	0%	0%	0%
$0.18	0%	2%	0%	1%	2%	0%	0%	0%	0%	0%	0%	0%	0%	0%	0%	0%	0%	0%	0%	0%
$0.20	2%	0%	0%	1%	2%	2%	0%	0%	0%	1%	1%	0%	0%	0%	0%	0%	0%	0%	0%	0%
$0.22	1%	0%	3%	3%	3%	2%	1%	1%	0%	1%	1%	0%	0%	0%	0%	0%	0%	0%	0%	0%
$0.24	2%	0%	8%	2%	2%	0%	0%	0%	0%	0%	0%	2%	2%	0%	0%	0%	0%	0%	0%	0%
$0.26	2%	5%	5%	2%	0%	0%	0%	0%	0%	2%	0%	0%	0%	0%	0%	0%	0%	0%	0%	0%
$0.28	0%	5%	0%	0%	0%	2%	0%	0%	1%	0%	0%	0%	0%	2%	2%	1%	0%	1%	0%	1%
$0.30	0%	1%	0%	0%	0%	1%	1%	1%	0%	0%	0%	0%	0%	0%	0%	0%	0%	0%	0%	0%
$0.32	0%	0%	0%	0%	0%	0%	0%	1%	0%	0%	0%	0%	0%	0%	0%	0%	0%	0%	0%	0%
$0.34	0%	0%	0%	0%	0%	0%	2%	1%	0%	0%	0%	0%	0%	0%	0%	0%	0%	0%	0%	0%

Percent of the time the combination of soybean oil expense and biodiesel selling price produced a profit based on monthly averages was 33%

* Shaded area indicates a 30 mm gpy biodiesel plant operating at a loss with not including any government incentives
** Based on only 4 months
*** Biodiesel price is based on No. 2 Diesel annual average price plus $1.00, which is equivalent to 1 cent per percent biodiesel content

Table 41. Biodiesel and Soybean Oil Historical Price Frequency as a Percent for 30 mm gpy Plant.

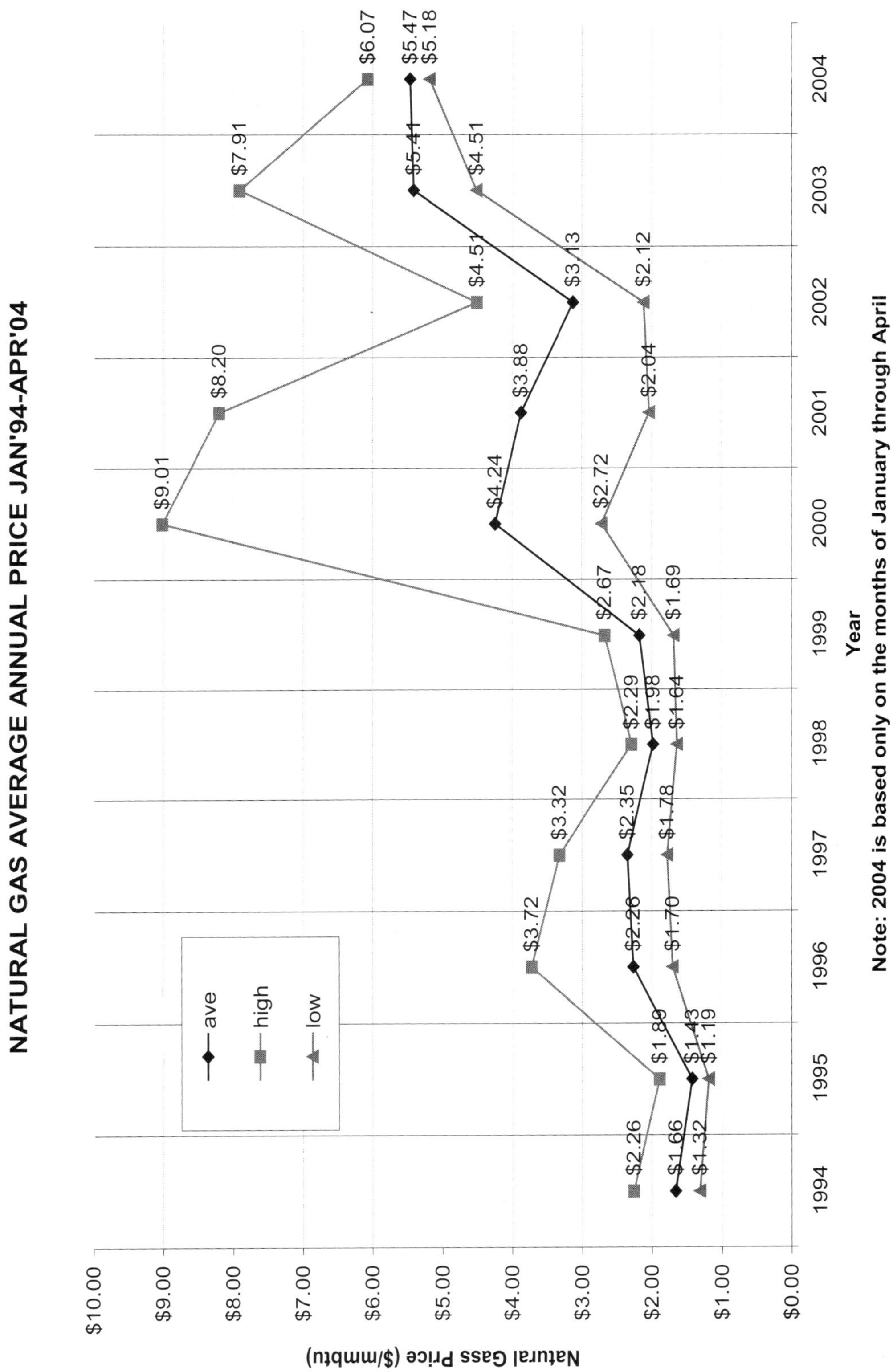

Figure 61. Natural Gas Average Annual Prices.

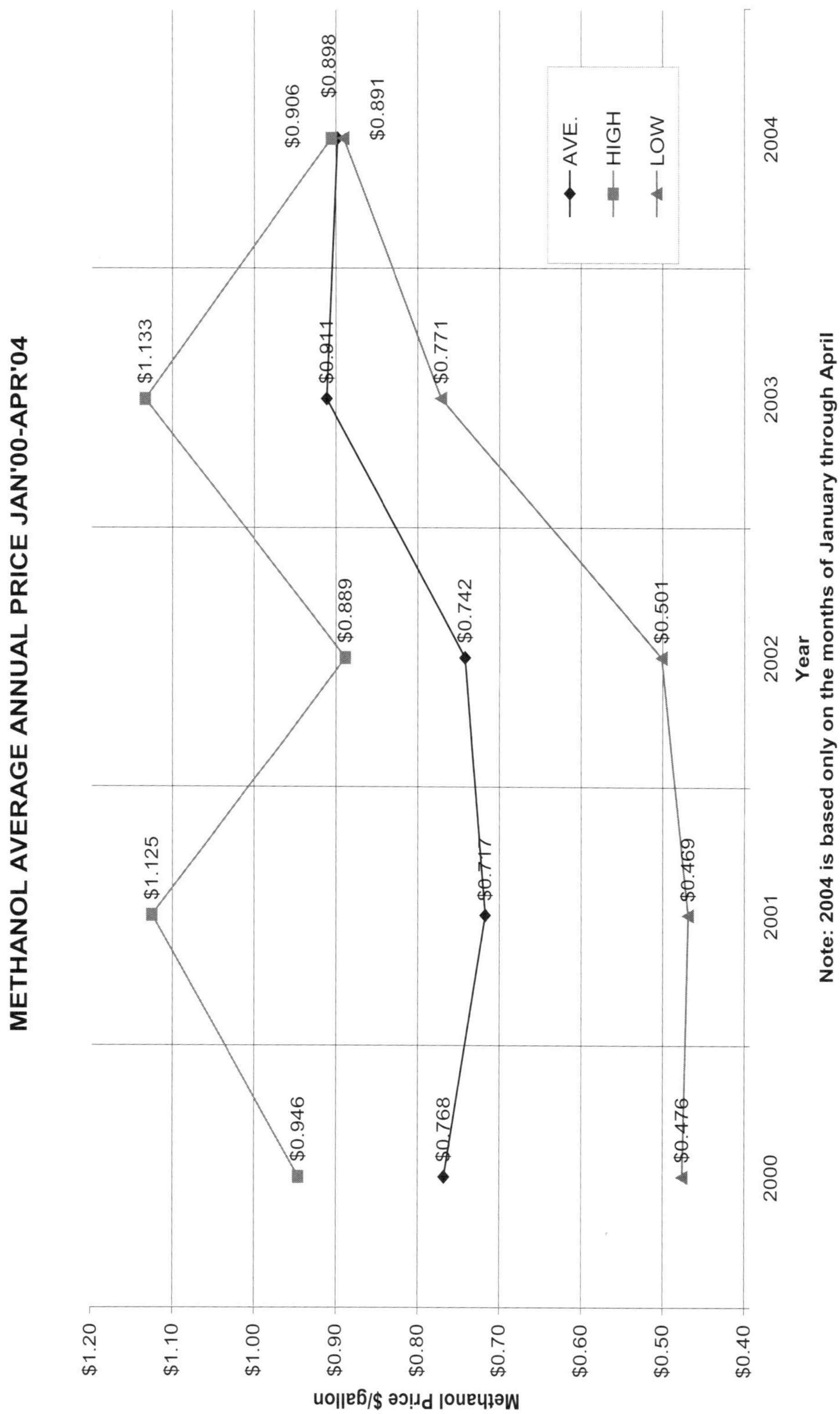

Figure 62. Methanol Average Annual Prices.

Feedstock still has the greatest effect on production cost; therefore finding lower cost feedstocks is an important goal for a biodiesel plant. If a different feedstock is being considered for use, a similar sensitivity analysis should be conducted to understand pricing cycles for that commodity. As an example, consider everyone's surprise at the cost of soybean oil at the start of 2004. It was almost that high about 10 years ago, especially if you adjust the cost of money to today's values.

KEY TERMS	DEFINITION
Reseller	A firm (other than a refiner) that is engaged in a trade or business that buys refined petroleum products and then sells them to a purchaser who is not the ultimate consumer of those refined products.
Retailer	A firm (other than a refiner, reseller or reseller/retailer) that carries on the trade or business of purchasing refined petroleum products and reselling them to ultimate consumers without substantially changing their form.
Price (unit price)	Total revenue derived from the sale of product during the reference month divided by the total volume sold; also known as the weighted average price. Total revenue should exclude all taxes but include transportation costs that were paid as part of the purchase price.
Sale	The transfer of title from the seller to a buyer for a price. Excludes intra-firm transfers, products consumed directly by the reporting firm or sales of bonded fuel. Also excludes products delivered/loaned to exchange partners, except where the amount supplied exceeds the amount received and the differential is invoiced as a sale during the reference month.
Sales to end users	Sales made directly to the consumer of the product. Includes bulk consumers, such as agriculture, industry and utilities, as well as residential and commercial consumers.
Residential	An energy-consuming sector that consists of living quarters for private households. Common uses of energy associated with this sector include space heating, water heating, air conditioning, lighting, refrigeration, cooking and running a variety of other appliances. The residential sector excludes institutional living quarters.
Commercial/institutional	An energy-consuming sector that consists of service-providing facilities and equipment of: businesses; federal, state, and local governments; and other private and public organizations, such as religious, social or fraternal groups. The commercial sector includes institutional living quarters. It also includes sewage treatment facilities. Common uses of energy associated with this sector include space heating, water heating, air conditioning, lighting, refrigeration, cooking and running a wide variety of other equipment. Note: This sector includes generators that produce electricity and/or useful thermal output primarily to support the activities of the above-mentioned commercial establishments.
Industrial	An energy-consuming sector that consists of all facilities and equipment used for producing, processing or assembling goods. The industrial sector encompasses the following types of activity: manufacturing (NAICS codes 31-33); agriculture, forestry, fishing and hunting (NAICS code 11); mining, including oil and gas extraction (NAICS code 21); natural gas distribution (NAICS code 2212); and construction (NAICS code 23). Overall energy use in this sector is largely for process heat and cooling and powering machinery, with lesser amounts used for facility heating, air conditioning and lighting. Fossil fuels are also used as raw material inputs to manufactured products. Note; This sector includes generators that produce electricity and/or useful thermal output primarily to support the above-mentioned industrial activities.
Retail outlet	Any company-owned outlet (e.g., service station) selling gasoline, on-highway low-sulfur diesel fuel or propane for on-highway vehicle use which is under the direct control of the firm filing the EIA-782 by virtue of the ability to set the retail product price and directly collect all or part of the retail margin. This category includes retail outlets: (1) being operated by salaried employees of the company and/or its subsidiaries and affiliates, and/or (2) involving personnel services contracted by the firm.
Other end users	For motor gasoline, all direct sales to end users other than those made through company outlets. For No. 2 distillate, all direct sales to end users other than residential, commercial/institutional, industrial sales and sales through company outlets. Included in the "other end users" category are sales to utilities and agricultural users.

Sales for resale	Sales of refined petroleum products to purchasers who are other-than-ultimate consumers; wholesale sales.
No. 2 distillate	A petroleum distillate that can be used as either a diesel fuel (see No. 2 diesel fuel) or a fuel oil (see No. 2 fuel oil).
No. 2 diesel fuel	A fuel that has distillation temperatures of 500 degrees Fahrenheit at the 10-percent recovery point and 640 degrees Fahrenheit at the 90-percent recovery point and meets the specifications defined in ASTM Specification D 975. It is used in high-speed diesel engines that are generally operated under uniform speed and load conditions, such as those in railroad locomotives, trucks and automobiles.
No. 2 diesel fuel, low sulfur	No. 2 diesel fuel that has a sulfur level no higher than 0.05 percent by weight. It is used primarily in motor vehicle diesel engines for on-highway use.
No. 2 diesel fuel, high sulfur	No. 2 diesel fuel that has a sulfur level above 0.05 percent by weight.
No. 2 fuel oil (heating oil)	A distillate fuel oil that has distillation temperatures of 400 degrees Fahrenheit at the 10-percent recovery point and 640 degrees Fahrenheit at the 90-percent recovery point and meets the specifications defined in ASTM Specification D 396. It is used in atomizing type burners for domestic heating or for moderate capacity commercial/industrial burner units.
Sulfur	A yellowish nonmetallic element, sometimes known as "brimstone." It is present at various levels of concentration in many fossil fuels whose combustion releases sulfur compounds that are considered harmful to the environment. Some of the most commonly used fossil fuels are categorized according to their sulfur content, with lower sulfur fuels usually selling at a higher price.
Petroleum Administration for Defense District (PADD)	**PADD I** (East Coast): **PADD IA** (New England): Connecticut, Maine, Massachusetts, New Hampshire, Rhode Island, Vermont. **PADD IB** (Central Atlantic): Delaware, District of Columbia, Maryland, New Jersey, New York, Pennsylvania. **PADD IC** (Lower Atlantic): Florida, Georgia, North Carolina, South Carolina, Virginia, West Virginia. **PADD II** (Midwest): Illinois, Indiana, Iowa, Kansas, Kentucky, Michigan, Minnesota, Missouri, Nebraska, North Dakota, Ohio, Oklahoma, South Dakota, Tennessee, Wisconsin. **PADD III** (Gulf Coast): Alabama, Arkansas, Louisiana, Mississippi, New Mexico, Texas. **PADD IV** (Rocky Mountain): Colorado, Idaho, Montana, Utah, Wyoming. **PADD V** (West Coast): Alaska (North Slope and Other Mainland), Arizona, California, Hawaii, Nevada, Oregon, Washington.

Sources: Energy Information Administration, Forms EIA-782A, "Refiners'/Gas Plant Operators' Monthly Petroleum Product Sales Report" and EIA-782B, "Resellers'/Retailers' Monthly Petroleum Product Sales Report" (Form/Instructions, Background, Survey Methodology and Statistical Details).

Explanatory Notes: 1. Values shown in the current month for No. 2 distillate at the U.S. and PADD levels in the residential, retail outlets, and sales for resale categories are initial estimates calculated using prior history of the series as well as present and past values of other related time series, such as spot prices and heating degree days. **2.** For all other data, the most current values shown are preliminary while values shown for previous months are revised. Data for all data categories are final upon publication in the Petroleum Marketing Annual. **3.** For purposes of this survey, No. 2 diesel fuel and No. 2 fuel oil are classified in accordance to what the product was sold as, regardless of the actual specifications of that product (i.e., if a No. 2 distillate was sold as a heating oil or fuel oil, the volume and price would be published in the category "No. 2 fuel oil" even if the product conformed to the higher specifications of a diesel fuel. **4.** "Through retail outlets" category includes low-sulfur diesel fuel only. **5.** The 4th quarter of 1993 was a transitional period between the predecessor EIA-782 survey and the revised EIA-782 survey system. The revised survey system contains additional product and sales categories, which may not be consistent with categories derived from the predecessor survey system. Beginning January 1994 all data are from the revised survey system and are consistent. **6.** Map of PAD Districts and federal/state motor fuel tax table from the Petroleum Marketing Monthly.

Table 42. Key Terms and Definitions.

CHAPTER 37 | Financing

Financing is one of the most important facets of starting or expanding a business. It will fund the manpower, equipment purchases and day-to-day expenses. There are a number of financing resources. Some of those outlined in this chapter will relate to Iowa, because of resources available in preparing this manuscript. Although listed as Iowa resources, other states and government agencies have similar resources available.

Financial resources given here can be broken down into the following categories:

- Federal sources
- State sources
- Local/community sources
- Other sources of monies
- Resources for further information

Federal Sources of Financing

United States Department of Agriculture (USDA)

There are many federal sources of financing, but for biodiesel, which falls under the umbrella of value added agriculture and rural development, the United States Department of Agriculture (USDA) is one of the best. Following is a list of the major loan and grant programs that USDA has to offer.

- Business and Industry Guaranteed Loans (B&I Guar.)
- Intermediary Relending Program (IRP)
- Rural Business Enterprise Grants (RBEG)
- Rural Business Opportunity Grants (RBOG)
- Rural Economic Development Loans (REDL)
- Rural Economic Development Grants (REDG)
- Renewable Energy and Energy Efficiency Program (REEEP)

The amounts allocated for the fiscal year 2004 from the USDA are listed below.

Fiscal year 2004 program levels:

- Business and Industry Guaranteed - $839.98 million
- Intermediary Relending Program - $39.76 million
- Rural Business Enterprise Grant - $43.77 million
- Rural Economic Development Loan - $14.91 million
- Rural Economic Development Grant - $4 million
- Rural Business Opportunity Grant - $3.05 million
- Renewable Energy Grant Program - $22.86 million

A brief description of each of these programs follows. Each of the programs can be used for financing biodiesel if you meet the program requirements for eligibility. The USDA is constantly updating its business programs, and you are encouraged to visit its website to obtain the up-to-date information on each of these programs or new ones being introduced.

Business and Industry Guaranteed Loans

The Business and Industry (B&I) Guaranteed Loan Program helps create jobs and stimulates rural economies by providing financial backing for rural businesses. This program provides guarantees up to 80 percent of a loan made by a commercial lender. Loan proceeds may be used for working capital, machinery and equipment, buildings and real estate, and certain types of debt refinancing. The primary purpose is to create and maintain employment and improve the economic climate in rural communities. This is achieved by expanding the lending capability of private lenders in rural areas, helping them make and service quality loans that provide lasting community benefits. This program represents a true private- public partnership.

B&I loan guarantees can be extended to loans made by recognized commercial or other authorized lenders in rural areas (this includes all areas other than cities of more than 50,000 people and the contiguous and urbanized area of such cities or towns). Generally, authorized lenders include federal or State chartered banks, credit unions, insurance companies, savings and loan associations, Farm Credit Banks or other Farm Credit System institutions with direct lending authority, a mortgage company that is part of a bank holding company, and the National Rural Utilities Finance Corporation. Other loan sources include eligible Utilities Program electric and telecommunications borrowers and other lenders approved by Business and Cooperative Programs who have met the designated criteria.

Assistance under the B&I Guaranteed Loan Program is available to virtually any legally organized entity, including a cooperative, corporation, partnership, trust or other profit or non-profit entity, Indian tribe or federally recognized tribal group, municipality, county or other political subdivision of a state. Applicants need not have been denied credit elsewhere to apply for this program.

The maximum aggregate B&I Guaranteed Loan(s) amount that can be offered to any one borrower under this program is $25 million. A maximum of 10 percent of program funding is available to value-added cooperative organizations for loans above $25 million to a maximum aggregate of $40 million.

Intermediary Relending Program

The purpose of the Intermediary Relending Program (IRP) is to finance business facilities and community development projects in rural areas. This is achieved through loans made by the Rural Development, Business and Cooperative Programs (BCP) to intermediaries. Intermediaries relend funds to ultimate recipients for business facilities or community development. Intermediaries establish revolving loan funds so collections from loans made to ultimate recipients in excess of necessary operating expenses and debt payments will be used for more loans to ultimate recipients.

Rural Business Enterprise Grants

The Rural Development, Business and Cooperative Programs (BCP) makes grants under the Rural Business Enterprise Grants (RBEG) Program to public bodies, private nonprofit corporations and federally-recognized Indian tribal groups to finance and facilitate development of small and emerging private business enterprises located in any area other than a city or town that has a population of greater than 50,000 inhabitants and the urbanized area contiguous and adjacent to such a city or town. The public bodies, private nonprofit corporations and federally recognized Indian tribes receive the grant to assist a business. Grant funds do not go directly to the business.

Rural Economic Development Grants

Purpose

- Provides grant funds to electric and telephone utilities financed by the Utilities Programs, an agency of the USDA, to promote sustainable rural economic development and job creation projects through the operation of a revolving loan fund program.
- Reference: Section 313 of the Rural Electrification Act of 1936 and 7 CFR 1703, Subpart B.

Rural Economic Development Loans

Purpose

- Provides zero-interest loans to electric and telephone utilities financed by the Utilities Program, an agency of the USDA, to promote sustainable rural economic development and job creation projects.
- Reference: Section 313 of the Rural Electrification Act of 1936 and 7 CFR 1703, Subpart B.

Rural Business Opportunity Grants

Purpose

The purpose is to promote sustainable economic development in rural communities with exceptional needs. This is accomplished by making grants to pay costs of providing economic planning for rural communities, technical assistance for rural businesses or training for rural entrepreneurs or economic development officials.

Section 9006: Renewable Energy and Energy Efficiency Program

The Farm Security and Rural Investment Act of 2002 (the Farm Bill) established the Renewable Energy Systems and Energy Efficiency Improvements Program under Title IX, Section 9006. This program currently funds grants and loan guarantees to agricultural producers and rural small businesses for assistance with purchasing renewable energy systems and making energy efficiency improvements.

New for 2006, the program offers both grants and guaranteed loans for eligible projects. In addition, projects with total eligible costs under $200,000 can apply under a Simplified Application Process designed to streamline the application process for small projects. (Program is described in more detail later in this chapter.)

Small Business Administration (SBA)

The SBA offers numerous loan programs to assist small businesses. It is important to note, however, that the SBA is primarily a guarantor of loans made by private and other institutions.

To learn more about SBA financing, see **www.sba.gov/financing/**.

Basic 7(a) Loan Guaranty

FUNCTION: Serves as the SBA's primary business loan program to help qualified small businesses obtain financing when they might not be eligible for business loans through normal lending channels. It is also the agency's most flexible business loan program, since financing under this program can be guaranteed for a variety of general business purposes.

Loan proceeds can be used for most sound business purposes including working capital, machinery and equipment, furniture and fixtures, land and building (including purchase, renovation and new construction), leasehold improvements and debt refinancing (under special conditions). Loan maturity is up to 10 years for working capital and generally up to 25 years for fixed assets.

CUSTOMER: Start-up and existing small businesses, commercial lending institutions

DELIVERED THROUGH: Commercial lending institutions

SBA offers multiple variations of the basic 7(a) loan program to accommodate targeted needs.

Certified Development Company (CDC), a 504 Loan Program

FUNCTION: Provides long-term, fixed-rate financing to small businesses to acquire real estate, machinery or equipment for expansion or modernization. Typically a 504 project includes a loan secured from a private-sector lender with a senior lien, a loan secured from a CDC (funded by a 100 percent SBA-guaranteed debenture) with a junior lien covering up to 40 percent of the total cost and a contribution of at least 10 percent equity from the borrower.

CUSTOMER: Small businesses requiring "brick and mortar" financing

DELIVERED THROUGH: Certified development companies (private, nonprofit corporations set up to contribute to the economic development of their communities or regions)

Microloan, a 7(m) Loan Program

FUNCTION: Provides short-term loans of up to $35,000 to small businesses and not-for-profit child-care centers for working capital or the purchase of inventory, supplies, furniture, fixtures, machinery and/or equipment. Proceeds cannot be used to pay existing debts or to purchase real estate. The SBA makes or guarantees a loan to an intermediary, who in turn, makes the microloan to the applicant. These organizations also provide management and technical assistance. The loans are not guaranteed by the SBA. The microloan program is available in selected locations in most states.

CUSTOMER: Small businesses and not-for-profit child-care centers needing small-scale financing and technical assistance for start-up or expansion

DELIVERED THROUGH: Specially designated intermediary lenders (nonprofit organizations with experience in lending and in technical assistance)

Loan Prequalification

FUNCTION: Allows business applicants to have their loan applications for $250,000 or less analyzed and potentially sanctioned by the SBA before they are taken to lenders for consideration. The program focuses on the applicant's character, credit, experience and reliability rather than assets. An SBA-designated intermediary works with the business owner to review and strengthen the loan application. The review is based on key financial ratios, credit and business history, and the loan-request terms. The program is administered by the SBA's Office of Field Operations and SBA district offices.

CUSTOMER: Designated small businesses

DELIVERED THROUGH: Intermediaries operating in specific geographic areas

Export Working Capital

The Export Working Capital Program (EWCP) was designed to provide short-term working capital to exporters.

FUNCTION: Supports export financing to small businesses when that financing is not otherwise available on reasonable terms. The program encourages lenders to offer export working capital loans by guaranteeing repayment of up to $1.5 million or 90 percent of a loan amount, whichever is less. A loan can support a single transaction or multiple sales on a revolving basis.

Designed to provide short-term working capital to exporters, the EWCP is a combined effort of the SBA and the Export-Import Bank. The two agencies have joined their working capital programs to offer a unified approach to the government's support of export financing. The EWCP uses a one-page application form and streamlined documentation with turnaround usually 10 days or less. A letter of prequalification is also available from the SBA. Maximum 7(a) Loan Amounts.

Small Business Innovation Research Grants (SBIR)

Description of the Small Business Innovation Research Program (SBIR):

SBIR is a highly competitive program that encourages small businesses to explore their technological potential and provides the incentive to profit from its commercialization. By including qualified small businesses in the nation's R&D arena, high-tech innovation is stimulated and the United States gains entrepreneurial spirit as it meets its specific research and development needs.

Competitive Opportunity for Small Business

SBIR targets the entrepreneurial sector because that is where most innovation and innovators thrive. However, the risk and expense of conducting serious R&D efforts are often beyond the means of many small businesses. By reserving a specific percentage of federal R&D funds for small business, SBIR protects the small business and enables it to compete on the same level as larger businesses. SBIR funds the critical start-up and development stages and encourages the commercialization of the technology, product or service, which, in turn, stimulates the U.S. economy. Since its enactment in 1982, as part of the Small Business Innovation Development Act, SBIR has helped thousands of small businesses to compete for federal research and development awards. Their contributions have enhanced the nation's defense, protected our environment, advanced health care and improved our ability to manage information and manipulate data.

SBIR Qualifications

Small businesses must meet certain eligibility criteria to participate in the SBIR program.

- American-owned and independently operated
- For-profit
- Principal researcher employed by business
- Company size limited to 500 employees

The SBIR System

Each year, 10 federal departments and agencies are required by SBIR to reserve a portion of their R&D funds for award to small business. They are:

- Department of Agriculture
- Department of Commerce
- Department of Defense
- Department of Education
- Department of Energy
- Department of Health and Human Services
- Department of Transportation
- Environmental Protection Agency
- National Aeronautics and Space Administration
- National Science Foundation

These agencies designate R&D topics and accept proposals.

Three-phase Program

Following submission of proposals, agencies make SBIR awards based on small business qualification, degree of innovation, technical merit and future market potential. Small businesses that receive awards or grants then begin a three-phase program.

- Phase I is the start-up phase. Awards of up to $100,000 for approximately six months support exploration of the technical merit or feasibility of an idea or technology.
- Phase II awards of up to $750,000, for as many as two years, expand Phase I results. During this time, the R&D work is performed and the developer evaluates commercialization potential. Only Phase I award winners are considered for Phase II.
- Phase III is the period during which Phase II innovation moves from the laboratory into the marketplace. No SBIR funds support this phase. The small business must find funding in the private sector or other non-SBIR federal agency funding.

SBA Role

The US Small Business Administration plays an important role as the coordinating agency for the SBIR program. It directs the 11 agencies' implementation of SBIR, reviews their progress and reports annually to Congress on its operation. SBA is also the information link to SBIR. SBA collects solicitation information from all participating agencies and publishes it quarterly in a Pre-Solicitation Announcement (PSA). The PSA is a single source for the topics and anticipated release and closing dates for each agency's solicitations.

For more information on the SBIR Program, please contact:

US Small Business Administration Office of Technology
409 Third Street SW
Washington, DC 20416
(202) 205-6450

All of SBA's programs and services are extended to the public on a nondiscriminatory basis.

DSIRE Federal Incentives Programs

DSIRE tracks federal financial incentives that promote renewable energy and energy efficiency – those designed primarily for residents, businesses and other end-users rather than funding opportunities for research and development, outreach or inter-governmental programs. New federal programs and incentives for renewable energy will be added to DSIRE as they become available.

More information on research, development and demonstration funding opportunities is available from the U.S. Department of Energy's Office of Energy Efficiency and Renewable Energy at **http://www.eere.energy.gov/financing/types_assistance.html**.

For more information on other departments and their programs, visit the websites of the following agencies:

- U.S. Department of Agriculture (USDA)
- U.S. Department of Housing and Urban Development (HUD)
- U.S. Department of Transportation (DOT)
- U.S. Department of Veterans Affairs (VA)
- U.S. Environmental Protection Agency (EPA)
- U.S. Small Business Administration (SBA)

Energy Efficient Commercial Buildings Tax Deduction

Last DSIRE Review: 02/27/2006

Incentive type:	Corporate deduction
Eligible efficiency technologies:	Equipment insulation, water heaters, lighting, lighting controls/sensors, chillers, furnaces, boilers, heat pumps, air conditioners, CHP/cogeneration, caulking/weather-stripping, duct/air sealing, building insulation, windows, doors, siding, roofs
Applicable sectors:	Commercial, construction
Amount:	\$0.30-\$1.80 per square foot, depending on technology and amount of energy reduction
Maximum incentive:	\$1.80 per square foot
Equipment requirements:	Must meet certification requirements
Authority 1:	26 USC § 179D
Date enacted:	8/8/2005
Effective date:	1/1/2006
Expiration date:	12/31/2007

The Energy Policy Act of 2005 established a tax deduction for energy efficient commercial buildings applicable to qualifying systems and buildings placed in service from January 1, 2006 through December 31, 2007. A tax deduction of \$1.80 per square foot is available to owners of new or existing buildings who install (1) interior lighting; (2) building envelope, or (3) heating, cooling, ventilation or hot water systems that reduce the building's total energy and power cost by 50 percent or more in comparison to a building meeting minimum requirements set by ASHRAE Standard 90.1-2001. Energy savings shall be calculated using qualified computer software approved by the IRS.

Note that the eligible technologies listed above are provided as examples and do not represent an official list specified in the statute.

Deductions of \$0.60 per square foot are available to owners of buildings in which individual lighting, building envelope or heating and cooling systems meet target levels that would reasonably contribute to an overall building savings of 50 percent if additional systems were installed. The U.S. Department of Energy has yet to make recommendations to the IRS regarding appropriate target levels for each relevant technology. However, interim rules for lighting systems are in effect (detailed in 26 USC § 45M) until and unless the IRS issues a different final regulation. The interim provision allows prorated deductions from \$0.30 to \$0.60 per square foot for lighting systems that meet certain criteria.

The deductions are available primarily to building owners, although tenants may be eligible if they make construction expenditures. In the case of energy efficient systems installed on or in government property, tax deductions will be given to the person primarily responsible for the systems' design. Deductions are taken in the year when construction is completed.

Modified Accelerated Cost-Recovery System (MACRS)

Last DSIRE Review: 05/05/2005

Incentive type:	Corporate depreciation
Eligible renewable/other technologies:	Solar water heat, solar space heat, solar thermal electric, solar thermal process heat, photovoltaics, wind, geothermal electric
Applicable sectors:	Commercial, industrial
Amount:	Varies
Terms:	5 years
Authority 1:	26 USC § 168
Effective date:	1986

Under the Modified Accelerated Cost-Recovery System (MACRS), businesses can recover investments in solar, wind and geothermal property through depreciation deductions. The MACRS establishes a set of class lives for various types of property, ranging from three to 50 years, over which the property may be depreciated. For solar, wind and geothermal property placed in service after 1986, the current MACRS property class is five years.

For more information, see *IRS Publication 946, IRS Form 4562: Depreciation and Amortization*, and *Instructions for Form 4562*. The IRS website provides a search mechanism for forms and publications. Enter the relevant form, publication name or number, and click "GO" to receive the requested form or publication.

Renewable Energy Production Incentive (REPI)

Last DSIRE Review: 10/25/2005

Incentive type:	Production incentive
Eligible renewable/other technologies:	Solar thermal electric, photovoltaics, landfill gas, wind, biomass, geothermal electric, livestock methane, tidal energy, wave energy, ocean thermal, fuel cells (renewable fuels)
Applicable sectors:	Tribal government, municipal utility, rural electric cooperative, state/local gov't that sell project's electricity
Amount:	1.5 cents per kWh (indexed for inflation)
Terms:	10 years
Website:	http://www.eere.energy.gov/wip/program/repi.html
Authority 1:	42 USCS § 13317
Date enacted:	1992, amended 2005

The Renewable Energy Production Incentive (REPI) provides financial incentive payments for electricity produced and sold by new qualifying renewable energy generation facilities. Qualifying facilities are eligible for annual incentive payments of 1.5 cents per kilowatt-hour (1993 dollars and indexed for inflation) for the first 10-year period of their operation, subject to the availability of annual appropriations in each Federal fiscal year of operation.

REPI was originally authorized under section 1212 of the Energy Policy Act of 1992 and had expired for new projects as of 9/30/03. However, Section 202 of the Energy Policy Act of 2005 (H.R. 6) reauthorized appropriations for fiscal years 2006 through 2026 and expanded the list of eligible technologies and facilities owners. See 42 USCS § 13317 above for the new REPI statute. New regulations established as a result of the new law will be posted when they become available.

Eligible electric production facilities include not-for-profit electrical cooperatives, public utilities, state governments, commonwealths, territories, possessions of the U.S., the District of Columbia, Indian tribal governments or a political subdivision thereof or native corporations that sell the project's electricity to someone else.

Qualifying facilities must use solar, wind, geothermal (with certain restrictions as contained in the rulemaking) or biomass (except for municipal solid waste combustion), landfill gas, livestock methane, and ocean (including tidal, wave, current and thermal) generation technologies. Fuel cells using hydrogen derived from eligible biomass facilities are also considered an eligible technology.

If there are insufficient appropriations to make full payments for electric production from all qualified facilities for a fiscal year, 60 percent of appropriated funds are to be assigned to facilities that use solar, wind, ocean (including tidal, wave, current and thermal), geothermal or closed-loop biomass technologies; and 40 percent of appropriated funds for the fiscal year to other projects.

REPI complements sections 1914 and 1916 of the Energy Policy Act of 1992, which provide tax incentives to certain private sector entities for certain types of new renewable energy generation facilities.

The point of contact for questions concerning REPI policy issues and the availability of appropriations for the REPI program is Dan Beckley. The point of contact on REPI implementation (facility qualifications, applications and payments) is Christine Carter.

USDA Renewable Energy Systems and Energy Efficiency Improvements Program

Last DSIRE Review: 03/30/2006

Incentive type:	Federal grant program
Eligible efficiency technologies:	Yes; specific technologies not identified
Eligible renewable/other technologies:	Solar water heat, solar space heat, photovoltaics, wind, biomass, geothermal electric, geothermal heat pumps, hydrogen, direct-use geothermal, anaerobic digestion, renewable fuels, fuel cells (renewable fuels)
Applicable sectors:	Commercial, agricultural
Amount:	Grants: 25 percent of eligible project costs; Guaranteed loans: 50 percent of eligible project costs (pending)
Max. limit:	Grants: $500,000 per renewable-energy project; Guaranteed loans: $10 million
Website:	http://www.rurdev.usda.gov/rd/farmbill/9006resources.html
Authority 2:	Farm Security And Rural Act of 2002 (Sec. 9006)
Date enacted:	5/13/02
Effective date:	FY 2003
Expiration date:	FY 2007
Authority 3:	Renewable Energy Systems and Energy Efficiency Improvements Program (Final Rule)
Effective date:	7/18/05

Section 9006 of the 2002 Farm Bill requires the U.S. Department of Agriculture (USDA) to create a program to make direct loans, loan guarantees, and grants to agricultural producers and rural small businesses to purchase renewable-energy systems and make energy-efficiency improvements. This program is known as the Renewable Energy Systems and Energy Efficiency Improvements Program.

The maximum grant award is 25 percent of eligible project costs up to $500,000 for renewable energy projects and up to $250,000 for energy efficiency improvements. Assistance to one individual or entity is not to exceed $750,000. The minimum grant request is $2,500 for renewable energy projects and $1,500 for efficiency projects. Eligible renewable energy projects include wind, solar, biomass and geothermal; and hydrogen derived from biomass or water using wind, solar or geothermal energy sources. Applications must be submitted to the appropriate Rural Development State Office.

Under the guaranteed loan option, funds up to 50 percent of eligible project costs (with a maximum project cost of $10 million) are available. The minimum amount of a guaranteed loan made to a borrower is $5,000. A combined grant and guaranteed loan under this program cannot exceed 50 percent of eligible project costs, and the applicant or borrower is responsible for having other funding sources for the remaining funds. The maximum percentage of guarantee ranges from 70 to 85 percent depending on the loan value; the percentage for a given project will be negotiated between the lender and the Rural Business-Cooperative Service. The interest rate will be negotiated between the lender and the applicant and the repayment term must not exceed 30 years for real estate, 20 years for machinery and equipment and seven years for working capital.

The USDA has implemented this program through a Notice of Funds Availability (NOFA) for each of the last three years. The latest round of funding was made available in February 2006. The USDA will determine each year if ***direct loan*** funds are available. If funds are available, a NOFA will appear in the *Federal Register*.

State Sources of Financing

The state sources of financing vary by state and category. One such category to research is renewable fuel. Organizations like DSIRE track such state sources and information can be found on its website **http://www.dsireusa.org**. Financial incentives that apply to renewable fuels are described here.

Corporate Tax Incentives

Corporate tax incentives allow corporations to receive credits or deductions ranging from 10 to 35 percent against the cost of equipment or installation to promote renewable energy equipment. In some cases, the incentive decreases over time. Some states allow the tax credit only if a corporation has invested a certain dollar amount into a given renewable energy project. In most cases, there is no maximum limit imposed on the amount of the deductible or credit.

Grant Programs

States offer a variety of grant programs to encourage the use and development of renewable energy technologies. Most programs offer support for a broad range of renewable energy technologies, while some states focus on promoting one particular type of renewable energy such as wind technology or alternative fuels.

Grants are available primarily to the commercial, industrial, utility, education and government sectors. Some grant programs focus on research and development, while others are designed to help a project achieve commercialization. Programs vary in the amount offered – from \$500 to \$1,000,000 – with some states not setting a limit.

Industry Recruitment Incentives

This category focuses on special efforts and programs designed to attract renewable energy equipment manufacturers to locate within a state or city. Renewable energy industrial recruitment usually consists of financial incentives like tax credits, grants or a commitment to purchase a specific amount of the product for use by a government agency.

The recruitment incentives are designed to attract industries that will benefit the environment and create jobs. In most cases, the financial incentives are temporary measures that will help support the industries in their early years but include a sunset provision to encourage the industries to become self-sufficient within a number of years.

Loan Programs

Loan programs offer financing for the purchase of renewable energy equipment. Low-interest or no-interest loans for energy efficiency are a very common strategy for demand-side management by utilities. State governments also offer loans to assist in the purchase of renewable energy equipment. A broad range of renewable energy technologies are eligible. In many states, loans are available to residential, commercial, industrial, transportation, public and nonprofit sectors. Repayment schedules vary; while most are determined on an individual project basis, some offer a seven- to 10-year loan term.

Personal Income Tax Incentives

Many states offer personal income tax credits or deductions to cover the expense of purchasing and installing renewable energy equipment. Some states offer personal income tax credits up to a certain percentage or predetermined dollar amount for the cost or installation or renewable energy equipment. Allowable credit may be limited to a certain number of years following the purchase or installation or renewable energy equipment. Eligible technologies may include solar and photovoltaic energy systems, geothermal energy, wind energy, biomass, hydroelectric and alternative fuel technologies.

Production Incentives

Production incentives provide project owners with cash payments based on electricity production on a \$/kWh basis, as is the case with the Federal Renewable Energy Production Incentive, or based on the volume of renewable fuels produced on a $/gallon basis, as is the case with a number of state ethanol production incentives. Payments based on performance rather than capital investments can often be a more effective mechanism for ensuring quality projects.

Property Tax Incentives

Property tax incentives typically follow one of three basic structures: exemptions, exclusions and credits. The majority of the property tax provisions for renewable energy follow a simple model that provides that the added value of the renewable device is not included in the valuation of the property for taxation purposes. For example, if a renewable energy heating system costs $1,500 to install versus $1,000 for a conventional heating system, and then the renewable energy system is assessed at $1,000.

Property taxes are collected locally, so some states allow the local authorities the option of providing a property tax incentive for renewable energy devices. Six states have such provisions: Connecticut, Iowa, Maryland, New Hampshire, Vermont and Virginia.

Rebate Programs

Rebate programs are offered at the state, local and utility levels to promote the installation of renewable energy equipment. The majority of the programs are available from state agencies and municipally-owned utilities and support solar water heating and/or photovoltaic systems. Eligible sectors usually include residents and businesses, although some programs are available to industry, institutions and government agencies as well. Rebates typically range from $150 to $4,000. In some cases, rebate programs are combined with low or no-interest loans.

Sales Tax Incentives

Sales tax incentives typically provide an exemption from the state sales tax for the cost of renewable energy equipment.

A sampling of maps for some of the states' financing for renewable fuels tracked by DSIRE follows in **Figures 63 through 70**.

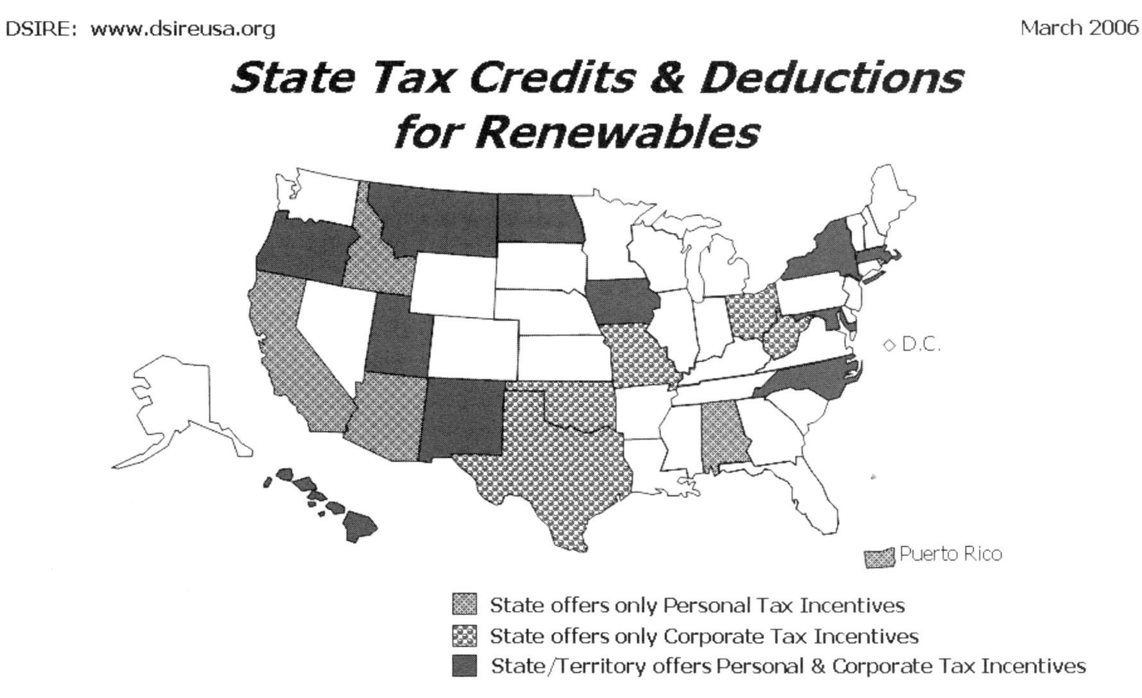

Figure 63.

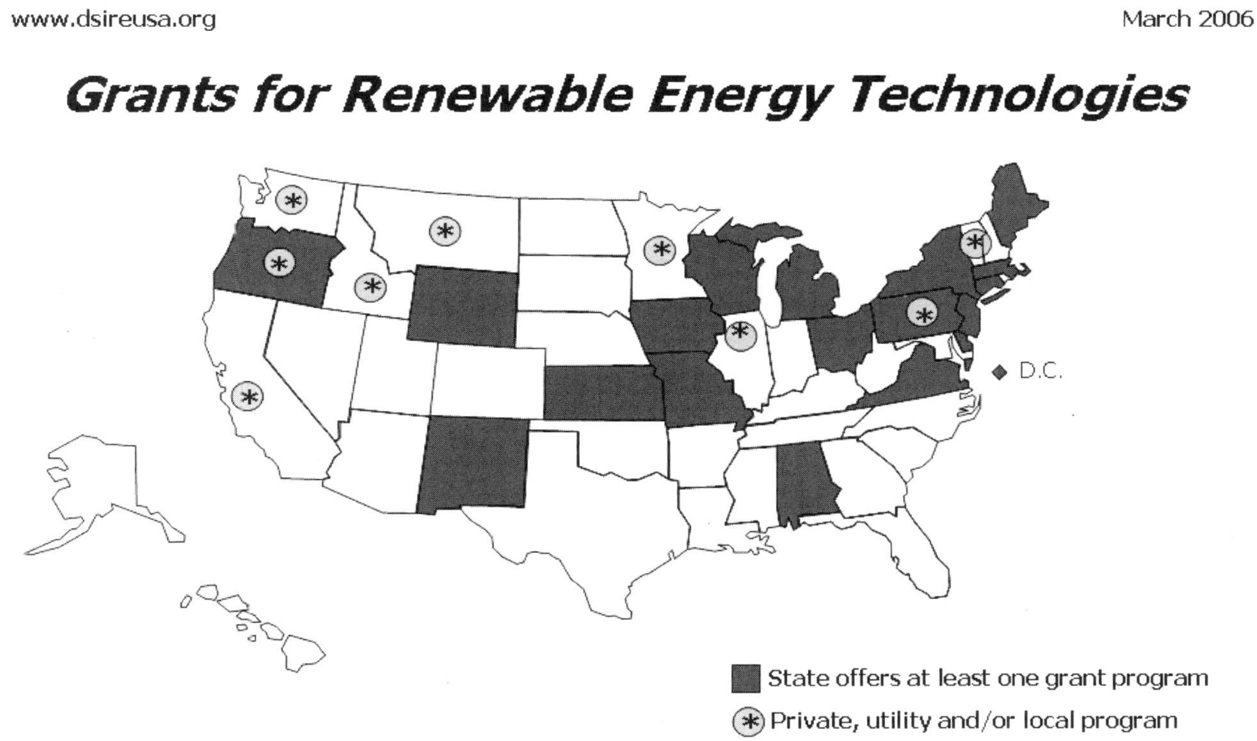

Figure 64.

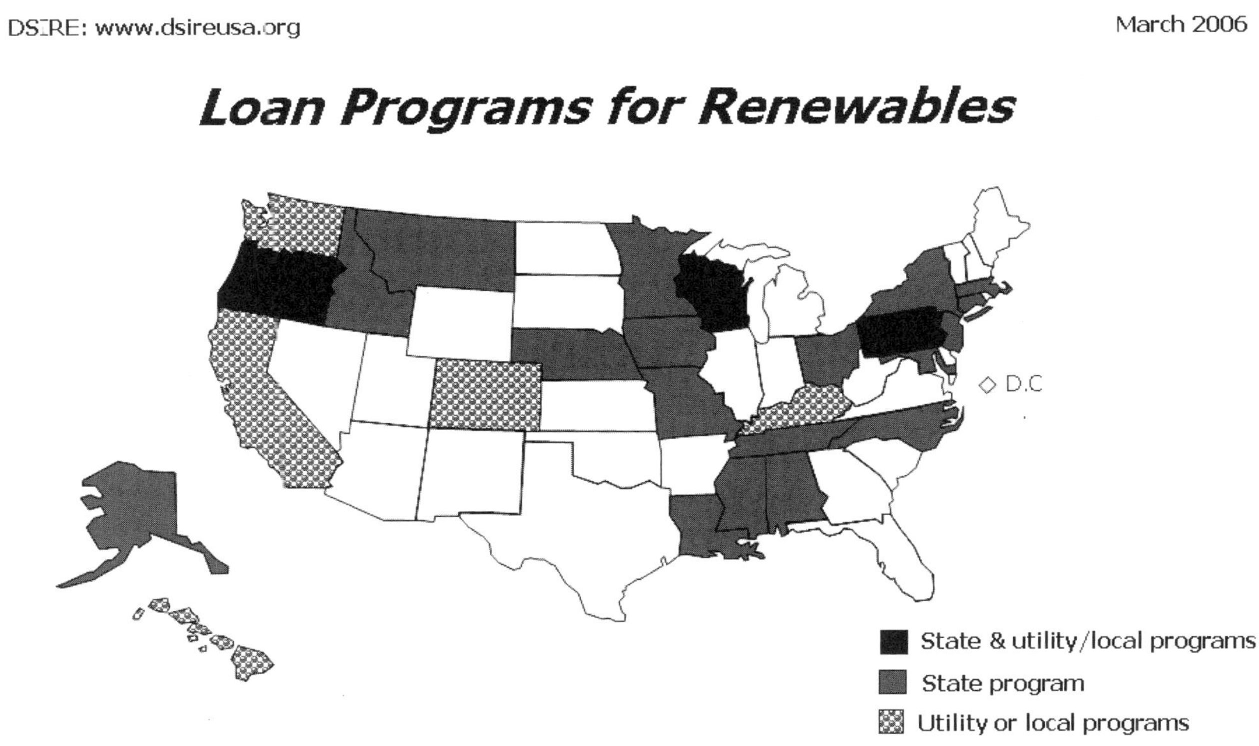

Figure 65.

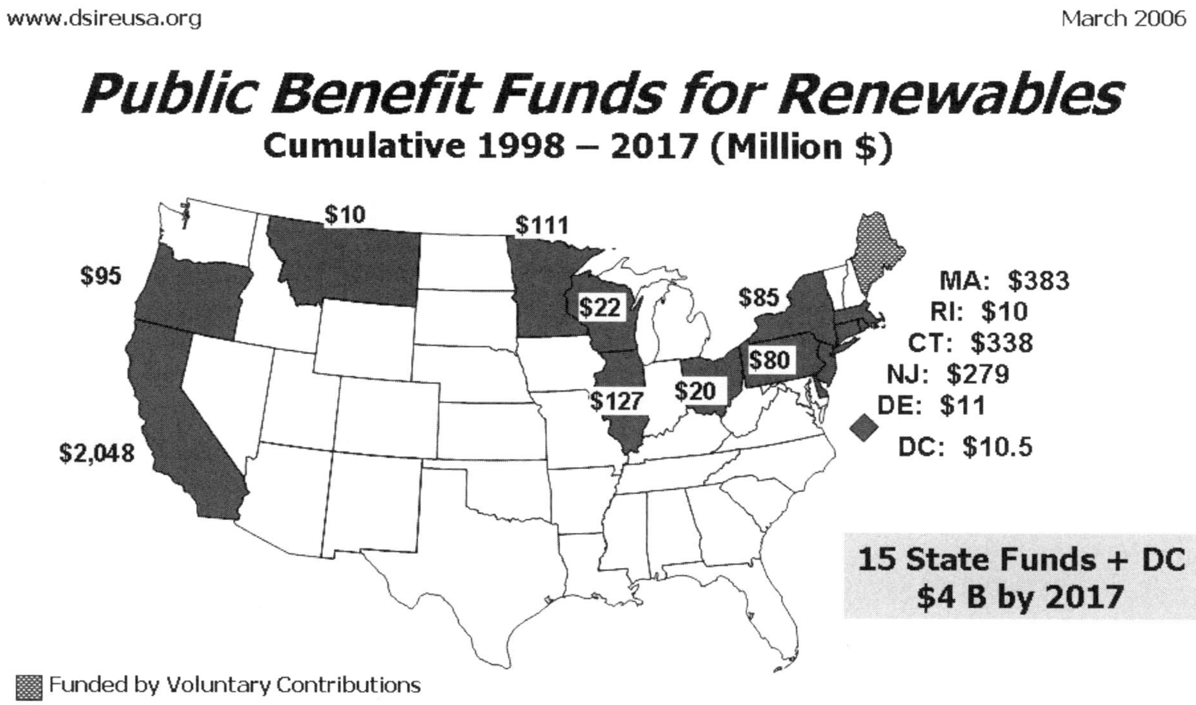

Figure 66.

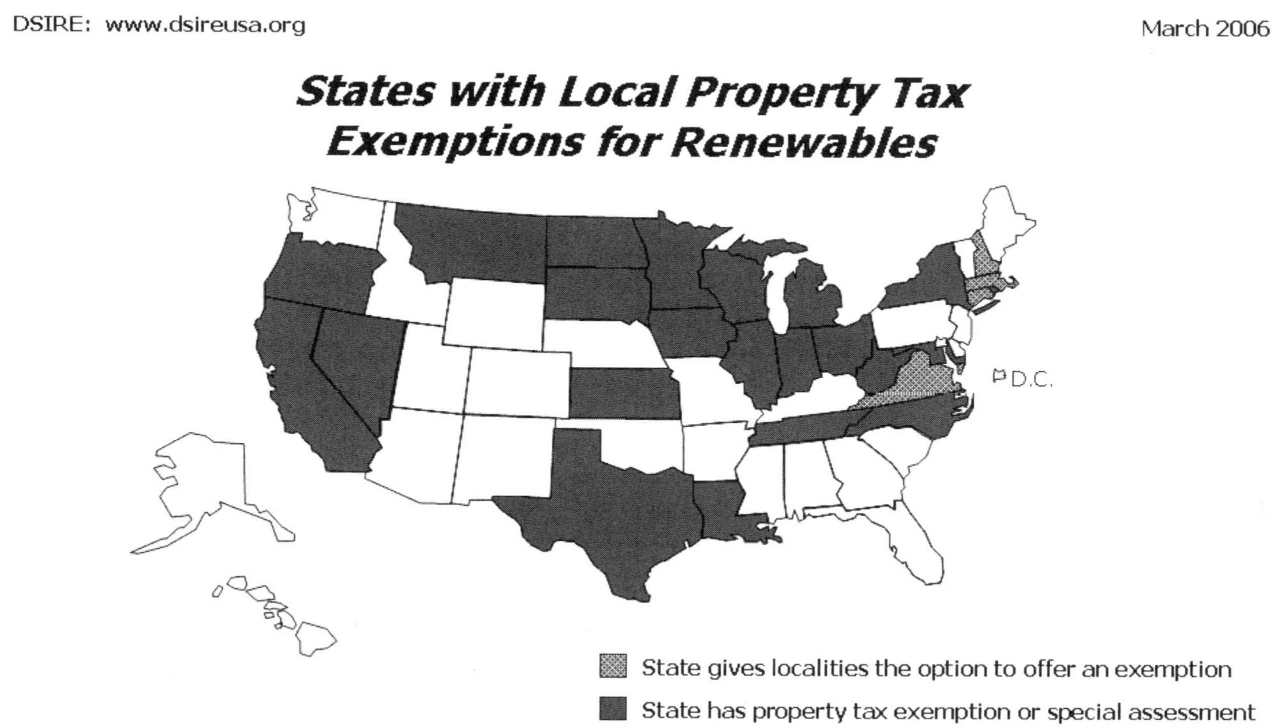

Figure 67.

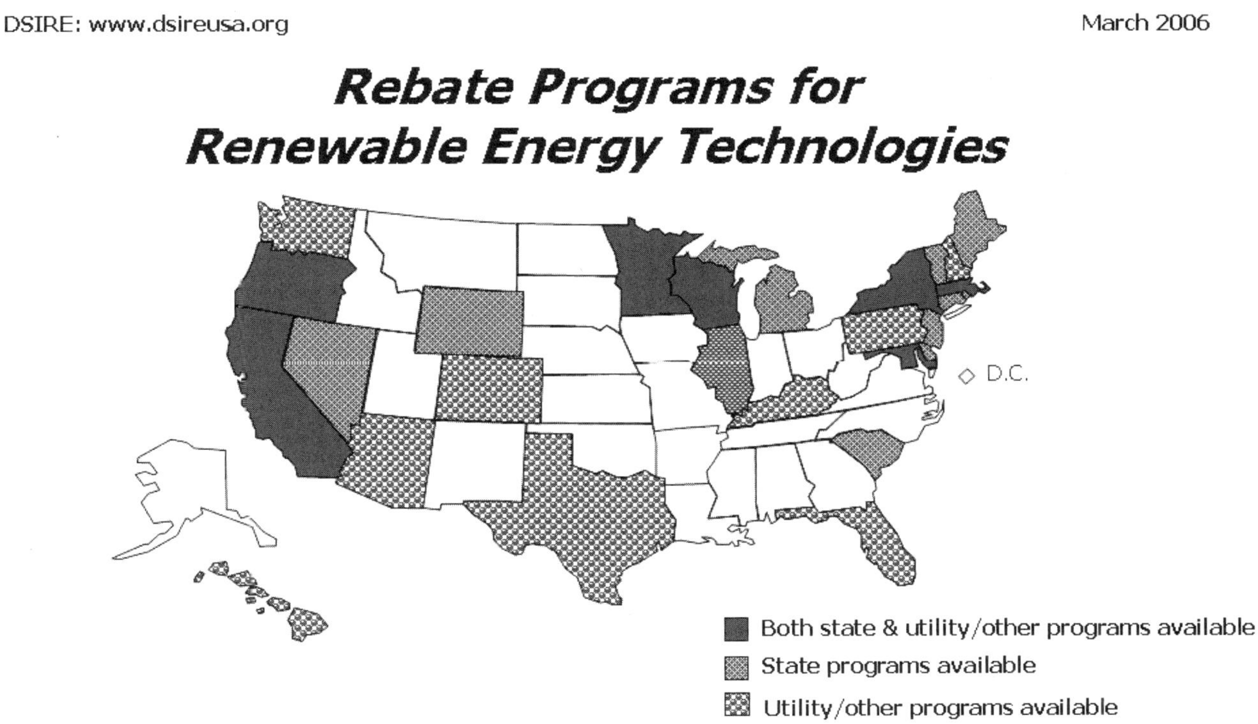

Figure 68.

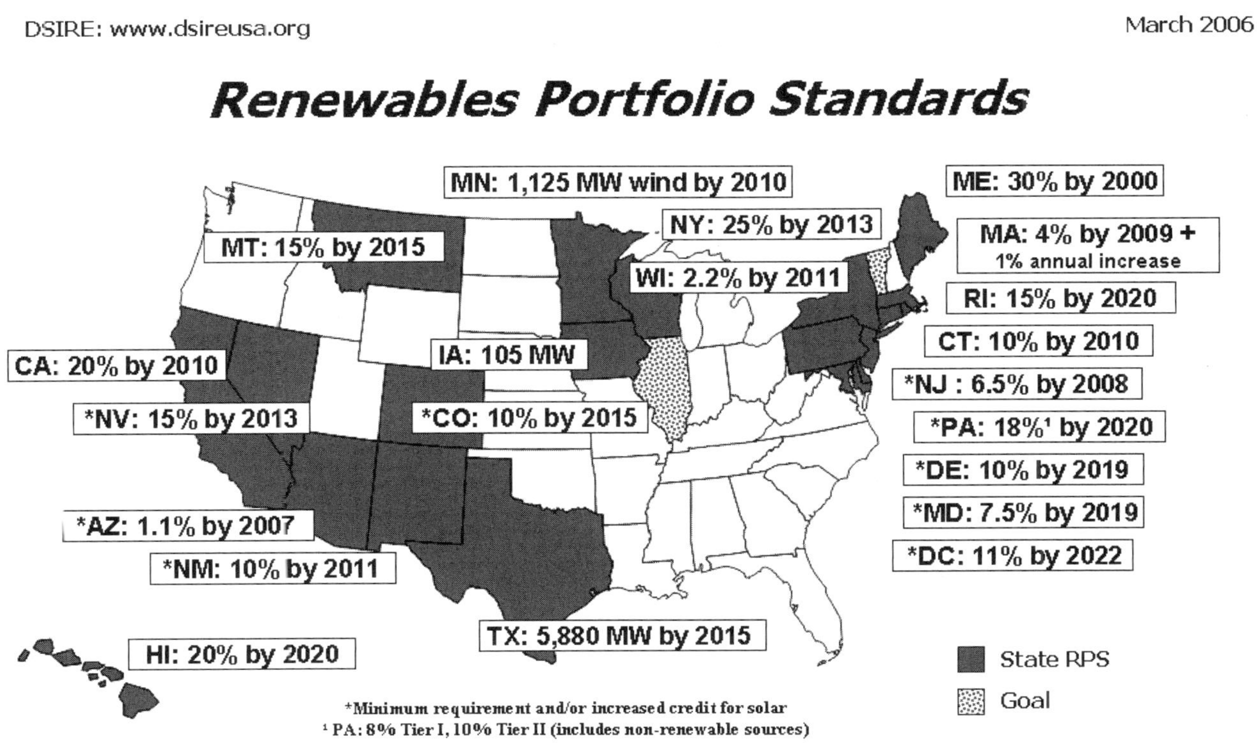

Figure 69.

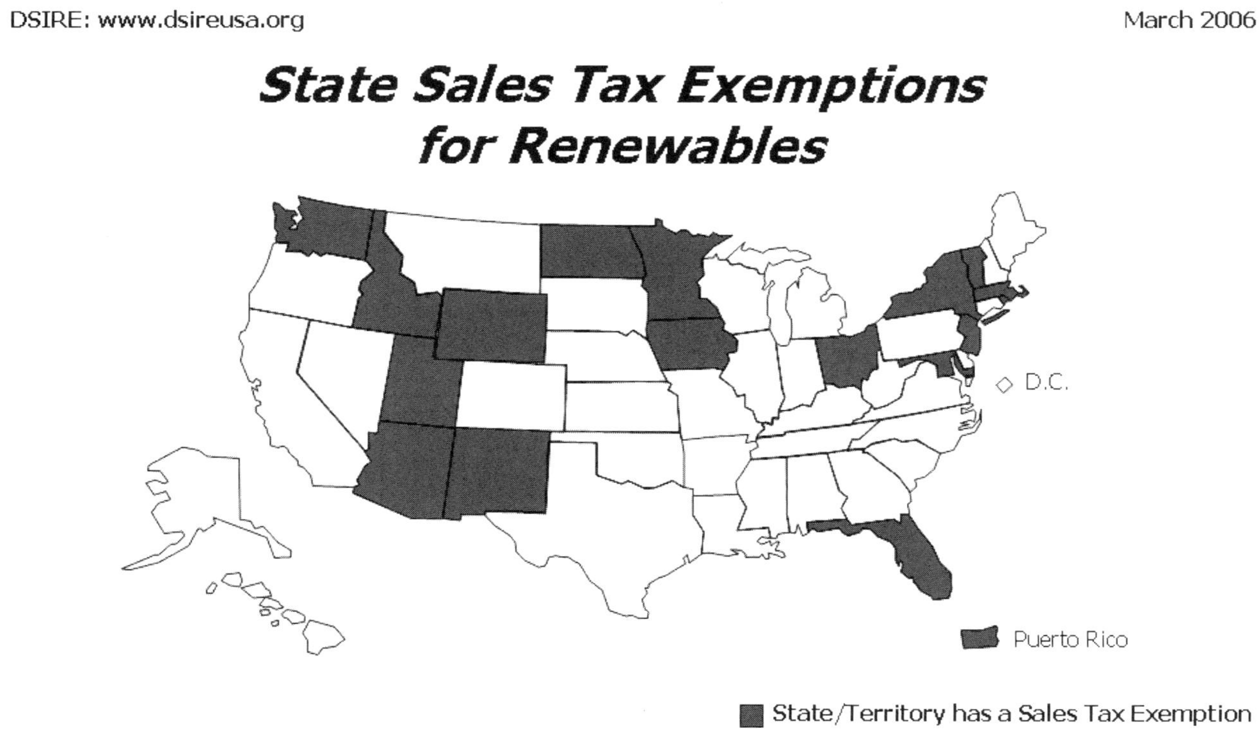

Figure 70.

There are many other possible incentives and programs available on the state level, such as those from the Department of Economic Development Financial Assistance office (examples from the State of Iowa). Each state has different programs and agencies to administer programs. Companies need to research what is available in their state. Following is a partial list of some programs available at the state level. It gives a sampling of possible financing resources.

One possible program is the Venture Project Component of the Community Economic Betterment Account (CEBA). The Venture Project Component of CEBA makes equity-like investments of up to $250,000 to start-up and early-stage businesses.

Specifics on the CEBA program can be obtained from the local economic development agency or organization. The purpose of the CEBA program is to increase employment opportunities for Iowans by increasing the level of economic activity within the state. The program is structured to provide financial assistance to businesses and industries that require assistance in order to create new job opportunities. The Venture Project Component allows for longer-term job creation and investment performance periods than offered through CEBA.

Another Department of Economic Development Financial Assistance sponsored program is the Economic Development Set-Aside (EDSA) program. It provides financial assistance to companies that create new employment opportunities and/or retain existing jobs and make new capital investment in Iowa. The amount of funding is based, in part, on the number of jobs to be created/retained. Funds are provided in the form of loans and forgivable loans. Projects eligible for EDSA funding include, but are not limited to, the following:

- Building construction or reconstruction
- Land or building acquisition
- Equipment purchases
- Operating and maintenance expenses
- Site development – clearance, demolition and building removal

EDSA investments can be used for working capital but should not be considered a sole funding source. The program leverages other financial support, such as bank financing and private investment.

The EDSA program assistance is targeted toward business projects located in communities of fewer than 50,000 in population. At least 51 percent of the created/retained employment opportunities must be made available to individuals presently earning wages defined as low- and moderate-income. The EDSA program can provide assistance up to $500,000.

The Entrepreneurial Ventures Assistance (EVA) program offers technical and financial assistance to eligible start-up and early-stage enterprises. To qualify, applicants must have completed or be participating in entrepreneurial training from a John Pappajohn Entrepreneurial Center or have comparable training or experience.

EVA provides financial and technical assistance to early-stage technology companies. An eligible business must be located in Iowa and in an industry sector offering the greatest start-up and growth potential for the state, including but not limited to:

- Biotechnology
- Recyclable materials
- Software development and computer related products
- Advanced materials
- Advanced manufacturing
- Medical and surgical instruments

Businesses engaged in retail sales, the provision of health care or professional services and distributors of products or services are not eligible for EVA funds. Financial assistance of up to $50,000 may be awarded for the purchase of machinery and equipment and other business acceleration expenses. (The maximum investment available may change; please contact the EVA program manager for more information.)

Applicants must identify a specific project and associated costs. EVA funds may be used to finance up to 50 percent of the total project costs. The preferred investment is in the form of a royalty arrangement; terms of the royalty arrangement are flexible and designed to meet the needs of the company. Other investment tools, such as low-interest loans, may be available.

A single applicant may also be awarded up to $10,000 for technical assistance. Technical assistance funds may be used for such activities as product testing and design improvement or the purchase of technical or professional expertise. Technical assistance funding is considered a grant, and repayment is not required.

In Iowa (and other agricultural states), the Department of Economic Development Financial Assistance has developed a Value-Added Agricultural Products and Processes Financial Assistance Program (VAAPFAP). This program seeks to increase the innovative utilization of Iowa's agricultural commodities.

VAAPFAP may assist projects in two primary categories and a third, secondary category:

- Innovative Products and Processes encourages the processing of agricultural commodities into higher-value products not commonly produced in Iowa, or utilizing a process not commonly used in Iowa to produce new and innovative products from agricultural commodities.
- Renewable Fuels and Co-products encourage the production of renewable fuels, such as soy diesel and ethanol, and co-products for livestock feed.
- A third category, Project Creation Assistance, is also available for projects that may be ineligible under the regular program due to their "pre-commercial" status. Project Creation Assistance is limited to projects within certain designated strategic initiatives.

Program eligibility requirements:

- The proposed project must be located in Iowa.
- The business must have a business plan demonstrating a viable market and managerial and technical experience.
- The business should also have completed a feasibility study documenting the viability of the proposed start-up business.

Projects also are evaluated according to the following criteria:

- The degree to which the facility will increase the utilization of agricultural commodities produced in the state
- Being new and innovative
- Feasibility
- The proportion of local match to be contributed to the project
- The level of need of the region where the existing facility is or the proposed facility is to be located
- The degree to which the facility produces a co-product that is marketed in the same locality as the facility

Any single project may apply for up to $525,000 in assistance. Financial assistance is provided in the form of loans and forgivable loans. Assistance of $20,000 or more is usually awarded as a combination of loans and forgivable loans, with the forgivable portion decreasing as the award size increases.

Another Department of Economic Development Financial Assistance program is the Export Trade Assistance Program (ETAP). ETAP provides financial assistance to eligible businesses wishing to enter new markets by participating in foreign trade shows and trade missions.

This program is available in many other states and also in Europe. It is best to check with your local or international economic development department.

Department of Economic Development Financial Assistance provides the Self-Employment Loan Program (SELP), which offers low-interest loans of up to $10,000 to low-income owners of new or expanding small businesses. This program is designed to assist in the creation and expansion of businesses owned, operated and managed by women, minorities or persons with a disability.

To qualify for a SELP loan, applicants must have an annualized family income that does not exceed current income guide-lines for the program. An applicant is automatically eligible for SELP if he or she is receiving Family Investment Plan (FIP) assistance or other general assistance such as disability benefits. The applicant can also qualify for SELP if determined eligible under the Job Training Partnership Act or is certified as having a disability under standards established by the Iowa Department of Education, Division of Vocational Rehabilitation Services.

SELP loans of up to $10,000 are available. The interest rate is 5 percent, and the loan is to be repaid in monthly installments over a five-year period. The first installment can be deferred for three months.

SELP applicants must obtain a local sponsor for the application process. Examples of a local sponsor include the Institute for Social and Economic Development (ISED), Small Business Development Center (SBDC), the local Job Training Partnership Act (JTPA) representative or a representative from an area Chamber of Commerce or local community college.

Department of Economic Development Financial Assistance has targeted small businesses with the Targeted Small Business Financial Assistance Program (TSBFAP). It is designed to create and expand minority- or women-owned businesses through direct loans of up to $25,000 and loan guarantees of up to $40,000. It also is designed to create and expand minority- or women-owned businesses through direct loans of up to $25,000 and loan guarantees of up to $40,000.

The TSB Financial Assistance Program is designed to assist in the creation and expansion of Iowa small businesses that have annual gross sales of $3 million and are at least 51 percent owned, operated and managed by women, minorities or persons with a disability. The business must be certified as a Targeted Small Business by the Iowa Department of Inspections and Appeals before applying for or receiving TSB funds.

Awards may be obtained in one of the following forms of assistance:

- Low-interest loans. Loans of up to $25,000 may be provided at interest rates of 0-5 percent, to be repaid in monthly installments over a five- to seven-year period. The first installment can be deferred for three months for a start-up business and one month for an existing business.
- Loan guarantees are available up to $40,000. Loan guarantees can cover up to 75 percent of a loan obtained from a bank or other conventional lender. The interest rate is at the discretion of the lender.
- In limited cases, equity grants – to be used to leverage other financing (SBA or conventional) – are available in amounts of up to $25,000.

TSB funds may be used to purchase equipment, acquire inventory, provide operating capital or to leverage additional funding.

The State Treasurer's Office in each state can be a source of financial assistance. In your state financial assistance could be available through the Small Business Loan Program. This program provides financing to new and expanding businesses through the sale of tax-exempt bonds. The program is sponsored by the State Treasurer's Office and the details can be found on the state's website.

Another program available in Iowa through the State Treasurer's Office is Link Investments for Tomorrow (LIFT). LIFT assists with rural small-business transfer, horticulture and alternative agricultural crops.

LIFT Value-added Agriculture was started in 1999 to help stimulate existing businesses or encourage the establishment of new businesses that add value through the processing of agricultural commodities.

Other characteristics of the Iowa program include:

- Value added agriculture includes agricultural commodities raised in Iowa that are processed into a more highly-valued state. This program is available for qualifying production agriculture projects. The Treasurer's office usually works closely with the Department of Economic Development to ensure that the projects approved for funding under this program meet the strategic goals of the state in value added agriculture.
- Any project that has been approved or that is otherwise eligible for participation in the regular VAAPFAP program in either the Innovative Products and Processes or Renewable Fuels category.
- Any value adding agricultural processing enterprise, regardless of whether or not its product or process is innovative, provided that Iowa farm operators have gained or are to gain significant equity ownership in the enterprise. Significant equity ownership should mean not less than a 30 percent ownership share.
- If the level of activity in the LIFT program does not use the funds available in a timely manner, eligibility may be expanded to include farm operator investment in capitol facilities, other than livestock, which are used for the care, feeding or production of livestock products.
- Agricultural commodities, which qualify for this program, are corn, soybeans, oats, hay, hogs, cattle, dairy cattle, sheep, lambs, chickens, turkeys and eggs.
- The borrower must be at least 18 years of age.
- The borrower is not eligible if he/she is currently participating in any LIFT program or has previously participated in any LIFT program other than Traditional Livestock.
- The maximum amount an owner may borrow is $250,000.
- The maximum amount all owners of a value added project or business may borrow is $1,000,000.
- The Quality Beef project has been granted a waiver to $10 million.
- The maximum eligibility for all new borrowers is five years.

State of Financial Assistance could include the Entrepreneurs with Disabilities_(EWD) assistance program. The EWD program helps qualified individuals with disabilities establish, acquire, maintain or expand a small business by providing technical and financial assistance. To be eligible for the program, applicants must be active clients of the Department of Education Division of Vocational Rehabilitation Services or the Department for the Blind.

Technical Assistance Grants of up to $10,000 may be used to pay for any specific business-related consulting service, such as developing a feasibility study or business plan, or accounting and legal services.

Financial Assistance Grants of up to $10,000 may be used to purchase equipment, supplies, rent or other start-up, expansion or acquisition costs identified in an approved business plan. Total financial assistance provided to an individual may not exceed 50 percent (maximum of $10,000) of the financial package. EWD financial assistance must be fully matched by funding from other sources.

Another source of financial assistance on the state level is in the form of state tax relief. In the states, tax relief comes in the form of many different programs. Some of them are discussed here.

Corporate Income Tax in which the state allows 50 percent deductibility of federal taxes from state corporate income tax, and the corporate income tax may be reduced or eliminated by the New Jobs Tax Credit. This and the following information are examples of state tax relief financing that are available in the various states. Similar programs are offered in other states. Check with economic development offices or the state department of taxation.

The New Jobs and Income Program provides a package of tax credits and exemptions to businesses making a capital investment of at least $10.38 million and creating 50 or more jobs meeting wage and benefit targets.

Pollution Control or Recycling Property Tax Exemption_may cover improvements to real property that are primarily used to control pollution of air or water, or primarily used for recycling. An application must be filed for exemption.

A business increasing its workforce by at least 10 percent may qualify for the New Jobs Tax Credit. This credit is equal to 6 percent of the state unemployment insurance taxable wage base. The credit for 2001 is up to $1,074 per new employee and can be carried forward up to 10 years.

Local/Community Sources of Financing

On the city and county level financing can come in the form of Local Tax Abatement. Iowa law allows cities and counties to abate local property taxes for value added to industrial real estate.

This is one of the most widely used incentives for new businesses on a local level. By abating local property taxes, the local community can reduce the tax burden on a new start-up company in the beginning years when money is short. Contact your local city council or economic development group for the process to secure this abatement.

Enterprise Zones are another financing program on the local level. Businesses locating or expanding in an Enterprise Zone may receive the following tax benefits:

- Property tax exemptions on all or part of the costs of improvements to land and buildings for up to 10 years. Certain areas within a state can be designated as an enterprise zone, which allows financing to be available for businesses locating in one of these areas. The economic development website will have the current enterprise zones listed.
- An investment tax credit of up to 10 percent on corporate income taxes for investments in machinery and equipment, land, buildings and improvements to existing buildings.
- Refund of sales, services or use taxes paid to contractors or subcontractors during construction.
- A 13 percent research and development activities credit (refundable) on corporate income taxes.
- Supplemental new jobs training withholding credit of 1½ percent of the gross wages. This credit is in addition to, and not in lieu of, the withholding credit of 1½ percent authorized for the Iowa New Jobs Training Program.

Additional city and county financial sources include Tax Increment Financing (TIF). Tax Increment Financing can offset the cost of public improvements and utilities, finance direct grants or loans to a company and provide the local match for federal or state economic development assistance programs. It often helps with the infrastructure needed in a new development and business start-up, and it is commonly seen with the first couple of businesses that locate in industrial parks. Often, the local match for other development programs and grants is just as valuable as the initial monies.

Other Sources of Financing

Venture capital groups can be sources for financing for new start-up businesses. In Iowa, Venture Network of Iowa (VNI) is a statewide forum using the Iowa Communications Network in which Iowa entrepreneurs, investors and business advisors interact, network and find financial and intellectual capital forge profitable, long-term business relationships. VNI, established in 1990, has been a valuable catalyst in the formation of new, viable businesses throughout Iowa.

VNI extends a unique opportunity – the Five Minute Forum – for entrepreneurs to present ideas to an audience of many of Iowa's most adept and eager investors. VNI offers professionals a resource for reviewing numerous investment opportunities on behalf of clients. In addition, resources are available to any individual or firm interested in funding or managing a business. VNI services the needs of both emerging and existing businesses.

Another venture capital source is the TecTERRA Food Capital Fund, through which equity is available for Iowa food processors and biotechnology companies with growth potential, especially those offering ownership opportunities and other benefits to agricultural producers.

Cybus Capital is a unique, national, private equity fund manager, advisor and investment banker specializing in helping companies in the North American value added food chain to:

- Invest or obtain growth capital.
- Buy or sell companies.
- Otherwise increase or protect stakeholder value through other capital structure events.

Its food and agricultural background includes approximately $2.0 billion in transactions. Companies represented include public and private corporations with revenues typically exceeding $5 million, producer groups, pools of capital and government and academic groups.

Cybus is headquartered in Des Moines, Iowa, with field and affiliate offices in Sacramento, California, and Denver, Colorado.

Access Capital Electronic Network (ACE-NET) is another venture capital source. ACE-NET is an Internet-based listing service providing information to angel investors on small, dynamic, growing businesses seeking equity financing. This will apply to any state and is a good resource to find venture capital.

A seldom thought of venture capital source is the Small Business Investment Companies (SBICs). SBICs were created in 1958 by Congress to fill the gap between the availability of venture capital and the needs of small businesses in start-up and growth situations. SBICs are privately owned investment firms that provide capital plus funds borrowed at favorable rates with an SBA guarantee.

SBICs, licensed by the SBA, are privately owned and managed investment firms. They are participants in a vital partnership between government and the private sector economy and they use their own capital and funds borrowed at favorable rates through the federal government. All SBICs are profit-motivated businesses. A major incentive for SBICs to invest in small businesses is the chance to share in the success of the small business if it grows and prospers.

Small businesses that qualify for assistance from the SBIC program are able to receive equity capital, long-term loans and expert management assistance. Venture capitalists participating in the SBIC program can supplement their own private investment capital with funds borrowed at favorable rates through the federal government.

Non-traditional sources can include local power and energy sources. In Iowa, there is the Iowa Energy Center. It has an Alternate Energy Revolving Loan Program (AERLP) to meet needs of companies focusing on wind, solar, biomass and hydro technologies. Commercial, industrial and residential applications are accepted in amounts set at half of financed cost, up to $250,000. There is 0 percent interest on AERLP's portion of the loan.

The AERLP provides loans to any individual or organization wanting to build renewable energy production facilities in Iowa. Successful applicants receive a single, low-interest loan that consists of a combination of AERLP funds and lender-provided funds. The AERLP provides 50 percent of the total loan up to a maximum of $250,000 at 0 percent interest. The remainder of the loan is made by the lender at a negotiated interest rate. The maximum loan term allowed for the AERLP funds is 20 years. As the loans are paid back to the Iowa Energy Center, those funds revolve back into the program and are made available to new applicants.

Another source available for all states is the Database of State Incentives for Renewable Energy (DSIRE). It contains data categorized by state, federal incentives, national trends in renewable energy incentives and additional resources.

DSIRE is a comprehensive source of information on state, local, utility and selected federal incentives that promote renewable energy. This database has an easy map to locate the incentives in your state. The link is provided in website resources.

A sample list of some of the Federal Incentives for Renewable Energy that DSIRE tracks is listed here:

<u>Corporate Deduction</u>

- Deductions for Clean-Fuel Vehicles and Refueling Property
- Corporate Depreciation
- Job Creation and Worker Assistance Act of 2002 - Special Depreciation
- Solar, Wind and Geothermal Modified Accelerated Cost Recovery System (MACRS)

<u>Corporate Exemption</u>

- Federal Excise Tax Exemption for Gasohol

<u>Corporate Tax Credit</u>

- Alcohol Fuel Credit
- Renewable Electricity Production Credit
- Solar and Geothermal Business Energy Tax Credit

<u>Federal Grant Program</u>

- Renewable Energy Systems and Energy Efficiency Improvements Program
- Tribal Energy Program Grant

<u>Federal Loan Program</u>

- Energy Efficient Mortgage
- Energy Star Financing and Mortgages
- Deductions for Clean-Fuel Vehicles and Refueling Property
- Renewable Energy Production Incentive (REPI)

<u>Personal Deduction</u> (see **http://www.dsireusa.org**)
<u>Production Incentive</u> (see **http://www.dsireusa.org**)

Resources

- Department of Economic Development (DED)
- State Treasurer's Office
- U.S. Small Business Administration (SBA)
- John Pappajohn Entrepreneurial Center
- United States Department of Agricultural (USDA)
- Iowa Energy Center
- Internet

Many of these resources are available in other states, nationwide or are federal organizations. Additional resources can be obtained from the Internet or through local and state economic development agencies. The following websites also provided large amounts of information for this chapter:

- Iowa Department of Economic Development (IDED) **http://www.iowa.gov/state/main/index.html**
- State Treasurer's Office **http://www.treasurer.state.ia.us/index.cfm**
- U.S. Small Business Administration (SBA) **http://www.sba.gov**
- John Pappajohn Entrepreneurial Center **http://www.pappajohn.com/**
- United States Department of Agricultural (USDA) **http://www.rurdev.usda.gov**
- Iowa Energy Center **http://www.energy.iastate.edu**
- Database of State Incentives for Renewable Energy (DSIRE) **http://www.dsireusa.org/**

Acknowledgement

USDA and SBA financing programs were copied from the respective public websites.

CHAPTER 38 | Taxation

Taxation and environmental permitting are two of the basic and most important facets of starting or expanding a biodiesel business. They involve meeting state and federal requirements associated with operating a business and having a payroll.

Many of the taxes and permitting regulations stated in this chapter are standard from state to state, but can differ slightly. Some example requirements are listed; they may apply to other states and businesses. Using the given resources as a base, you may find similar resources available in your own location.

Federal Taxes

There are many federal taxes or requirements of which you must be aware that apply to your type of business structure. They are Employers Identification Number (EIN), withholding tax, Social Security tax, self-employment tax, personal income tax and corporate income tax. They are discussed in further detail in this chapter; complete information can be obtained from the Internal Revenue Service (IRS).

The first step that must be taken is to obtain an EIN. The EIN is issued by the IRS; obtain an application from the IRS, Form SS-4, and you will receive information on the business liability associated with it. Included with the EIN, is information on Social Security withholding, federal income tax and other applicable federal tax information.

Partnerships, corporations and, in some cases, sole proprietorships are required to have an EIN. Sole proprietorships are required to have an EIN if any one of the following apply: they pay wages to one or more employees; file any type of federal excise tax return; sell alcoholic beverages, tobacco, firearms or ammunition; own a truck and transport goods or materials for others on public roads; or pay $6,000 or more annually to an independent contractor.

State Taxes

State taxes include the following: withholding tax, unemployment tax, sales tax and property tax. We will discuss these taxes in more detail starting with payroll taxes.

Payroll taxes include a host of state and federal taxes; for now, we will look at the state taxes. Employers in Iowa and most states are required to withhold estimated income tax payments from employees' wages. The business must register with state and federal tax collecting agencies and collect both state and federal taxes. Federal payroll taxes include Social Security and withholding taxes. Ask to receive information about withholding procedures for state and federal tax collecting agencies when you register.

State income tax withholding must be collected by the business if there is a state income tax. This applies if the business has one or more employees or if the business is a corporation. The business must register with the Department of Revenue as a state withholding agent or with the appropriate agency in your state. The business' EIN will be needed for registration and the Department of Revenue will send you information about state income tax withholding policies.

The unemployment tax covers benefits the state offers to employees that lose their jobs. In most states, employers must contribute to an unemployment insurance fund for the state. Obtain a copy of the *Employer's Handbook on Unemployment Insurance* from the Job Service of the appropriate state.

Specific questions on regulations effecting business operations can be obtained from the state; in Iowa, the Department of Economic Developments' CALL ONE Program 1-800-532-1216.

What Is Next?

Most businesses can start operating once state and federal taxes are satisfied and local regulations are met. However, certain businesses and occupations require permits and/or licenses to be issued. Engineers, architects and land surveyors are just a few of the occupations for which a state may require licensing. Commercial feed product sales are an example of a business that requires special permits/licenses from the Department of Agriculture and Land Stewardship or similar agency in your state. If your biodiesel plant will be buying and selling feedstock, this might be required.

Local Requirements

Visit the city clerk's office and county auditor's offices in your areas of governance to determine what local requirements might apply to your biodiesel plant. You need to research information on local permits, licensing requirements and procedures to follow to obtain the necessary permits and licenses. Many requirements are common; however, regulations may vary from place to place.

Zoning Regulations

Check the local zoning regulations to find out what areas are restricted to commercial activities. This is one of the first steps in selecting a site or plant. Determine if there are guidelines on regulating activities within the area you will be operating.

Make sure you know the local ordinances, road and bridge limitations and sewage requirements for the location. Address issues related to the surrounding community. This may involve attending a zoning or city council meeting to obtain input about your project. Local ordinances may have limitations on advertising in specific areas. School and other community sensitive building may have specific requirements to be addressed. There are a variety of considerations.

Business Permits

Some locales may require a general business permit before operating. In addition, special permits may be required for certain businesses and/or operations. Check with your local government agencies and offices. You will have to abide by the local fire codes and have an inspection before starting operation.

Building Permits

Business or construction permits are needed if your business is erecting or remodeling a building. A building inspector from either the county or city will inspect to make sure local building codes are met before opening. Often, the engineering or construction firm you use will obtain the necessary permits in your business' name.

Business Registration

Before you can register your business, you have to decide on its organizational structure. It can be a partnership, sole proprietorship, corporation or a cooperative. There are many different forms of corporations and coops, each with its own tax and registration requirements.

Partnerships need to file a Partnership Certificate with the Secretary of State in the state in which it will be registered. A small fee may be required.

Sole proprietorships need to file a Certificate of Assumed or Fictitious Name at the State Recorder's Office. Again, a small fee may be required.

Corporations and coops involve much more in terms of documentation and procedure. If you plan to incorporate a new or existing business, you need to file articles of incorporation with the Secretary of State. Legal documents are best prepared by a lawyer in most cases. Usually, there is an initial fee to file plus a fee to register with the county.

Labor Laws
Be aware of both state and federal laws for employees and the workplace. Know about the minimum wage laws and be aware of child labor standards. This information can be obtained from the state's labor offices.

Workers' Compensation
Employers are required by the state to provide medical and disability benefits to employees in the case of a possible job-related injury or disease. Employers can provide insurance through a private insurance company or be self-insured. But to be self-insured, the employer must be qualified by the state.

Patents and Copyrights
The federal government is the only source to obtain patents for inventions. Trademarks and copyrights are also obtained through the federal government.

The following table may help you work through some of the necessary tax issues.

Federal taxes	
Employers Identification Number (EIN)	Obtained from the IRS. Used for federal and state tax purposes.
Withholding tax	Comes from employees' wages and held by the employers. Employee's W-4 Form is used to determine the amount of tax to be withheld.
Social Security tax	Must be paid by all employers with a net earning of $400 or more.
Self-employment tax	Provides Social Security coverage for self-employed individuals in place of Social Security tax.
Personal income tax	Paid by sole proprietors and members of partnerships on business earnings. Also paid by corporate shareholders on the corporate dividends.
Corporate income tax	Paid by corporations except those who qualify as "Subchapter S" corporations according to IRS. Filed estimate quarterly tax report with IRS.
State taxes	
Withholding tax	Comes from employees' wages and held by the employers just as federal income taxes. Payments made quarterly to Iowa Department of Revenue and Finance.
Unemployment tax	Benefits employees who have been laid off. Paid by the employer. Collected by the state and federal government.
Sales tax	Sales tax permits are required for retailers conducting business in Iowa. Iowa Department of Revenue and Finance provides applications.
Property tax	Businesses that own property are taxed for the property.

Table 43. Tax Examples

Biodiesel Tax Issues
You will need to pay federal excise taxes since biodiesel is a fuel; also road or transportation taxes, if applicable.

In general, excise taxes are not owed to the federal, state or local agencies until the fuel enters into a customer's tank. If a biodiesel producer has a retail pump on site, he/she will have to collect and pay all relevant excise taxes. If the producer sells the B100 to a second party that distributes (blended or not) to the end customer, then the second party is responsible for the taxes. The IRS has debated making the biodiesel producers liable for all taxes, but at this time, the firm that distributes the fuel to the end customer is responsible for collecting and remitting the taxes. Since state or local taxes may be different, the producers need to be aware of local tax regulations and should check with local, state and federal taxing authorities.

Environmental Permits

The environmental permitting issues must be addressed at the start of the project. This subject is covered in more detail in another chapter.

Other Issues

The general taxes, both federal and state, and biodiesel taxes are issues that must be addressed in your business. Most of these taxes and permits are basic to many other business ventures. There are other issues, however, that need to be considered and vary from state to state and from country to country. These include taxes and permits related to areas like insurance, equipment, fire, liability, product quality, distribution related permits, and weights and measurements. You need to research these in your specific location to learn which ones apply.

CHAPTER 39

Inventory and Management

In this chapter, we provide you an overview of some of the decisions and determinations you will have to consider in building and operating a biodiesel plant. It is just the opening of the door to issues you will encounter. Issues include:

- Storage requirements
- Transportation needs
- Branding recognition
- Inventory control
- Contract issues
- Management philosophy
- Quality

Many of these are to be addressed in your written, formal, business plan and will be influenced by the type of biodiesel plant you are building or operating and the markets you will be supplying.

Storage Requirements

Storage requirements are determined by the size of your biodiesel operation and customer requirements. Also regulations, such as zoning and fire, can dictate the number and sizes of your tanks and whether they are above or below ground.

The major areas you should address when looking at the storage requirements are:

- Storage of products and raw materials in tanks
- Type – vertical or horizontal
- Location – above ground or below
- Size – amount of capacity
- Design – material of construction
- Regulations – federal, state and local
- Heating requirements

If using feedstocks with variable properties, such as yellow grease, more storage will be required so that different feeds can be mixed to standardize the FFA level entering the plant. Since biodiesel is often sold under contract, sufficient finished product storage capacity will be needed to meet contract delivery requirements.

Transportation Needs

Transportation needs will be somewhat determined by your customers, suppliers of your raw materials and the size of your facility. Some customers can only accept drums, totes or tank trucks; on the other hand, small biodiesel facilities may be able to provide fuel only in small volumes best suited to drum, totes or tank trucks. Other customers have facilities to unload rail cars or access to a pipeline, which are better suited to producing larger volumes of biodiesel. Access to steam to melt railcars or tank trucks of biodiesel may be important for customers receiving shipments in those ways; even in southern climates as railcars or tank trucks can freeze in transit. If possible, you should have the resources to switch from one type of transportation to another as needed.

Some transportation needs depend on the volume of product you will be shipping at one time, such as with rail or pipelines. A secondary consideration is location of a rail spur or pipeline. Cost is always a factor in determining the best mode of transportation and the frequency of shipments. It will be cheaper to move large volumes infrequently compared to small volumes frequently. The availability of customer storage capacity will influence these decisions.

The decisions to own versus lease and handling internally vs. outsourcing are mainly financial ones. Other factors will come into play depending on whether or not you have the traffic to warrant having someone on your payroll or to purchase a tank truck or rail car with all the expenses required to maintain it.

Regardless of the form or frequency of transportation, a program needs to be implemented to ensure that the container used to transport the biodiesel fuel is kept clean and free of vegetable oils, dirt and other contaminants. The ability to clean these containers before use should always be available on site and staff should be trained to inspect and monitor container cleanliness.

Depending on some customer requirements, you may be required to filter the product before it enters the transportation container or when it leaves the container.

Brand Recognition

Brand recognition is important where there are many choices for the consumer of your products. Many times the decision to buy from a company is determined by the recognition of the buyer with all other factors equal. A customer will even pay more for a brand that he recognizes over one he has never heard of, because of perceived quality or familiarity.

Establishing brand recognition can be expensive. It takes a lot of advertising, marketing and often many years to establish your product or company name as a household word. A quality assurance program will be required to protect your brand and respond to consumer problems in a timely and professional manner. Most problems arise from failures in the chain of custody that result in contamination of the fuel. A liability bond that protects the consumer's equipment is just one part of an entire quality system that provides consumer security.

A quicker method in obtaining brand recognition is to buy someone else's brand name. This can be difficult because many companies do not want to sell their brand name. They may still be viable companies relying on the brand recognition. Also, it can be hard to find a brand name that would be appropriate for your product. The cost of purchasing a brand name can be very expensive, especially if it is a nationally known brand name rather than just regional.

Another way to go about the branding issue is to partner with a company with a well know brand name. Perhaps you bring the product and some money, and the other firm brings the brand name. If the right partnership can be formed, this can be a win-win situation for the partners. Biodiesel firms have successfully partnered with petroleum firms in this manner.

The last way to capitalize on brand name recognition is to sell your product to a national company that has a brand name established. That company will then use its brand name to promote your product. In return for the added market exposure, it may ask for a lower price on your product. Another advantage to selling to a national company is that it has the resources to create a new brand name for your product under that company's name.

Inventory Control

Management often asks the important question, "Why does inventory need to be controlled?" A few of the most frequently given answers are:

- Cost savings on equipment
- Limited space or tanks availability
- Cash flow
- Taxes
- Product quality

Inventory has to be stored, so you must have tank capacity for your entire inventory, both finished and work-in-process. In addition, you need piping, pumps, valves, heaters and a host of auxiliary equipment. There are daily costs associated with maintaining excess inventories.

Real estate and buildings cost money; when you have excess inventory you must have more of both these resources available. If your plant has limited space available for the storage of inventory, you must minimize the amount of inventory you have on-hand.

When a biodiesel business is starting out or hits slow times, cash is tight. This means cash flow needs to be managed very carefully. Inventory ties up your cash flow and does not produce profit until it is sold and the money is collected. Therefore, inventory is directly tied to cash flow; when inventory is high, cash flow is low. In addition, stability can be a problem with large inventories of finished product of biodiesel.

Income taxes and some other federal and state taxes are assessed on inventory of finished goods, including the biodiesel and glycerol.

Holding large quantities of biodiesel products and raw materials on-hand also can compromise product quality. The longer you have material in storage, the greater the risk of contamination, for example, by water. Stability can change if the products are held for a long time in the absence of stability additives.

The second most important question in this area that management must answer is, "Is inventory an asset or a liability?" It is a difficult question.

There are two opposing views, one from a financial perspective and one from a manufacturing perspective. In the financial world, inventory is an asset. It is seen as a source of money to repay a loan if it is sold. Accounting practices also list it as an asset.

You, as a manufacturer, should consider it as a liability because it ties up money that can be used in other parts of the operation. Inventory can become obsolete and therefore cannot be sold for normal price, again becoming a liability. In many cases, obsolete inventory cannot be sold at all.

In the case of biodiesel, excess inventory in the form of a mixed product may not be what a customer wants; therefore it can sit for a long time tying up storage tanks and preventing storage of the products in demand. This would be considered a liability, because it would affect production.

Another example would be producing biodiesel to a certain set of specifications and having the regulations change, making this product undesirable and obsolete. Quality can be affected for reasons already given.

In the real-world view, inventory is both an asset and a liability. Storing more than is needed to supply the customer demand or orders on-hand can be considered a liability. Keeping enough on-hand to supply the customer demand or orders can be considered an asset, because it protects the market share and customer base.

In the case of biodiesel, the quicker you can sell your inventory, the faster you can have the profit or cash from the sale in-hand. Therefore, you want to make your product to fill orders as they come in. In many cases, you must hold on-hand a certain amount of inventory to make sure the customer can get product when needed.

Sometimes, the amount you keep on-hand is dictated by the contract with your customer or by the means of transportation you are using to deliver it to the customer. For example: if you are using a tank truck for transportation, you would want to have on-hand at least one tank truck amount in storage when the time to deliver arrives. Not having that amount available in tanks could lead to a late shipment and a dissatisfied customer. This could cause you to lose your contract; therefore, this inventory would be an asset to your operations and future.

It can be an asset to have a small amount of excess inventory to supply the spot local market that can demand higher prices resulting in higher profits. Too much excess inventory, however, would become a liability, because it would tie up cash flow since inventory has costs associated with it.

In summary, too little inventory can be a liability in satisfying customer contracts and future business, and therefore inventory can be considered as an asset by the manufacturer. Too much inventory is a liability from the cash flow being tied up and chance for loss of product or obsolescence. You want to operate somewhere in between, and that can be determined by looking at inventory turns and the flow of product from your tanks to the customer.

You control inventory through various steps. You must establish procedures, a tracking system (MRP) and purchasing limits and controls. You also need to conduct a financial analysis and take customer considerations (lead time and customer understanding of production or process) into account.

Set up standard operating procedures (SOP) for your inventory control. These rules will help others keep the inventory amount to a minimum and make sure you are producing what the customer needs.

Establish some type of Material Resource Planning (MRP) system to track your inventory and raw materials used in your operation. This, coupled with sales orders, can give you an indication of the status of your inventory levels.

Your procedures or SOP can help set limits on raw material you purchase. Work with your supplier of raw materials or feedstock to limit the inventory you have on-hand at any one time. If possible, work out a deal where you receive your raw materials on a Just-In-Time basis in order to minimize the money you have tied up in inventory, but not at the risk of shutting down production. Consignment is another way to minimize your cost of raw material inventory.

Conduct a financial analysis of the cost of holding inventory and compare that to the loss of sales if you can not supply the product. Financial considerations are only one part of the equation in determining the amount of inventory to hold on-hand, but they are important.

Discuss customer needs with them to help define what level of inventory you must have. Explain that inventory costs you money, and ease of availability comes at a price. This will help customers take a look at their real product demands. Market norms can also help determine an estimate of lead times, but you should set your goal for shorter lead times than the industry norms. Help customers understand the biodiesel process and how the reliability of the process and equipment will dictate the safety stock of finished product and raw materials you keep in inventory. Once all risks are evaluated, you can determine a starting point for inventory levels and adjust them as time goes on or conditions change.

Inventory Control

The following items need to be controlled:

- Raw materials
- Intermediate products
- Finished products
- Supplies
- Contracts

Raw materials and supplies are purchased, while intermediate and finished products are a result of production and customer demand. In most cases, contracts dictate the amount of inventory and the determining factors for the inventory levels. Contracts can be broken down into the following categories: supplies, raw materials, finished products and other purchases.

Contracts specify the quantity being purchased or sold. This is directly related to the level of inventory in stock. These contracts should be flexible, if possible, on the delivery of the products as well as the quantity being delivered. In this way, you can limit the amount of excess inventory and spell out procedures to be used by the suppliers and customers. Many management philosophies can be used in a successful biodiesel plant and operation. They should be considered by management and used where determined applicable.

A few of these management philosophies are:

- Just in time
- Kanban
- Replenishment
- Safety stock or buffer

Just in time means material needed for production is available right before it is needed and not stocked before-hand in any appreciable quantities.

Kanban is a visual system to indicate when a raw material or intermediate product used in production needs to be produced or procured. For example: when an item is used up, let's say hydrochloric acid, the employee using the last of the acid takes the card from the container and passes it to the person in charge of supplying the item or ordering the acid. That person then orders the quantity indicated on the card and passes it on to production to be used. This visual card dictates the batch size to be produced or ordered.

Replenishment differs slightly from Kanban. In replenishment, when an item is used in production, it is replaced at a set time of production and only in the quantity that was used in production. This means the time at which the product is produced or ordered is set, and the quantity produced or ordered is variable, unlike Kanban where the quantity produced or ordered is set and time of producing or delivery of the item is variable.

In a biodiesel plant with limited variability in feedstock and products, Kanban and replenishment probably work the same and provide the same benefits. When variety or variability occurs, replenishment is perhaps a better inventory control system.

Safety stock or buffer allows you to keep amounts on-hand to cover the variability in demand of a product or raw material and to protect against Murphy Law (If anything can go wrong, it will!).

Contract Issues

Following are a few issues that should be addressed in a contract. By agreeing on these issues up front, you can minimize the problems that can arise later in the relationship between your company and the customer or suppler.

- Standard clauses for trucking and rail transportation
- Mediation clause for disputes between supplier and customer
- Quality standard and sampling procedure
- Testing methods used

In most cases you would want to have your attorney help with the language in the contract to protect your interests.

Management Philosophies

In today's economic climate it is important to strive to be the best among your competitors. Adopting management philosophies that will take your company to better productivity and quality is very important.

The most popular management philosophies being used today are listed here. Each one can benefit your organization and add profit to the bottom line. One is not necessarily better than another; often parts of each can be applicable in your company.

- Lean manufacturing
- Constraint management (Theory of Constraints)
- Six Sigma
- Malcolm Baldrige criteria
- BQ 9000 certification

Your management team should review what each one has to offer and see what would work best in your organization.

Quality

Quality involves testing requirements and customer acceptance of your product.

Here is a list of quality topics you need to address:

- ASTM test methods
- ISO certification or BQ 9000 certification
- Policies and procedures
- Quality manual

ASTM test methods are used to evaluate the quality of your product. The ones applicable to the biodiesel industry are discussed in other parts of this publication.

ISO certification (ISO 9001) is an international certification for quality accepted around the world. It is becoming more accepted as a requirement for conducting business and can expand your market to those who require this certification. The BQ 9000 is a quality management system requirement specific to the biodiesel industry developed by the National Biodiesel Accreditation Commission on November 12, 2002. It defines quality standards that need to be met to insure production and distribution of a quality product. Both ISO certification and BQ 9000 certification processes and requirements are very similar.

Policies and procedures need to be determined for your plant and process. This will help standardize the evaluation and testing of your product and process and help reduce operator error in sampling and testing. It will also spell out what is acceptable for the customer and what is not.

The quality manual is the place where all your quality procedures and policies are kept. It needs to be available to all employees, and the contents kept up to date and current. All documents are controlled and dated.

ISO 9001

The Center for Industrial Research and Service (CIRAS) at Iowa State University has developed extensive information and a variety of programs on the ISO quality system. For further information, contact CIRAS at **http://www.ciras.iastate.edu.**

In this publication, we talk about ISO 9000 as a quality system you may consider for a soy diesel facility. Many of you may recognize the ISO name and also may have preconceived ideas about ISO. The new standard is focused on processes. The documentation requirements have been minimized.

BQ 9000

More information about BQ 9000, Quality Management System Requirements for the Biodiesel Industry, can be obtained from the National Biodiesel Accreditation Commission, an autonomous committee formed by the National Biodiesel Board. Additional information can be found at the National Biodiesel Board website at **http://www.biodiesel.org**.

This set of standards is intended for use in a cooperative and voluntary program for the accreditation of producers and marketers of biodiesel. Compliance with the requirements is a minimum requirement of the accreditation process of the NBB. This certification process has received a lot of attention since the problems with off-spec biodiesel fuel entering the Minnesota diesel fuel market by mistake happened in late 2005. It is thought that the BQ 9000 certification will be requested by the customer as a part of doing business with them.

CHAPTER 40

Government Incentives

As discussed elsewhere, biodiesel is currently too expensive (at least from seed oils) to compete directly with diesel fuel, except in those cases where environmental or other advantages justify the added cost. Recent (2005) federal legislation and historically high petroleum-prices have created some cost equity. In Europe, biodiesel has been greatly aided by the fact that it is not taxed like petroleum-based fuels. In the United States, a number of state and federal programs have been developed to help make biodiesel more competitive to allow the industry to develop. Other programs are currently in the discussion phase. These programs are justified for many reasons, such as reducing imports of foreign oil, rural economic development and environmental improvement. This chapter will review some of the current and proposed programs.

Commodity Credit Corp.

Starting in 2000, the Federal Government authorized the USDA through the Commodity Credit Corporation (CCC), to make direct payments to producers of ethanol and biodiesel to offset part of their cost to buy commodities to produce these fuels. The current program will expire in 2006. At the time of this writing (April 2006), it is unclear whether anything will replace it. The program is described in the following website: **http://www.fsa.usda.gov/daco/Default.htm.**

The CCC program will pay biodiesel producers for 40 percent of the cost of purchasing soybeans (or other oil seeds such as corn, grain sorghum, sunflower seeds, rapeseed, canola, cottonseed and others), if the beans are processed to produce biodiesel. In the case of soybeans, the producer can still sell the soybean meal and the 40 percent CCC payment is approximately equal to the value of the oil in the beans. Current versions of the CCC program also include greases from rendered animal fats and recycled restaurant frying oils. However, the payment on these products depends on their price relative to soybean oil.

The original CCC program only applied to increases in production over the previous year. To continue to take advantage of the subsidy, a biodiesel producer had to continually grow and increase its production level. The current program allows partial payments on existing production (50 percent for base production in the second year, 30 percent in the third year and 15 percent in the fourth year).

Example 1: Suppose as a biodiesel producer you produce 1,000 gallons of biodiesel. The CCC assumes that one bushel of soybeans can produce 1.4 gallons of biodiesel. Thus, to produce the 1,000 gallons of biodiesel, you must have purchased 1000/1.4 = 714.3 bushels of soybeans. If soybeans were selling for $5.59/bushel, these beans cost you $3,992.90. The CCC will return to you 40 percent of this cost or $1,597. Now if these beans weighed 57 lb/bushel and contained 18 percent oil, they should yield 7,328.7 lbs of oil. At 22.59 cents per pound, this oil was worth $1,656. So, the payment from the CCC was roughly equivalent to the value of the oil. You still have the meal from the 714.3 bushels of soybeans, which would be (assume 5 percent moisture loss) 714.3 x 57 x 0.82 x 0.95 / 2000 = 15.9 tons. At $165/ton, this meal is worth $2,617. So, the value of the meal and the subsidy ($1,597 + $2,617 = $4,214) is more than the original cost of the soybeans.

Example 2: Suppose you produced 1,000 gallons of biodiesel from yellow grease that you purchased for $0.10/lb. The CCC figures that if you had produced this biodiesel from soybean oil it would have taken 714.3 bushels, and at the current price of beans, that would have cost $3,992.90, and you would have expected a check for 40 percent of that or $1,597 (see Example 1). However, since yellow grease was only selling for 10 cents/lb when soybean oil was at 22.59

cents/lb, CCC will decrease the size of the payment by the ratio of the two prices: 0.10/0.2259 = 0.443, which means the amount of the CCC payment would be 0.443 x $1,597 = $707. Now to produce 1000 gallons of biodiesel from yellow grease requires about 7,725 lb of grease (assumes 5 percent loss during processing). At 10 cents/lb, this would cost $772.50. Again, the CCC payment almost entirely pays for the cost of the feedstock.

For both of these cases, it should be remembered that this program was only funded for a total of $150 million per year; this was later dropped to $100 million. Since the program applies to both biodiesel and ethanol producers, there is more demand for the funds than funding available. Producers are required to submit estimates of their production levels; and then based on the total estimated production, the amount that each producer is eligible to receive is pro-rated so that the total payments do not exceed the funding level.

No producer can receive more than 5 percent of the total allocation. At the current subsidy rate, $1.59/gallon in the earlier example, a producer would exhaust his or her share of the $100 million after producing about 3.1 million gallons of biodiesel, a relatively small plant. Producers are cautioned against basing the economics of their plant on the CCC program. The subsidy is reduced every year on existing production; and the program may not be renewed after 2006.

EPAct

The Energy Policy Act (EPAct) is federal legislation to reduce the consumption of petroleum-based fuels for transportation. It is administered by the Department of Energy (DOE) and primarily focuses on replacing petroleum-based fuels with alternative non-petroleum fuels. It originated in 1992 with very aggressive goals of replacing 10 percent of petroleum-based motor fuels by 2000, and 30 percent by 2010. Its actual progress to-date has lagged these goals by a considerable amount. The primary strategy for increasing the use of alternative fueled vehicles has been to require state and federal fleets to purchase these vehicles. Alternative fuel providers have also been required to purchase alternative fuel vehicles, and starting in 2002, some municipal and private fleets are provided with purchasing guidelines. Although these fleets have provided a market for thousands of new alternative fuel vehicles, and all of the major automakers have alternative fuel models available, they only consume about 1-2 percent of the total fuel supply. Even when ethanol (E10) is included the total alternative fuel consumption is only about 3 percent of the supply.

Biodiesel was not originally included in the list of EPAct-defined alternative fuels, but B100 was quickly added. However B20 has a special exemption and is handled on a pro-rated basis. Essentially, five conventional diesel-powered heavy-duty vehicles fueled with B20 can be counted as one alternative fuel vehicle. The regulations require that the vehicles must be fueled with a B20 or higher blend and consumption equal to 450 gallons to be considered to be equal to one vehicle. More information about the EPAct program is available at **http://www1.eere.energy.gov/vehiclesandfuels/epact/**.

Federal Excise Tax Credit

Legislation was proposed in 2001 to reduce the federal fuel excise tax on biodiesel. Diesel fuel is currently subject to a federal excise tax of 25.0 cents/gallon. States generally add a similar amount. Two U.S. Senators introduced S. 1058 that would have reduced this tax by 1.5 cents per percent of biodiesel made from vegetable oil up to 2 percent and 1 cent per percent of biodiesel up to 20 percent. So, a 2 percent blend of biodiesel could receive a tax reduction of $0.03, which would correspond to a cost reduction of $1.50/gallon for the biodiesel. A 20 percent blend would receive a reduction of $0.20 and correspond to a cost reduction of $1.00/gallon of biodiesel. To make up for the reduction in tax income to the Highway Trust Fund caused by this subsidy, an equivalent amount of funds would be transferred from the CCC program to the Highway Trust Fund. This legislation did not pass but was reintroduced in subsequent years.

Starting in January 2005, provisions of the JOBS bill provide a federal excise tax credit of $0.01 per percent of biodiesel in a blended fuel. This is equivalent to an incentive of $1.00/gallon for the fuel blender. The full tax credit is limited to biodiesel produced from first-use vegetable oils and animal fats, so-called *agri-biodiesel*. Biodiesel produced from recycled products such as restaurant waste is only eligible for $0.50/gallon credit. If the feedstock consists of a mixture of both first-use vegetable oils or animal fats mixed with waste oil or second-use oil, then the entire mixtures is treated as second-use oil and is only eligible for $0.50/gallon credit, no matter what the percentages in the feedstock mixture is first-use oil. The $1.00/gallon tax credit must first be applied to the $0.25 excise tax liability for the biodiesel. It can then be applied to excise tax liability for other fuel sales. Finally, if the producer has excess tax liability, he can request a refund from the Internal Revenue Service (IRS) at the end of each quarter. This excise tax credit is authorized through 2008.

In order to qualify for the credit, some diesel fuel must be blended with the biodiesel even if the amount is as small as 1 part in a thousand. This fuel blend is sold as B99.9. Biodiesel producers have the choice of selling fuel as B100 to a blender and passing the opportunity to collect the credit downstream or blending the fuel with 0.1 percent diesel fuel, collecting the credit, and then selling the fuel as B99.9 at a price that is discounted by approximately $1.00/gallon. This tends to be a business decision. Since most biodiesel producers do not have excise tax liability beyond the fuel they produce, if they sell the fuel as B99.9, they must wait until the end of the quarter to receive payment. If they sell the fuel as B100, they might pass the $1.00/gallon credit on to a fuel distributor who can immediately use the credit against his excise tax liability from other product sales. This allows the distributor to receive the benefit of the credit immediately.

Biodiesel producers who produce under 60 million gallons annually are also eligible for a 10-cent per gallon tax credit up to 15 million gallons of agri-biodiesel produced. This is known as the Small Producer Tax Credit and is paid directly to the producer. This provision is authorized to the end of 2008.

B2 in Minnesota

The State of Minnesota currently has a state-wide mandate that requires the use of 2 percent biodiesel in all diesel fuel sold in the state for use in internal combustion engines (with a few exceptions). This mandate took effect in the fall of 2005. The law includes a reimbursement program for distributors who make capital expenditures to adapt or add equipment in order to comply with the mandate, if the mandate is repealed within eight years of becoming effective. This could make it very expensive for the state to consider repeal of the mandate.

A moratorium has been placed on the mandate twice since it was implemented. Both times were due to fuel quality issues. The first instance was associated with biodiesel that was slightly below the flash point specification in ASTM D 6751. This was probably due to biodiesel that still contained some residual methanol. The second instance was motiviated by widespread problems with fuel filter plugging. The cause of the plugging is still unclear, but high levels of monoglycerides appear to have been the primary issue. This is almost certainly due to incompletely reacted fuel entering the system.

The Minnesota mandate is being used as a model for many states (Washington passed a similar mandate in 2006). The moratoriums have reinforced the importance of fuel quality to the industry, and many efforts are underway to ensure that fuel quality is maintained.

Renewable Fuel Standard

The legislation mentioned earlier – the JOBS Bill that established the federal excise tax credit – contains a Renewable Fuels Standard (RFS) that would increase the nation's consumption of renewable fuels to 7.5 billion gallons by 2012.

The first target for this legislation is 4 billion gallons consumed in 2006. This would be primarily corn-based ethanol, although biodiesel is included. Since the Environmental Protection Agency 9EPA) was not expected to have sufficient time to develp a comprehensive plan for implementation of the Standard by 2006, the law mandates that ethanol and biodiesel must comprise 2.78 percent of the U.S. fuel supply in 2006.

The implementation plan from the EPA is expected to introduce new incentives for use of ethanol and biodiesel to raise their share of the transportation fuel market.

CHAPTER 41 Making a Profit

To make a profit today, businesses have to be more efficient, make a quality product and be better than the competition. They have to manage smarter and involve the workforce in the profit equation.

Two sets of guiding principles are included in the pages that follow. One group can be called Generally Accepted Business Principles. It is a partial list. The other is called Six Steps to Success, a collection of areas needing improvement as observed with new business start-ups.

Generally Accepted Business Principles

Do your homework or research before you start a biodiesel business. Find out as much as you can about the industry. Conduct a feasibility study of the biodiesel industry to obtain the necessary information to make an informed decision about entering the business.

Look at the financing. Determine the cost of starting a biodiesel business or adding it onto you current business. Determine your financial resources.

Understand the markets. This will allow you to assess the profit potential of the business and what is needed to penetrate the market.

Meet all regulations and permit requirements. You must make sure you understand what the regulations apply to a biodiesel business and what permits are required.

Write and use a business plan. By writing a business plan you will be forced to put your whole business plan on paper and make decisions about the specific areas that a business plan addresses. You should follow the plan like a road map and update it as your situation changes.

Research

- Read about biodiesel and its processes.
- Talk to the experts in the industry.
- Visit someone already in the biodiesel industry and ask questions.
- Understand why you want to go into this business.

In your research, you should receive a full understanding of biodiesel manufacturing and all the possible process variations. This will entail talking to experts and researchers. The Internet is another source of information, but talking to the experts is more reliable.

Visit people already in the business. You can learn from the problems they had and what they found to be the most difficult part of entering the biodiesel business. Keep in mind that every business is different. Also, you may run into some who will not want to share information because you will be a competitor.

Understand why you want to go into the biodiesel business. Know your motives for picking this business over another. Ask yourself if it will be worth the money and trouble.

Look at the Financing

Determine how much money or capital is needed to start a biodiesel business. This will require compiling all the costs associated with the project. This is the same information you will need for your business plan. Understand the pricing structure for the biodiesel industry and the selling price of the products. This should include the current and proposed legislation on incentive and tax credits.

Spend time researching all the available resources. This will help offset some of the financial cost of starting the business and involve the state and local community in the process. Find a local or regional bank that funds loans to start-up businesses and develop a relationship with it. You will want a bank that has some experience with commercial or value added loans. It can be a valuable resource on funding opportunities and provide you with help in developing a budget and financial statements.

Last, but not least, review the financial status of you, your group or your existing company and determine if the resources are available at this time to enter into a new business. The more solid you are financially, the easier you will be able to borrow money and the higher potential for success you can expect.

Understand the Markets

You need to understand the local, regional, national and global markets for biodiesel; determine the markets for byproducts from manufacturing; determine how you will find your customers; and look at the future markets.

The marketing of your byproducts is becoming more important to making a profit in value added agriculture industries like biodiesel and ethanol. Many times the sale of byproducts makes the difference between losing money, breaking even or making a profit. Find the market for your byproducts and determine what design changes in your plant will produce a byproduct that will satisfy it.

Spend some time determining how you are going to market your products and what distribution channels you will be using. Are you going to sell to wholesalers, the end user or both?

Take out your crystal ball and look into the future at how you expect the market to evolve. Read about the trends and forecast for the biodiesel market and what legislative bills are on the horizon that might affect the purchasing of biodiesel.

Meet All Regulations and Permit Requirements

The regulations and permits associated with your business may differ from state to state and in other countries. Find out what is applicable to your location and required by the local community.

You will need to involve a lawyer sometime in the business planning process. Make sure you address all the legal questions that might arise from your research. The legal and tax aspects of the type of business structure you use has to be addressed very early.

Make sure you meet the environmental regulations that apply. Consider the disposal of byproducts that you cannot sell. Make sure you obtain all the permits necessary for operations and address the regulation in the design of the plant.

Establish a dialog with local community and/or state officials at the beginning of your planning process.

Write and Use a Business Plan

The process of building a business plan has been extensively covered in this publication. It may be the document that persuades a lender or lenders to provide capital for your venture. A business plan also is an essential management tool for your business and may serve as the implementation tool for a strategic plan. The business plan outline in this publication applies to an entrepreneur or a businessperson seeking money for a new business start-up or a business expansion.

Six Steps to Success

- Entrepreneurial spirit
- Productivity
- Quality
- Management
- Innovation
- Employees

Entrepreneurial Spirit

The entrepreneurial spirit is what makes a good company great. The entrepreneurial spirit has to start at the top of the organization and trickle down to the rest of the employees. You want your management team to think like an entrepreneur.

If you expect great ideas from your management and employees, you have to reward them for successful ideas. You should also reward them for the failures as long as they are few and they learn from the mistakes. If people do not make a mistake or wrong decision every now and then, they are not trying new innovative ideas or suggestions.

All employees should think of the company as if it was their own and imagine how they would improve it if they could. They should believe that spending or wasting a dollar of company's money is like taking a dollar out of their pocket. In the long run, it is.

Know your limitations and don't try to be everything. There will be areas where you will have to bring in expertise. Fostering "thinking outside the box" is another way of expressing entrepreneurial spirit. You should never hear employees say, "We have always done it this way." They should know why the task should be done and what benefit it is to the company.

Productivity

Productivity and technology (a form of productivity improvement) are ways to improve bottom-line profits. Throughput is a way of measuring profitability and productivity. Throughput is the rate at which the system generates money through sales. Another way of stating it is: throughput equals the revenue from selling your product minus raw materials. In the case of biodiesel, throughput is the dollars received from the sale of biodiesel minus the cost of the raw materials. This is the money you have to pay your operating expenses including labor; what is left over is profit. This management philosophy is called constraints management or Theory of Constraints.

Minimizing wasted or non-value added work is a measurement of time used to make your product. You should be continually asking the question, "Does this add value to the finished product?" Most companies find out that there is only 10 percent value added work used to produce a product and the other 90 percent is waste. This management philosophy is call Lean Manufacturing.

If you don't see improvement in the profit of the company when a suggestion is made or a change takes place, then question why you need to do it. All improvements should lead to improved bottom-line results. The goal of all companies is to make money, now and in the future.

Reduction of inventory increases cash flow and throughput and maintains maximum flexibility in production.

Quality

You can insure quality in your product and manufacturing process by writing manufacturing procedures with quality in mind. Also use visual means to help the operator know if a product is within specifications or not. For example, have a vial of what the in-process biodiesel product should look like every step along the way. This way the operator can compare an in-process sample with the standard and have a quick initial indication that the product might be in spec.

Implement a quality system. Design quality into your biodiesel plant by making it difficult for the operator to make a mistake. As examples: One easy way to help reduce contamination is by using back flow prevention devices; and another way of reducing the possibility of adding the wrong material to the manufacturing process is to color code the pipes for the raw materials.

Management

Be sure the people who fill the management roles in your company have the right skills and knowledge and share your management philosophies.

Once you have managers in positions, continually evaluate them to see that they are meeting your need for that position. Make sure that the evaluations are conducted by their superiors as well as their peers and subordinates.

Keep management small and delegate responsibility downward. In many manufacturing operations today, there is only one level of management and the employees make most of their own production decisions.

Utilize the individual strengths of your management team and don't be afraid of creating job descriptions that maximize the knowledge and experiences of the managers. Sometimes the traditional job titles will not fit the job responsibilities your management team members have.

Share the profits of the company with your management and employees. A bonus system or profit sharing tie performance to personal profit or pay.

Innovation

Often we have heard that we shouldn't re-invent the wheel. Sometimes, it would be a good thing to do. Too often we do something because it has been done a certain way for as long as we can remember. In the competitive climate of today, it is the company that re-invents a better wheel or finds a better way to produce a product that ends up making the profit.

Don't be satisfied with current technology; always be seeking out new technology or research. Talk to the experts and find out where the technology is heading or what is on the horizon.

Devote some of your resources to developing new technology and conduct your own research. This can challenge the employees' innovative thinking and increase their understanding of the process.

Protect your technology if it is new and innovative. Patents are one means of protection and trade secrets are another.

Employees

- Treat them fairly and consistently.
- Involve them in the business decision process.
- Communicate the business's well-being and performance, including the finances.
- Consider them a resource not a cost.
- Empower them to make decisions.
- Create a win-win situation and share the profits.

Most of these suggestions are self-explanatory, although few companies take advantage of them. The more you involve your employees in the business and delegate responsibility to them, the more likely they will feel a part of the company. The more they feel involved in the decision making process of the company, the more they will be willing to help it succeed, especially if they can share in the profits or have a steady and rewarding job.

Communication is the key to any successful relationship with employees. They are the most valuable resource a company has, and the return on investment in training and developing them will pay off in the future.

SECTION FIVE

Regulations and Environment

CHAPTER 42

Environmental Requirements and Permitting

This chapter reviews the major federal laws that are the basis for setting environmental protection requirements across the U.S. These requirements can be implemented in the form of imposing limits to emissions, restricting or banning use of harmful materials or requiring permits for discharge of pollutants. Here we will:

- Review the major legislation defining environmental regulations in the U.S.
- Review the general requirements under the environmental titles.
- Describe the impact of the legislation on biodiesel production and use.
- Describe specific code requirements to be addressed in permitting a biodiesel facility.

Environmental laws are developed by the U.S. Congress and signed by the President. The responsibility for developing, interpreting and enforcing these laws is delegated to the U.S. Environmental Protection Agency (EPA). The EPA inspects and applies penalties for compliance violations and is the lead environmental research and development agency for the federal government.

Each state and locality also has the authority to enact and enforce its own environmental regulations. State and local requirements may be more restrictive than federal requirements, but they cannot be less restrictive. In the cases where permits are required, review and enforcement are usually at the state or local level. Also, many of the permitting and enforcement responsibilities at the federal level have been delegated to the state level.

Environmental law is a complex and rapidly changing area, and detailed requirements and interpretations can vary by locale. This chapter is an introduction to major acts and acronyms. It is recommended that you work with specialists as you develop your plans and responses to the requirements that must be met by your facility.

The review of general code and permitting requirements is based upon international code requirements. Local code requirements can be much more restrictive. It is strongly urged that the facility management work with a local consultant in gaining the required permits and in monitoring compliance.

Resource Conservation and Recovery Act (RCRA)

The RCRA addresses solid and hazardous waste management activities including waste management and underground storage tanks. The framework is a "cradle-to-grave" system governing hazardous waste from generation to disposal.

Facilities that generate hazardous waste are subject to accumulation (amount present), manifesting (tracking) and record keeping standards. The facilities must have a permit either from the EPA or the delegated state agency. Various provisions of the RCRA are delegated to state agencies in 46 states.

Important regulatory requirements under the RCRA include:

- Identification of solid and hazardous wastes.
- Standards for generators of hazardous wastes.
- Land disposal restrictions.
- Used motor oil storage and disposal regulations.
- Emission standards for tanks and containers.
- Underground storage tank design and release detection standards.
- Design and operating standards for boilers and industrial furnaces that burn fuel containing hazardous wastes.

The EPA publishes a detailed listing of materials and specific chemicals that fall under the following categories: "P," "U," "K," "F" and "D" shown in **Table 44**. In the case of biodiesel production, methanol, base and/or acid catalysts are listed materials.

P or U	Commercial products
K	Hazardous wastes from specific industries
F	Hazardous wastes from non-specific industries
D	Materials that exhibit hazardous characteristics such as ignitibility, corrosivity, reactivity or toxicity

Table 44. EPA Categories.

The key issue in working under the RCRA, however, is the designation of "waste." Raw materials in storage are not waste and materials recovered as a part of the process, such as methanol, are not waste unless so designated. Recovered co-products, such as glycerol, are not waste if they are to be used elsewhere in processing or sold to another entity. A material must be designated as "waste," to fall under the RCRA. However, the operation cannot stockpile materials to avoid the RCRA. There must be a "reasonable expectation" of use within the facility within a reasonable period of time (generally six months), or the facility could be cited as an unlicensed waste disposal or holding operation.

There is an incentive to keep generation of hazardous waste materials to a minimum both for plant safety and for RCRA reporting. The extent of required reporting and the requirements for detail in managing these materials grows in proportion to the amount generated.

Waste generators are categorized on the basis of the amount of hazardous waste generated. The amounts are further broken down as hazardous and acutely hazardous. Each category of generator has specified limits regarding the amount of hazardous material allowed on the property. The categories are:

- Conditionally Exempt – No more than 220 lbs. of hazardous waste (HW) and no more than 2.2 lbs of acutely hazardous waste (AHW) per month. There is a maximum of 2,200 lbs HW on the property.
- Small Quantity Generator (SQG) – Produce between 220 and 2,200 lbs. HW and no more than 2.2 lbs. of AHW per month.
- Large Quantity Generator (LQG) – Produce more than 2,200 lbs. of HW or more than 2.2 lbs. of AHW per month.

There are a number of operating and reporting requirements for generators. The requirements vary somewhat in proportion to the quantity of waste generated. These requirements include:

- Inventory – Identify all HW on-site, pounds generated per month and maximum accumulated.
- State and EPA ID number (Conditionally Exempt not required to have EPA ID).
- Accumulation limits – Specific rules for each level generator.
- Labeling – RCRA HW labels for LQG and SQG; DOT labels for Conditionally Exempt.
- Inspection – Storage areas weekly, tanks daily; spill and emergency prevention/detection equipment for LQG and SQG.
- Transport – Detailed instructions regarding packaging, manifests, exception reports, holding times and approved treatment/disposal facility (TSD).

The following requirements apply to LQG, with lesser requirements for smaller generators.

- Waste minimization – Certify program in place; annual reports (LQG).
- Training – Employees trained in handling and procedures.
- Emergency response – Plans and incident reporting.
- Reporting – Exception reports, quarterly reports, periodic survey.
- Recordkeeping – Manifests for three years, exception reports for three years, analyses for three years, inspection logs, biennial reports and land disposal restrictions.

Hazardous materials training is required under OSHA provisions, and MSDS materials are required in the workplace. There are a number of professionally-developed training materials available to meet these requirements for training and recertification. This is not only good practice, it is the law. The characteristics that define a hazardous material are:

- Ignitable – Flash point < 140°F.
- Corrosive – PH > 12.5 or < 2.
- Reactive – Rapid or violent reaction when exposed to heat, pressure, water or other materials.
- Toxic – Determined by acute and chronic response levels; carcinogens, teratogens, blood or kidney pathogens.

Comprehensive Environmental Response, Compensation, and Liability Act (CERCLA)

The CERCLA is also known as the Superfund Act. The EPA currently has about 1,300 active Superfund clean-up sites. The Superfund Act authorizes the EPA to respond to releases of hazardous substances and requires the reporting of releases to the National Response Center. Reportable quantities are listed in 40 CFR §302.4.

The National Oil and Hazardous Substances Pollution Contingency Plan (40 CFR Part 300) provides for permanent clean-ups of pollutants.

Emergency Planning and Community Right to Know Act (EPCRA)

The EPCRA is designed to improve community access to information about chemical hazards present in the community. It also is intended to facilitate the development of chemical emergency response plans by state and local governments. To implement the EPCRA, Congress required each state to appoint a State Emergency Response Commission (SERC). The SERCs were required to divide their states into Emergency Planning Districts and name a Local Emergency Planning Committee (LEPC) for each district.

Companies are required to notify their SERC and LEPC of the presence of "extremely hazardous substances" if they are in excess of the planning threshold quantity. Companies are also required to notify their SERC and LEPC in the event of a release exceeding the reportable quantity of a CERCLA hazardous substance or an EPCRA extremely hazardous sub-stance.

A facility having OSHA hazardous chemicals in excess of threshold amounts must submit to their SECR, LEPC and fire department copies of their MSDS and chemical inventory forms. Facilities in SIC Codes 20 – 39 with > 10 employees that manufacture or use more than the threshold amounts of TRI chemicals must submit an annual toxic chemical release report (Form R).

The Toxic Release Inventory (TRI) is a list of chemicals subject to reporting requirements under Title III of the 1986 Superfund Amendments and Reauthorization Act (SARA). The current TRI list contains 582 individually listed chemicals and 30 chemical categories. The list is published in EPA-260-B-01-001, which may be downloaded from the internet.

Clean Water Act (CWA)

The Water Pollution Control Act Amendments of 1972, as amended in 1977, are commonly known as the Clean Water Act. It established the basic structure for regulating discharges of pollutants into the waters of the U.S. It gave the EPA the authority to implement pollution control programs such as setting wastewater standards for industry. The CWA also set water quality standards for all contaminants in surface waters and made it unlawful to discharge any pollutants from a point source into navigable waters unless a permit was obtained.

The CWA is designed to restore and maintain the chemical and physical integrity of the nation's surface waters. To accomplish this, the Act provides for regulation of several categories of pollutants. The National Pollutant Discharge Elimination System (NPDES) controls direct discharges into navigable waters. Pollutant categories included are:

- Point source discharges from pipes or sewers.
- Storm water discharges.
- Discharges to public treatment works (pretreatment requirements).

In practice, essentially all surface running waters are considered navigable.

The CWA also distinguishes among "priority" pollutants, toxic pollutants and "conventional" pollutants (BOD, suspended solids, fecal coliform, oil and grease and pH). Some materials are also considered "non-conventional" pollutants.

Safe Drinking Water Act

This Act requires the EPA to sets standards for drinking water quality. It also provides for Underground Injection Control (UIC) permits to regulate five categories of injection wells. The Wellhead Protection Program is designed to protect areas adjacent to water wells from contamination by pollutants that could enter the water supply through the well bore.

Toxic Substances Control Act (TSCA)

The EPA has authority to regulate chemical imports and production to control unreasonable risks. Chemicals deemed to present unreasonable risks are entered into an inventory list. Non-inventory materials must be described in terms of health and environmental effects prior to import or manufacture. The EPA can require health and environmental testing before import or manufacture of specified chemicals. Also, the EPA can ban or limit chemicals that pose unreasonable risks. They currently include asbestos, CFCs and PCBs.

Clean Air Act (CAA)

The CAA sets emission standards for fixed operations in terms of "criteria pollutants" and specific chemical emissions. Of direct importance to the biodiesel community, the CAA has been the basis for increasingly strict vehicle emission standards that have created the market niches biodiesel addresses. The CAA is divided into a number of Titles that address specific areas of clean air concerns:

> ***Title I:*** Establishes national ambient air quality standards (NAAQS) to limit "criteria pollutants" (CO, Pb, NO2, PM, O3 and SO_2). The Title also establishes new source performance standards (NSPS) requiring best available technology for pollutant control. Very important for the chemical industry, the Title establishes the National Emission Standards for Hazardous Air Pollutants (NESHAP), which provides the authority for monitoring and permitting much of the chemical industry and a list of hazardous air pollutants (HAP).
>
> ***Title II:*** Provides for emissions regulation for mobile sources: cars, trucks, buses, rail, boats and airplanes. This Title is the basis for requiring measures such as reformulated gasoline, vapor recovery nozzles, catalytic converters and low-sulfur diesel fuel. The California Air Resources Board (CARB) is applying the strictest standards in the nation under this Title.
>
> This Title is driving the liquid fuels industry in the U.S. The emissions standards and the progressive tightening of these standards have an enormous impact on liquid fuels. They are the basis for the entry of biodiesel into the liquid fuel stream.
>
> The EPA has compiled a database and issued a report on biodiesel emissions under the Biodiedel Emissions Analysis Program. Also, there is health effect information for biodiesel emissions. Biodiesel has been approved under Tier I and Tier II requirements.

Title III: There are a number of very specific provisions that pertain to chemical process operations under Title III. They are covered in 40 CFR Parts 60, 61 and 63. The Act directed the EPA to develop a list of sources for the 189 HAP and a schedule for developing emission standards. The standards will be based on "maximum achievable control technology" (MACT). Importantly, permitting of existing and new sources will be based on MACT.

Title IV: Establishes a SO_2 emission reduction program to reduce acid rain. This rationale is a part of the "green power" effort for electrical power generation and a factor in the vehicle emissions programs.

Title V: Establishes a single permit program for all "major sources." The states are developing their programs with EPA guidance. Once approved, permits will be issued and monitored by the state.

Title VI: Requires phase-out of the manufacture of ozone-depleting chemicals and restrictions regarding their distribution. This is part of the U.S. enactment of the Montreal Protocol.

Oil Pollution Control Act

This Act, set up a trust fund to strengthen the EPA's ability to respond to large oil spills and for above-ground storage incidents, requires development of Area Contingency Plans for spill responses. It is not clear if/how biodiesel facilities will be affected, because some agencies treat fats and oils like petroleum, while others do not. Since there have been serious spills associated with the oleochemical industry, biodiesel producers should design their facilities in accordance with best practices for containment and mitigation of spills.

Energy Policy Act (EPAct)

The EPAct establishes alternative fuel goals for vehicles to replace petroleum with renewably-derived fuels. A January 11, 2001 ruling gave biodiesel fuel use credits for fleets *in lieu* of purchasing alternatively-fueled vehicles.

Code Requirements for Permitting Biodiesel Facilities

The authorization to operate a biodiesel production facility is secured by approval of a set of operating permits that regulate the design, construction, emissions and safe operation of the facility. The permits are based upon a number of specific construction codes, documented compliance with safety, emissions, training and house-keeping standards, and approval of the standing operating procedures of the company. In general, the permits are issued locally, but they are governed by a hierarchy of local, state, and federal requirements.

There are a number of primary code references that are used internationally that govern the basic design and construction of a permitable facility. These codes include:

- International Fire Code (IFC) – Deals with general precautions, fire-related building requirements, specific industry requirements and hazardous materials requirements. In the case of the biodiesel facility, the most important compliance section of the IFC is Section 2703 – Hazardous Materials-General Requirements. The key provisions in Section 2703 are listed in **Table 45**.

2703.1	Maximum allowable quantity of hazardous materials per control area
2703.2	Systems, equipment and processes
2703.3	Release of hazardous materials
2703.4	Material Safety Data Sheets (MSDS)
2703.5	Hazard identification signs
2703.6	Signs
2703.7	Sources of ignition
2703.8	Construction requirements
2703.9	General safety requirements
2703.10	Handling and transportation

Table 45. Key Titles in the IFC Section 2703.

Section 2703.1 is of particular importance for design of a biodiesel facility. This section establishes the maximum allowable quantities (indoors) of hazardous materials in a single control area. For a General Industrial (F-1 occupancy) building without an automatic sprinkler system, the allowable quantities are:

- Flammable liquid (Class FL I-B) - 120 gallons (methanol, methoxide solutions and reaction mixtures)
- Combustible liquid (Class CL III-B) - 13,200 gallons (yellow grease, virgin oils, esters, biodiesel, glycerol)
- Water reactive material (Class WR-2) - 4.5 gallons (sulfuric acid)

Exceeding the allowable quantities requires the installation of an automatic sprinkler system (doubles the allowable quantity) or the use of a building complying with an H-2 occupancy rating.

The impacts of an H-2 occupancy requirement include:

- Automatic fire extinguishing system required
- Ventilation required per IFC and IMC
- Spill control, drainage, and containment
- Stand-by power
- Emergency alarms
- Increased separations from other buildings, roads and property boundaries
- 3- to 4-hour fire barriers required between different class occupancies

- International Building Code (IBC) – Deals with building use and occupancy classifications, general construction requirements (fire resistance, fire protection systems, means of egress, interior and exterior walls, roofing design), structural requirements, foundations, and materials of construction. Key sections of the IBC for a biodiesel facility are listed in **Table 46**.

Chapter 3:	Use and occupancy classification
Section 414:	Hazardous materials
Section 415:	Groups H-1, H-2, H-3, H-4, and H-5
Section 503:	General height and area limitations
Section 720:	Prescriptive fire resistance (fire delay time/material requirements)
Chapter 10:	Means of egress
Chapter 11:	Seismic construction

Table 46. Key Sections in the International Building Code.

- International Mechanical Code (IMC) – Deals with the mechanical design of the facility including, structural requirements, heating, ventilation and air conditioning design and materials of construction.
- International Electrical Code (IEC) – Deals with the design and installation of the facility electrical system, including panels, wiring, conduits and grounding, etc.
- ASME B31 – Code for Pressure Piping – Sets forth requirements for materials of construction, wall thickness requirements for piping and vessels operating at elevated pressures and temperatures, fabrication standards, and pressure testing protocols.
- NFPA 30, Chapter 3 Piping Systems – Describes requirements for piping design and construction in facilities containing hazardous and flammable materials.

Hazardous Material Inventory Statement (HMIS)

Developing the HMIS is fundamental to the permitting of any process facility. The types and amounts of hazardous materials kept on-site affect the design of the facility from the basic layout to the materials of construction required. In some circumstances, the HMIS can be the basis for refusal of a permit application because the proposed facility does not meet the standards for the intended use.

Preparation of the HMIS begins with the facility's Chemical Inventory (required list of all chemicals on-site, their quantity and location). Typically, the facility is divided into several functional areas (chemical storage, process floor, quality control laboratory, etc.) that must be inventoried. The inventory includes not only process and laboratory chemicals, but also chemicals used for cleaning, maintenance and in the office areas. By law, there must be a notebook containing copies of the MSDS for every chemical in the inventory.

The facility can be divided into control areas based upon function, <u>if</u> the areas are separated by fire-rated walls. The use of control areas also is a means to increase the allowable amount of materials on-site. Each control area is inventoried separately, but the HMIS applies to the entire facility.

The inventory for the process floor is the most complicated. The inventory is the sum of the total volume of each chemical or mixture in:

1. Piping
2. Reactors and vessels
3. Day-tanks and feed and product tanks

In the case of biodiesel, the partially reacted mixtures in the direct esterification process step and the transesterification step must be inventoried as a Class I-B liquid. Flash points for the mixtures at the mid-point of the reaction are on the order of 60°F because of the excess methanol present.

Essentially all of the information needed to complete the chemical description part of the HMIS is available in the MSDS sheets for the chemicals. The information required for each chemical in the HMIS includes:

- Product — Name of chemical
- CAS # — Chemical Abstract Service Number
- Criterion — Primary physical and health hazards in order of severity
- Basis — HMEX (Hazardous Materials Expert) Fire Code Classification (1999)
- Storage — Total amount in pounds, gallons or cubic feet, as appropriate
- Use-closed — Amount in use-closed (vessel or system normally closed while in operation)
- Use-open — Amount in use-open (vessel or system normally open to the atmosphere during operation)

An example of the HMIS data for an acid esterification (sulfuric acid) and base transesterification (sodium methylate) system is shown in **Table 47**.

Product	CAS #	Criterion 1	Criterion 2	Criterion 3	Basis
Yellow grease	None	CL III-B			FP – 260
Virgin oil	None	CL III-B			FP – 260
Esters/biodiesel	None	CL III-B	IRR		FP – 260
Acid rxn. mix	None	FL I-B	TOX	IRR	FP – 58
Base rxn. mix	None	FL I-B	TOX	IRR	FP – 56
Methanol	67-56-1	FL I-B	TOX	IRR	FP – 52
Glycerol	56-81-5	CL III-B	IRR	OHH	FP – 320
NaOH as methoxide	121-41-4	FL I-B	COR	TOX	FR – 84
Sulfuric acid 98%	7664-93-9	WR-2	TOX	COR	HMEX

Table 47. HMIS Inventory Data for Biodiesel Production.

Summary

- Compliance with the international codes is imperative in the design, construction and permitting of a biodiesel facility.
- Use UL, CE or other similar listed components if you are building your own system.
- Select components such as heaters and pumps that are "Approved for Use" by the vendor.
- Require your fabricator to demonstrate UL, CE and "Approved for Use" listings for components and require a system that can be certified as "Code Approved" if you are purchasing a process system, tankage or other process components.
- Remember, changing your process system after it has been approved and permitted must be reported and may require a new permitting review.

CHAPTER 43

EPA Registration

In recent years, regulators and policy-makers have become concerned about the potential impact of exhaust emissions on human health. The Environmental Protection Agency (EPA) has promulgated strict regulations for the amount of carbon monoxide, unburned hydrocarbons, oxides of nitrogen and particulate matter that an engine is allowed to emit. This has resulted in large reductions in the amount of these compounds entering our atmosphere. In addition, the EPA enforces ambient air standards across the country and holds those regions that are not in compliance with these standards responsible for developing plans to correct the noncompliance. As an additional safeguard, the EPA requires that producers of fuels and fuel additives (F/FA) intended for use in motor vehicles register their products. The authorization for this registration process is provided in sections 211(b) and 211(e) of the Clean Air Act and the F/FA regulations are sometimes referred to as the 211 regulations. The regulations are described in detail in 40 CFR Part 79.

All fuel and fuel additive manufacturers are required to register each product they sell, offer for sale or introduce into commerce with the EPA. This registration must include the following [derived from 40 CFR 79.11 and 79.33]:

a) The commercial identifying name of each additive that will or may be used in a designated fuel.
b) The name of the additive manufacturer of each additive named.
c) The range of concentration of each additive named, as follows:
    - In the case of an additive which has been or is being used in the designated fuel, the range during any three-month or longer period prior to the sate of submission.
    - In the case of an additive which has not been used in the designated fuel, the expected or estimated range.
d) The purpose in-use of each additive named;
e) The description (or identification in the case of a generally accepted method) of a suitable analytical technique (if one is known) that can be used to detect the presence of each named additive in the designated fuel and/or to measure its concentration therein;
f) Such other data and information as follows:
    - Hydrocarbon composition (aromatic content, olefin content, saturate content), with the methods of analysis identified.
    - Polynuclear organic material content, sulfur content, and trace element content, with the methods of analysis identified.
    - Distillation temperatures (90 percent point, end point).
    - Cetane number or cetane index.
    - Mechanism of action of each additive reported.
    - Reactions between the additives and motor vehicle diesel fuel.
    - Identification and measurement of the emission products of such additives when used in motor vehicle diesel fuel.
    - Effects of such additives on all emissions.
    - Toxicity and any other public health and welfare effect of the emission products of such additives.
g) Assurances that the fuel manufacturer will notify EPA if any information changes.
h) An assurance that the manufacturer will not advertise that registration constitutes endorsement of its product by EPA.
i) Assurances that the fuel used is substantially similar to fuels used for certification of motor vehicles after 1974 (does not increase emissions).
j) The manufacturer shall submit or shall reference prior submissions, all of the test data and other information required in Subpart F (Tier 1, 2 and 3 testing).

The requirement given in j) is the most substantive and requires testing that can cost millions of dollars. These requirements can be organized into three tiers as shown in **Figure 71** (from **http://www.epa.gov/otaq/regs/fuels/additive/spring94.pdf**).

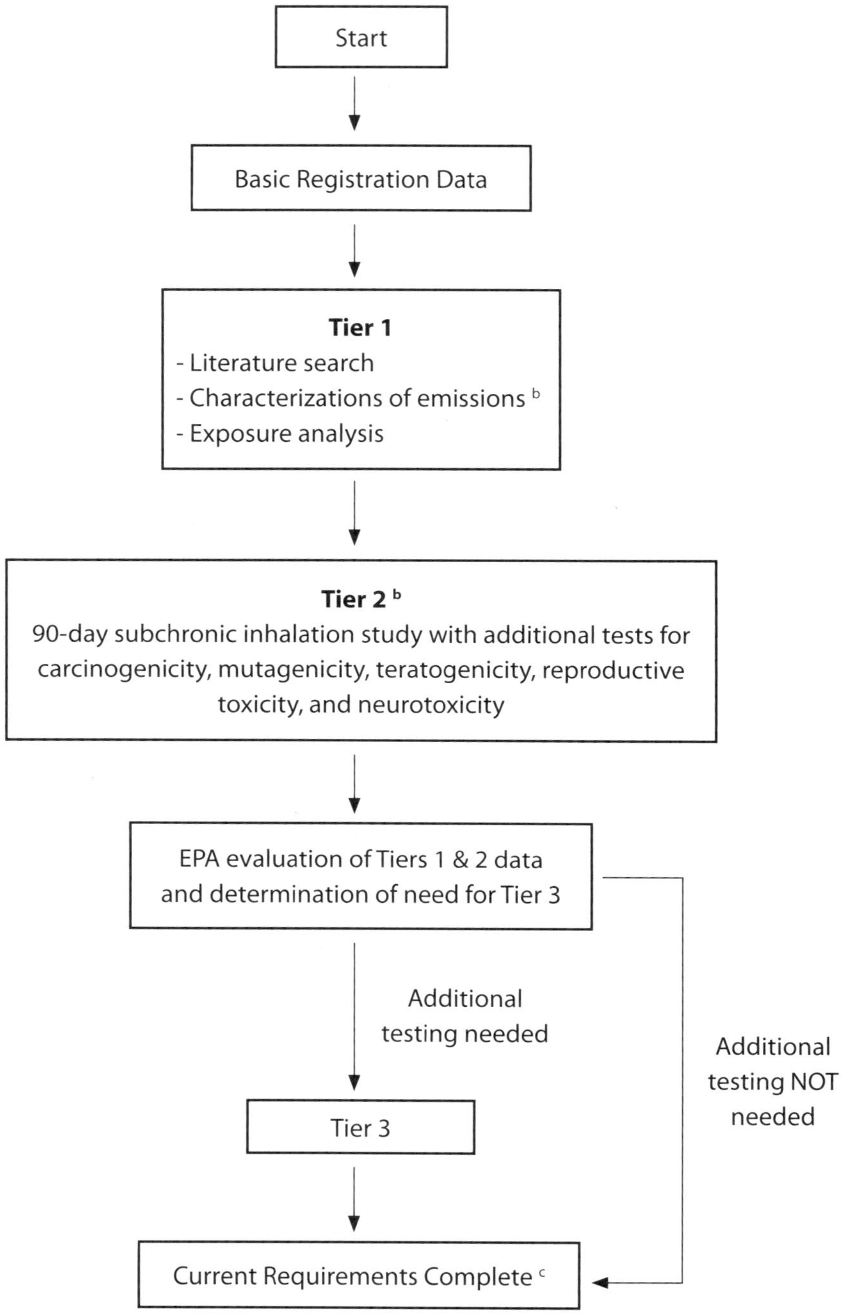

[a] This chart shows the general requirements for most F/FAs, and does not take into account any special provisions which may apply.

[b] Required unless adequate data exists.

[c] EPA retains the authority to require additional testing if new concerns arise.

Figure 71. Overview of Evaluation Tiers [a].

Testing Requirements

Tier 1 is the first step of data collection. It starts with a comprehensive literature review of the publications that have investigated the health effects of the exhaust from engines fueled with the fuel to be registered.

The second step of Tier 1 is an emissions characterization. This requires testing of engines following a protocol that goes well beyond the usual emissions certification test. The protocol requires speciation of hydrocarbon emissions (identification of individual chemical compounds) and measurement of the amounts of the polynuclear aromatic hydrocarbons (PAHs) and nitro-PAHs.

Tier I data in the public domain (generated outside of the National Biodiesel Board, or NBB) can be referenced so long as the engine, testing methods and other criteria involved in the test meet the EPA's standards. Anyone pursuing a Tier I data should check with EPA in person and read the 40CFR79 regulations to determine what is appropriate to submit.

To-date, a complete body of emission data that meets Tier I criteria has not been conducted independently from NBB and is not available for reference. The NBB is the only organization to submit Tier II data to the EPA. Only members of NBB can access this data for registration purposes. All other manufacturers will need to either join NBB for their coverage or conduct their own independent tests.

Tier 2 consists of a test of laboratory rats that are exposed for 90 days to the exhaust of engines fueled with the fuel (sub-chronic exposure). After 90 days, the rats are dissected and examined for evidence of carcinogenicity (tendency to cause cancer), mutagenicity (tendency to cause genetic mutations-conducted on diesel particulate extract), teratogenicity (tendency to cause fetal abnormalities), reproductive toxicity (inability to reproduce) and neurotoxicity (nervous system problems).

Tier 3 is additional testing that EPA may require after the results of Tier 1 and Tier 2 testing have been submitted. If EPA believes that additional testing is needed to confirm the results of Tier 1 and Tier 2 or if new testing is justified, it can require as part of Tier 3.

Special Provisions

The cost of the testing to satisfy Tier 1, 2 and 3 is very large. NBB has stated that its total cost was $2.2 million. In recognition of the high potential costs for this testing, the EPA included several provisions intended to ease the burden of the program. These provisions include the ability for manufacturers to group together and share costs and waivers for small businesses. A manufacturer may make use of jointly-submitted testing and analysis for a product that conforms to the same grouping criteria as the tested product. However, previous submitters (NBB in this case) are entitled to reimbursement for an appropriate portion of its costs incurred to report the information.

The grouping below is not solely intended for the evaluation of shared costs and waivers, but also dictates the testing required by the manufacturer. The grouping system established by the EPA divides the candidate fuels into three groups: baseline fuels, non-baseline fuels and atypical fuels. These fuels are defined as follows:

Baseline fuels – satisfy all of the following criteria:

1) Contains no elements other than carbon, hydrogen, oxygen, nitrogen and/or sulfur.
2) Contain less than 1.0 percent oxygen by weight.
3) Contain less than 0.05 percent sulfur by weight.
4) Possess the characteristics of diesel fuel as specified by ASTM standard D 975-93.
5) Derived only from conventional petroleum, heavy oil deposits, coal, tar sands and/or oil sands.

Biodiesel only contains carbon, hydrogen, and oxygen, so it satisfies requirement 1). However, B100 contains about 11 percent oxygen so 2) is not satisfied, and B100 cannot be considered to be baseline. Further, while B100 does have a sulfur content of less than 0.05 percent, it does not conform to ASTM D 975 and is not derived from conventional petroleum or the other sources listed.

Non-baseline fuels – These fuels satisfy the criteria for the baseline fuel except that they contain more than 1 percent oxygen and may be derived from sources that are not fossil-based.

Atypical fuels – The definition of an atypical fuel is that it contains elements other than carbon, hydrogen, oxygen, nitrogen and/or sulfur. However, EPA also indicates that fuels that do not fall into the baseline or non-baseline categories should also be considered to be atypical.

The non-baseline category would appear to be a better fit for biodiesel. It allows the fuel to contain more than 1 percent oxygen and to be from other than fossil sources. In fact, biodiesel (B100) is specifically identified as a non-baseline fuel in one area of the Code of Federal Regulations [40 CFR 79.56(e)(4)(ii)(B*)(2)(iv)(A)*], so it would appear that it was the intent of the original writers of the regulations that there is a biodiesel group that is considered as a non-baseline group. However, the EPA has subsequently determined that B100 does not qualify as non-baseline since it does not satisfy some of the parameters specified in ASTM D 975. The ASTM standard for diesel fuel oils specifies the requirements for diesel fuel shown in **Table 48**.

Property	Grade LS #1	Grade LS #2	Grade No. 1-D	Grade No. 2-D	Grade No. 4-D
Flash point °C, min.	38	52	38	52	55
Water and sediment, % vol, max.	0.05	0.05	0.05	0.05	0.50
Distillation temp., °C, 90%					
Min.	--	282	--	282	--
Max.	288	338	288	338	--
Kinematic viscosity mm²/s at 40°C					
Min.	1.3	1.9	1.3	1.9	5.5
Max.	2.4	4.1	2.4	4.1	24.0
Ramsbottom carbon residue On 10%, %mass max.	0.15	0.35	0.15	0.35	--
Ash, %mass max.	0.01	0.01	0.01	0.01	0.10
Sulfur, %mass max.	0.05	0.05	0.50	0.50	2.00
Copper strip corrosion, max. 3hrs at 50°C	No.3	No.3	No.3	No.3	--
Cetane number, min.	40	40	40	40	30
One of the following properties must be met:					
(1) cetane index	40	40	--	--	--
(2) aromaticity, %vol, max.	35	35	--	--	--
Cloud point, °C, max.	Determined by local climate				

Table 48. Requirements for Diesel Fuel Oils (ASTM D 975-97).

Excluding biodiesel from consideration as a non-baseline fuel based on the differences from ASTM D 975 is a decision that is likely to be challenged. The areas where biodiesel does not conform to D 975, such as the 90 percent distillation temperature, are primarily included in the standard to allow distinctions to be made between different grades of diesel fuel such as No. 1 and No. 2. They are not relevant to the question of whether or not a fuel is suitable for use in a diesel engine. The only property in D 975 that pertains to the suitability of the fuel is the cetane number, and biodiesel easily meets the minimum value of 40 given in the standard.

Due to the deviations with D 975, the EPA has categorized biodiesel as an *atypical* fuel. While biodiesel does not satisfy the primary requirement of atypical fuels, which is that it should contain elements other than carbon, hydrogen, oxygen, nitrogen, and/or sulfur, the EPA considers fuels to be atypical if they do not fit in the baseline or non-baseline categories [40 CFR 79.56 (e)(2)(iii)]. The significance of this decision is the impact it has on a company's eligibility for the small business waivers.

Small Business Provisions

To further ease the impact of the testing costs on small producers, the F/FA regulations state that fuel manufacturers of baseline and non-baseline fuels with annual sales of less than $50 million only need to submit the basic registration data. These companies are not required to submit Tier 1 and Tier 2 data. However, since EPA does not consider biodiesel to be a non-baseline fuel, small biodiesel producers are not eligible for this exemption.

Small producers of atypical fuels can also qualify for an exemption, but the limiting size is only $10 million. These producers are also still required to submit Tier 1 data although Tier 2 can be waived. Collecting Tier 1 data can still be quite expensive, requiring between $100,000 and $250,000.

NBB Health Effects Data

NBB, using funds from the National Soybean Checkoff Program, sponsored a program to develop Tier 1 data. NBB and the Department of Energy co-sponsored a program to develop Tier 2 data. This test program has been described in a number of publications [1-4].

The EPA has accepted the NBB data and registered many biodiesel producers. These producers are considered to be part of NBB's group and, with NBB's permission, are allowed to use its data.

The EPA regulations recognize that groups that submit health effects data are entitled to reimbursement from groups that submit later requests that reference the earlier data. Before the EPA will consider the later registration requests, the group must submit documentation that the original submitter has been notified and reimbursement has been arranged. The entitlement to compensation remains in effect for 15 years following the original submission. Any person who violates the registration requirements is subject to a civil penalty set up by the EPA, which may be up to $25,000 per violation, plus the monetary savings, which may have resulted from the violation. Each day of a continuing offense is a separate violation.

The NBB has made its health effects data available to all its members at no charge. Non-members must pay $100,000 plus twice the volume dues paid by NBB members on gallons produced for access to the data.

NBB Membership

NBB is a trade association that has been formed to promote the common business interests of its members and to promote the use of biodiesel as a fuel that meets ASTM standards.

NBB has two classes of Voting Director membership:

1) Feedstock producers or feedstock producer organizations.
2) Biodiesel processors or biodiesel marketers.

The board also has non-voting associate memberships available to individuals and organizations that do not fall into one of the categories for voting membership.

The NBB charges annual dues that depend on whether you are in the feedstock class or the biodiesel processor/marketer class. The dues for a feedstock producer or feedstock producer organization are currently \$10,000/year. Several organizations can split the dues but they only receive one Voting Director position. Dues for a biodiesel processor or marketer are a minimum of \$5,000/year but primarily depend on annual sales. **Table 49** shows how the actual dues are calculated from the sales volume.

The annual dues are billed at the start of each fiscal year (October 1). The volume dues must be paid quarterly based on quarterly sales reports submitted to NBB. NBB may audit members' books to confirm that they are correctly reporting their volumes.

Volume produced or marketed (gallons/yr)	Volume dues
0 to 1,000,000	\$0.01/ gallon (\$5,000 minimum annual dues)
1,000,001 to 5,000,000	\$0.0075/gallon
5,000,001 to 10,000,000	\$0.0050/gallon
10,000,001 to 15,000,000	\$0.0025/gallon
Amounts over 15,000,000	\$0.0010/gallon
Sample calculation: *Suppose a processor sells 65 million gallons of biodiesel.* *\$10,000 + \$30,000 +\$25,000 + \$12,500 +\$50,000 = \$127,500 annual dues*	

Table 49. Voting Director; Biodiesel Producer/Marketer - Volume Dues.

NBB has also offered a small producer membership for \$2,500 that provides access to the health effects data needed for registration but this is a non-voting membership. This membership is limited to producers with production of 250,000 gallons/year or less.

NBB Fuel Accreditation

NBB has initiated a fuel accreditation program, known as BQ 9000, that will allow fuel producers to claim that they are an "accredited" producer. This is intended to ensure that biodiesel producers follow good business practices and provide fuel that meets the ASTM standard. The accreditation process involves ISO 9000 – type inspections (paid for by producer) and periodic fuel testing. The latest guidelines can be obtained by contacting NBB through **http://ww.biodiesel.org**.

References

1. Koo-Oshima, S., N. Hahn, and J. Van Gerpen, "Comprehensive Health and Environmental Effects of Biodiesel as an Alternative Fuel," presented at the fall meeting of the Society for Risk Analysis, Paris, France, Oct. 11-14, 1998.
2. Sharp, C.A., S.A. Howell, J. Jobe, "The Effect of Biodiesel Fuels on Transient Emissions from Modern Diesel Engines, Part I Regulated Emissions and Performance," Society of Automotive Engineers Paper No. 2000-01-1967, 2000.
3. Sharp, C.A., S.A. Howell, J. Jobe, "The Effect of Biodiesel Fuels on Transient Emissions from Modern Diesel Engines, Part II Unregulated Emissions and Chemical Characterization," Society of Automotive Engineers Paper No. 2000-01-1968, 2000.
4. Finch, G.L., et al., "Effects of Subchronic Inhalation Exposure of Rats to Emissions from Diesel Engine Burning Soybean Oil-Derived Biodiesel Fuel," Inhalation Toxicology, 14:1017-1048, 2002.

About the Authors

All of the authors whose writings are included in this publication have served as faculty at the widely recognized Biodiesel Workshop Series, held in cooperation with Iowa State University, Ames, Iowa. The five-day workshops take place several times annually. Materials from those workshops, which cover technology, analytical methods and business management for biodiesel producers, is included in this book.

Jon Van Gerpen

Jon Van Gerpen is head of the Department of Biological and Agricultural Engineering at the University of Idaho, Moscow. He was an initial developer and leader in establishing the Biodiesel Workshop Series. Van Gerpen previously was a professor of Mechanical Engineering at Iowa State. He has been conducting biodiesel-related research for 10 years, including investigations into biodiesel-fueled engine emissions, fuel composition effects, thermal and oxidative stability and contaminant effects. While at Iowa State, he oversaw the operation of a biodiesel pilot plant at the Iowa Energy Center's Biomass Energy Conversion Center and directed research at the Internal Combustion Engines Laboratory in the university's Department of Mechanical Engineering.

Rudy Pruszko

Rudy Pruszko is a chemical engineer and project manager with the Center for Industrial Research and Service (CIRAS), which is part of Iowa State University Extension. He has more than 25 years of manufacturing and management experience in a broad range of industries. His experience includes design, construction, start-up and operations of five chemical plants, including business planning, cost analysis, marketing research and project development from concept to commercialization. Pruszko has assisted several biodiesel plant start-ups with feasibility studies and business planning.

L. Davis Clements

L. Davis Clements is founder and president of Renewable Products Development Laboratories, Inc. (RPDL), Lincoln, Nebraska, a process development and licensing company founded in 1999. He has been involved in chemical process design and process development activities in industry and academia, and as a consultant since 1966. The current focus of RPDL is processes and products based on the utilization of animal fats, vegetable oils and waste cooking grease as a source of energy and industrial chemicals.

Gerhard Knothe

Gerhard Knothe has a Ph.D. in organic polymer chemistry and has worked at the National Center for Agricultural Utilization Research of the U.S. Department of Agriculture in Peoria, Illinois, since 1989. In his position as research chemist, he has been involved with developing industrial uses of vegetable oils. His contributions include research on pre-combustion reactions of fatty compounds as they occur in vegetable oils and boidiesel, combustion-improving additives, additives for improving cold-flow properties and analytical methods for biodiesel production and fuel quality assessment.

Brent Shanks

Brent Shanks is an associate professor of chemical engineering at Iowa State University, Ames, joining the faculty in 1999. Prior to that, he worked for Shell Chemical Company, Houston, Texas, for 11 years as a research engineer in the area of heterogeneous catalysis and as department manager of the chemical catalyst department. Shanks' work includes leading training courses for process engineers and operators in manufacturing facilities. A current project involves the synthesis of novel, highly basic heterogeneous catalysts for the transesterification of triglycerides to biodiesel.